GASEOUS RADIATION DETECTORS

Widely used in high-energy and particle physics, gaseous radiation detectors are undergoing continuous development. The first part of this book provides a solid background for understanding the basic processes leading to the detection and tracking of charged particles, photons, and neutrons.

Continuing then with the development of the multi-wire proportional chamber, the book describes the design and operation of successive generations of gas-based radiation detectors, as well as their use in experimental physics and other fields. Examples are provided of applications for complex event tracking, particle identification, and neutral radiation imaging. Limitations of the devices are discussed in detail.

Including an extensive collection of data and references, this book is ideal for researchers and experimentalists in nuclear and particle physics.

This title, first published in 2015, has been reissued as an Open Access publication on Cambridge Core.

FABIO SAULI is Research Associate for the Italian TERA Foundation, responsible for the development of medical diagnostic instrumentation for hadrontherapy. Prior to this, he was part of the Research Staff at CERN in the Gas Detectors Development group, initiated by Georges Charpak, before leading the group from 1989 until his retirement in 2006. He has more than 200 scientific publications, and is an editor of several books on instrumentation in high energy physics. His achievements include inventing the Gas Electron Multiplier (GEM), which is widely used in advanced detectors.

CAMBRIDGE MONOGRAPHS ON PARTICLE PHYSICS, NUCLEAR PHYSICS AND COSMOLOGY

General Editors: T. Ericson, P. V. Landshoff

GASEOUS RADIATION DETECTORS

Fundamentals and Applications

FABIO SAULI

European Organization for Nuclear Research
CERN, Geneva, Switzerland

CAMBRIDGE
UNIVERSITY PRESS

Shaftesbury Road, Cambridge CB2 8EA, United Kingdom

One Liberty Plaza, 20th Floor, New York, NY 10006, USA

477 Williamstown Road, Port Melbourne, VIC 3207, Australia

314–321, 3rd Floor, Plot 3, Splendor Forum, Jasola District Centre, New Delhi – 110025, India

103 Penang Road, #05–06/07, Visioncrest Commercial, Singapore 238467

Cambridge University Press is part of Cambridge University Press & Assessment,
a department of the University of Cambridge.

We share the University's mission to contribute to society through the pursuit of
education, learning and research at the highest international levels of excellence.

www.cambridge.org
Information on this title: www.cambridge.org/9781009291187

DOI: 10.1017/9781009291200

First published 2015
Reissued as OA 2022

A catalogue record for this publication is available from the British Library.

ISBN 978-1-009-29118-7 Hardback
ISBN 978-1-009-29121-7 Paperback

Contents

Contents

Acronyms

ADC:	analogue to digital converter
ASIC:	application specific integrated circuit
ATLAS:	one of the LHC experiments at CERN
BNL:	Brookhaven National Laboratory, USA
CAT:	compteur à trous
CEN-Saclay:	Centre d'Etudes Nucléaires, Saclay, France
CERN:	European Organization for Nuclear Research, Geneva, Switzerland
CGS:	electrostatic units: centimetres, grams, seconds
COG:	centre of gravity
CRID:	Cherenkov ring imaging detector
CSC:	cathode strip chamber
CsI:	caesium iodide
CVD:	carbon vapour deposition
DC:	drift chambers
DME:	dimethyl ether $(CH_3)_2O$
FERMILAB:	Fermi National Laboratory, Batavia, Illinois, USA
FGLD:	field gradient lattice detector
FWHM:	full width at half maximum
GDD:	Gas Detectors Development group at CERN
GEM:	gas electron multiplier
GSPC:	gas proportional scintillation counter
HADC:	high-accuracy drift chamber
HBD:	hadron blind detector
HMPID:	high momentum particle identification detector
IHEP:	Institute of High Energy Physics, Protvino, Russia Federation
ILC:	International Linear Collider
ILL:	Institut Laue-Langevin, Grenoble, France

INFN:	Istituto Nazionale di Fisica Nucleare, Italy
ISIS:	identification of secondary particles by ionization sampling
IVI:	interaction vertex imaging
KEK:	High Energy Accelerator Research Organization, Kamiokande, Japan
LAr:	liquid argon
LBL:	Lawrence Berkeley Laboratory
LEM:	large electron multiplier
LEP:	Large Electron–Positron collider at CERN
LHC:	Large Hadron Collider at CERN
LNF:	Laboratori Nazionali Frascati, Italy
LNGS:	Laboratori Nazionali Gran Sasso, Italy
MDT:	monitored drift tubes
Micromegas:	micro-mesh gaseous structure
MIPA:	micro-pin array
MPGD:	micro-pattern gas detector
MRPC:	multi-gap resistive plate chamber
MSC:	multi-step chamber
MSGC:	micro-strip gas counter
MWDC:	multi-wire drift chamber
MWPC:	multi-wire proportional chamber
μPIC:	micro-pixel chamber
NSR:	nuclear scattering radiography
NTP:	normal temperature and pressure: 0°C, 1 atmosphere
PEP:	Electron Positron Collider at SLAC
PET:	positron emission tomography
PPAC:	parallel plate avalanche counter
PRR:	proton range radiography
PST:	plastic streamer tubes
P10:	mixture of 10% methane in argon
QE:	quantum efficiency
RHIC:	Relativistic Heavy Ion Collider, Brookhaven, USA
RICH:	ring imaging Cherenkov counter
RMS (rms):	root mean square (Gaussian standard deviation)
RPC:	resistive plate counter
SLAC:	Stanford Linear Accelerator Center
SLHC:	Super LHC at CERN
SPECT:	single photon emission computed tomography
SQS:	self-quenching streamer
SSC:	Superconducting Supercollider

STP:	standard temperature and pressure: 20°, 1 atmosphere
SWDC:	single-wire drift chambers
TEA:	triethyl amine $(C_2H_5)_3N$
TEC:	time expansion chamber
TERA:	Fondazione per Adroterapia Oncologica, Novara, Italy
TGC:	thin-gap chambers
TGEM:	thick gas electron multiplier
TMAE:	tetrakis dimethyl amino ethylene $C[(CH_3)_2N]_4$
TPC:	time projection chamber
TRIUMF:	Canada's National Laboratory for Particle and Nuclear Physics, Vancouver
TRT:	transition radiation tracker
UV:	ultra-violet
VUV:	vacuum ultra-violet
WIMP:	weakly interacting massive particle
WLS:	wavelength shifter

Preface

Major scientific advances are the result of interplay between ground breaking theoretical intuitions and experimental observations, validating or contradicting the predictions. In elementary particle physics, the commissioning of high-energy accelerators and colliders demanded the development of innovative detectors capable of recording increasingly complex events; in astrophysics, where the scope is to detect radiation from remote sources, or ubiquitously present in the Universe but with little if any interaction with ordinary matter, the focus is rather on the realization of large volume, low noise devices capable of revealing rare events obscured by diffuse backgrounds. In both cases, dedicated gas-filled detectors have demonstrated their flexibility of conception and excellent performances.

Starting with Ernest Rutherford's original development of the single-wire proportional counter in the early 1900s, through the multi-wire and drift chambers introduced by Georges Charpak in the late sixties, to the powerful new tracking devices collectively named micro-pattern gas chambers, the development of gaseous detectors has been a continuous story of success and, sometimes, disappointments.

While many textbooks exist on gaseous detectors (see the Further Reading section), most of the information on recent progress in the field is scattered in thousands of articles, conference records, doctoral theses and other documents. This book aims to collate selected information in an organized way, reproducing relevant data on the various developments, providing extended references to published material as well as links to useful web-based tools and databases. The content is largely based on the many courses given by the author at CERN and various universities and research laboratories worldwide, and greatly profits from constructive interactions with the students. Whenever possible, simplified, back-of-the-envelope calculation examples are provided as a complement to more rigorous algorithms.

After a recall of the major processes of interaction between charged particles, photons and neutrons with the medium, releasing detectable messages in matter, the first part of the book follows the fate of the ionisation yields, released in a counter's gas, under the effect of externally applied electric and magnetic fields, from simple collection to charge multiplication and breakdown. Depending on the counter geometry and field strength, a detector can be made then to operate in simple charge collection, in a regime of avalanche charge multiplication with the detected charge proportional to the primary ionisation, or in gain-saturated regimes providing conveniently large signals, almost independent from the original charge. Primary or field-enhanced photon emission can also be exploited for detection. Each mode has its own advantages and disadvantages, discussed in the subsequent chapters, which have to be thoroughly analysed to best cope with the experimental needs.

Associated for many years with Georges Charpak's research group at CERN, I was easily fascinated by his enthusiasm in searching new directions for the development and applications of detectors, mostly based on the use of a gas as sensitive medium; a short personal recollection of my participation to these research efforts is illustrated in the first chapter of the book. Over the years, the activity of the group attracted many young scholars, who contributed to the various developments before returning to their home institution, often subsequently creating their own research team while keeping friendly and constructive contacts with our group; this book is dedicated to them and their works, with apologies for any mistake or omission.

I am particularly grateful to Ugo Amaldi, who hosted me in the TERA Foundation premises at CERN during the final drafting of the book; a warm word of appreciation goes to many colleagues who provided scientific help and support in obtaining original documents and reprint permissions: Marcello Abbrescia, Ugo Amaldi, Elena Aprile, Tullio Basaglia, Malte Hildebrandt, John Kadyk, Salete Leite, Eugenio Nappi, Anna Peisert-Elliott, Archana Sharma, Emile Schins, Graham Smith, Jerry Va'vra, and many others.

Last but not least, warm thanks to my daughter Raffaella who undertook the strenuous task of improving the language on a subject rather extraneous to her field of expertise.

1

Introduction

1.1 Historical background

The foundations of modern gaseous detectors can be traced back to the works of Ernest Rutherford, 1908 Nobel Laureate in Chemistry. In the course of his studies of the atomic structure, he conceived an instrument capable of detecting individual ionization trails left in a gas by natural radioactivity. Knowledgeable of John Sealy Townsend's studies on collisional charge multiplication in gases at high electric fields, and with the help of Hans Geiger, he built a tool capable of amplifying the weak primary ionization signal. The device consisted of a thin metal wire, the anode, coaxial with a gas-filled cylindrical cathode; on application of a potential difference between the electrodes, electrons released in the gas drift towards the anode, undergo inelastic ionizing collisions in the fast increasing field and multiply in an appropriately named electron–ion avalanche. Restricted to a narrow region around the wire, the multiplication process amplifies the charge released in most of the gas volume and yields a signal proportional to the primary charge, hence its name 'proportional counter' (Rutherford and Geiger, 1908). Large multiplication factors, or gains, could be achieved, permitting the detection of small amounts of primary charge with the rudimentary electrical instrumentation available at the time. Further developments of the device by Geiger and Walther Müller permitted them to reach the ultimate goal of detecting single electrons released in the counter's gas (Geiger and Müller, 1928).

Proportional counters of various sizes and shapes were employed for decades in the detection of ionizing radiation; Geiger–Müller counters are still widely used for radiation monitoring. Arrays of proportional counters have been built to cover larger areas; however, limited in location to their physical size, they could hardly satisfy the tracking requirements of the emerging high-energy particle physics. Already in the 1930s, this goal was mainly achieved using photographic emulsions, capable of recording the trails left by the passage of charged particles.

1

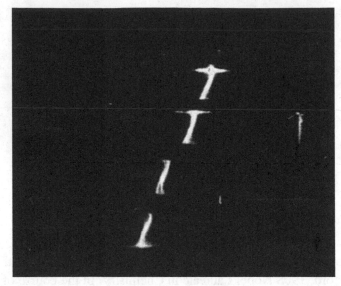

Figure 1.1   A cosmic ray detected in a four-gap spark chamber (Fukui and Miyamoto, 1959). By kind permission of Springer Science+Business Media.

The development of other types of detector having excellent imaging capability, such as the cloud chamber (Charles Thomson Wilson, 1927 Nobel Laureate) and the bubble chamber (Donald Arthur Glaser, 1960 Nobel Laureate), relegated the use of emulsions to specialized nuclear physics investigations. Bubble chambers, at the same time target and detector and providing accurate three-dimensional optical images of complex events, were successfully used for decades in particle physics and still powerful tools of investigation in the 1960s. However, these devices have a major drawback: they are made sensitive under the action of an external mechanical control only during a selected time interval, uncorrelated to the physical events under study. Well adapted to the analysis of frequent processes, they are less suited for the study of rare events.

A new type of gas counter that could be made sensitive in coincidence with selected events, the triggered spark chamber, was developed in the late fifties (Fukui and Miyamoto, 1959). On application, shortly after the passage of a charged particle, of a high voltage pulse across a thin gas layer between two electrodes, a detectable spark would grow along the ionization trails left in the gas. A system of external coarse devices, as a set of scintillation counters, provides a signal to trigger the chambers in coincidence with specific geometrical or energy loss requirements; the concept of selective event trigger was born.

Figure 1.1, from the reference above, is one of the first pictures of a cosmic ray track detected with a four-gap spark chamber. Stacks of thin-gap spark chambers could thus provide a sampled image of tracks crossing the detector within a short

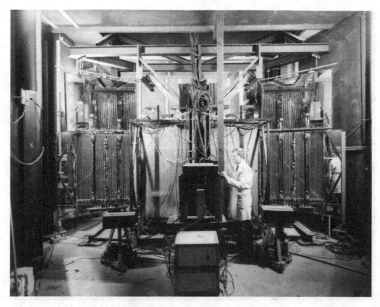

Figure 1.2 PS11, an optical spark chamber experiment at CERN's Proton Synchrotron, with accidental cosmic rays tracks. Picture CERN (1967).

time window, and were extensively used in particle physics experiments, cosmic ray studies and other applications. Recording of the events was done by photography or, in the later times, electronic video digitization. Figure 1.2 is an example of an experimental setup with several optical spark chamber stacks surrounding a target, operating in the 1960s at CERN.

Limited originally by the slow picture recording process, spark chambers evolved into faster electronic devices thanks to the development of methods capable of detecting the current pulse produced by a spark on electrodes made with thin wires. The most successful employed small ferrite core beads, used at the time in computer memories, interlaced with the wires and read out with a sequence of electrical pulses (Krienen, 1962). Simple to implement, the magneto-strictive readout method, introduced in the early sixties, relied on the detection of the sonic waves induced by a discharge on an external wire transducer, perpendicular to the wire electrodes; coordinates were then deduced from the time lapse between the spark and the detection of the sound pulse at the two ends of the pickup wire (Perez-Mendez and Pfab, 1965). Other methods included capacitive charge storage and direct detection of the spark sound with microphones located in strategic positions; for a review see for example Charpak (1970).

In thin-gap chambers, the applied high voltage pulse causes a discharge propagating from anode to cathode. Further developments of the technology led to the introduction of a more powerful family of devices, named streamer chambers: these

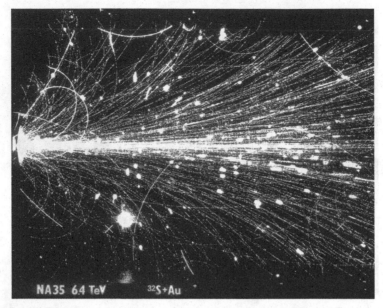

Figure 1.3   Heavy ion collision recorded with CERN's NA35 streamer chamber.
Picture CERN (1991).

are large volume detectors in which a very narrow and high voltage pulse induces
the formation of short local discharges following the ionized trails in the gas. While
having rate capability limited to a few hertz, needed to generate nanosecond-long,
hundreds of kV/m voltage pulses, the streamer chambers had an impressive multi-
track imaging capability, as shown by the example of Figure 1.3, recorded with the
NA35 streamer chamber at CERN on a relativistic heavy ion collision (Brinkmann
et al., 1995). In many ways, later developments with gaseous detectors, the main
subject of this book, have been inspired by the challenge to achieve similar image
qualities with faster, fully electronic devices. For a review of streamer chambers
development and performances see Rohrbach (1988).

1.2 Gaseous detectors: a personal recollection

In the late 1960s, as a post-doc at the University of Trieste (Italy), I contributed to
the realization of a detector system using wire spark chambers with magnetostrictive readout, used in an experiment at CERN. While a technical staff was in charge
of the chamber's construction, the delicate but tedious work of winding the
miniature coils used to detect the sonic pulse on the acoustic sensing wire was a
task for the young students. The results of the test beam measurements of efficiency and position accuracy with a set of detectors are described in my first

Figure 1.4 The author with the CERN–Trieste magnetostrictive spark chambers setup. Picture CERN (1967).

publication (Bradamante and Sauli, 1967); I can be seen in in Figure 1.4 working on the experimental setup, a fixed target experiment to study proton–proton and proton–deuteron interactions at (for that time) high energies.

Albeit selective and faster in response than previous generations of detectors, spark chambers are limited in operating rate to a few tens of hertz, due to the time needed to clear the excited and ionized species from the region of a spark before the application of another pulse, in order to prevent re-firing.

In the late 1960s, the need for large area and faster electronic detectors acquired paramount importance, motivated by the challenging requirements of the increasingly high-energy particle physics. The multi-wire proportional chamber (MWPC), invented in 1967 by CERN's Georges Charpak, revolutionized the field of position-sensitive detectors (Charpak *et al.*, 1968). In Figure 1.5 Charpak's technician, Roger Bouclier, stands next to the first MWPC, with 24 anode wires

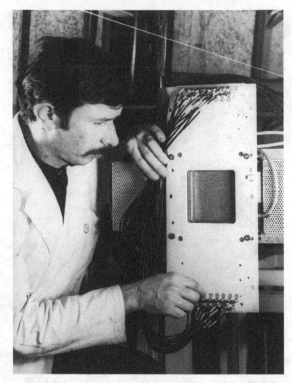

Figure 1.5   Roger Bouclier with the first multi-wire proportional chamber.
Picture CERN (1968).

and 10×10 cm^2 active area.[1] For his invention, and the contribution of the new
family of detectors to fundamental research, Charpak received the 1992 Nobel
Prize for Physics.

The outstanding innovative performances of the new device were soon recognized, despite the challenge posed at the time by the need of using individual
recording electronic channels on many wires a few mm apart: nanosecond time
resolution, sub-mm position accuracy, continuous sensitivity and high rate capability. The new detector technology, swiftly adopted by several experiments, gave
Charpak resources and support to continue and expand the research activity on
gaseous detectors. I joined Charpak's group in 1969, contributing for many years
to the development and applications of innovative gaseous detectors; after
Georges' retirement in 1989, I took the leadership of the group then named Gas
Detectors Development (GDD) until my own retirement in 2006. During all those
years, the continuing challenge posed by the increasing requirements of particle

---

[1] There is no known picture of Charpak himself with the early MWPCs; Figure 1.8, taken several years later,
is sometimes quoted to be one.

Figure 1.6　A large MWPC prototype, with (left to right) Georges Charpak, Fabio Sauli and Jean-Claude Santiard. Picture CERN (1970).

physics experimentation motivated the search for faster and more performing devices that exploit the properties of charge transport and multiplication in gases.

The original MWPC could attain avalanche gains around 10^5; detection of the signal released by fast particles (a few tens of electron–ion pairs) required the use of low noise amplifiers, which was possible but rather demanding for the electronics of the time. A major discovery by Charpak's group, and possibly a reason for the fast spread of the technology, was a gas mixture in which saturated gains above 10^7 could be reached, providing pulses of amplitude independent of the primary ionization release, thus leading to simpler requirements for the readout electronics. Quite appropriately, this mixture (argon-isobutane with a trace of freon) was named 'magic gas' (Bouclier *et al.*, 1970).

The first MWPC was only 10 cm on the side; soon, a large effort was put into developing the technology for manufacturing larger detectors. However, unexpected problems of electrostatic instability, discussed in Section 8.4, resulted in a dramatic failure of the early prototypes; the problem was solved with the introduction of internal insulating supports or spacers. Figure 1.6 shows one of the first large size working devices, about one and a half metres on the side, built by the group in 1970 (Charpak *et al.*, 1971).

Suitable for fixed target experiments, the heavily framed construction of the chamber seen in the picture was not optimal for use within a magnet, due to the unfavourable ratio of active to total area; a lighter design of the detector, which

Figure 1.7   The 40 MWPC array of the Split Field Magnet Detector spectrometer. Picture CERN (1973).

made use of self-supporting, light honeycomb panels holding the wires, was developed by the group to equip the multi-particle spectrometer of the Split Field Magnet experiment at CERN's proton–proton storage rings (Figure 1.7) (Bouclier *et al.*, 1974). Deploying 40 large MWPC modules, the instrument featured data taking rates of several kHz, a performance unthinkable when using older tracking devices, and operated for many years for systematic measurements of particle yields in proton–proton collisions. One of the searches, the quest for free quarks, yielded no results for fundamental reasons that become clear only later; however, it motivated one of my early works to estimate the detection efficiency of MWPC on charge 1/3 particles (Breidenbach *et al.*, 1973).

   In the initial conception of the MWPC, space accuracy was determined by wire spacing, a few mm at best. As anticipated in seminal work by Charpak and collaborators, sub-mm position accuracies could be achieved by exploiting the time lag, or drift time, of the detected charge in respect to an external trigger (Charpak *et al.*, 1970). Developed in the early seventies, and using several centimetres wire spacing, drift chambers provided position accuracies between 300 and 400 μm, while substantially reducing the number of electronics channels (Walenta, 1973). A thorough optimization of the electric field structure and detailed studies on the electrons' drift properties permitted them to reach position accuracies around 50 μm for fast particles perpendicular to the detector (Charpak *et al.*, 1973). Figure 1.8 shows Charpak with an early prototype of the High

Figure 1.8   Georges Charpak with the first prototype of the high accuracy drift chamber. By permission of SPL Science Photo Library (1982).

Accuracy Drift Chamber, a single cell 50 mm wide; curiously, in the absence of a picture of the inventor with the first MWPC, this picture is sometimes referred to as such. In Figure 1.9, Guy Schultz and Amos Breskin, former members of the group, are seen inserting a set of medium-size high accuracy drift chamber prototypes in a magnet for systematic measurements of performances in magnetic fields. As will be discussed in more detail in Chapter 9, each chamber provided two perpendicular coordinates, resolving the right–left ambiguity, intrinsic in a time measurement, thanks to the use of anode wire doublets mounted at a close distance.

The temperature dependence of the drift properties, crucial for a stable long-term use of the devices, was studied thoroughly with dedicated detectors, and led to the choice of operating gases having a saturated drift velocity, with minimal variation with temperature (see Section 4.7). Requiring the heating of the detectors while operating, these studies resulted often in spectacular failures due to the appearance of heavy discharges in flammable gas mixtures (Figure 1.10).[2]

---

[2] This event is colloquially named 'Breskin's thermodynamics experiment' from the name of the team member in charge of the measurement.

Figure 1.9   Guy Schultz (left) and Amos Breskin (right) inserting a set of high accuracy drift chambers in a magnet. Picture CERN (1972).

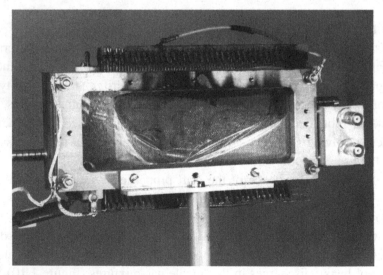

Figure 1.10   Burned-out drift chamber, the end of a temperature dependence study. Picture by the author at CERN (1972).

Figure 1.11   The JINR high-accuracy drift chamber setup at Fermilab. Picture by the author (Sauli, 1977).

The unique position accuracy properties of the detectors were soon exploited in experiments. I participated in the initial study of the kaon form factor in a K-e scattering experiment at Fermi National Laboratory in Batavia by a group of the Joint Institute of Nuclear Research (JINR Dubna) led by Edick Tsyganov (Filatova *et al.*, 1977); in Figure 1.11 I pose with the two drift chamber telescopes of the experiment installed in a beam line at Fermilab.

Another experiment exploiting the excellent space localization properties of our drift chambers was the study of channelling effects of fast charged particles in crystals, set up at CERN by a group from Aarhus University (Denmark). Figure 1.12 is a stereogram of the distribution of incidence angle of 1.35 GeV π^+ on a germanium crystal, selected for releasing less than average energy in the crystal, clearly showing an increase of the yield along the crystal axis and planes (Esbensen *et al.*, 1977). The study of channelling properties in crystals continued for many years, eventually focussing on the possibility of deflecting high-energy particle beams using bent crystals (Tsyganov, 1976) and is a subject of continuing investigation, see for example Chesnokov *et al.* (2013) and references therein.

In the mid-1970s we used a set of high-accuracy drift chambers to investigate the prospects of exploiting the reconstruction of the vertex of interaction of high energy protons within a target, thus obtaining a three-dimensional density map, with the so-called nuclear scattering radiography method (NSR)[3] (Saudinos *et al.*, 1975).

---

[3] Given its full 3-D imaging capability, the method would be more appropriately named nuclear scattering tomography.

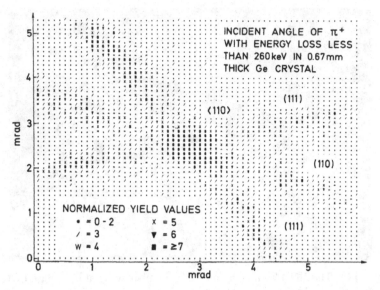

Figure 1.12   Angular distribution of 1.35 GeV pions releasing less than the average energy loss in a germanium crystal (Esbensen *et al.*, 1977). By kind permission of Elsevier.

The measurements demonstrated the feasibility of a diagnostic tool that provided good image quality at low patient irradiation doses compared to X-ray tomography; tagging of the proton–proton elastic scattering component, with a simple angular correlation cut, permitted the identification of the hydrogen content in the body (see Figure 1.13).

Requiring the use of ~GeV proton beams, available in only a few high energy physics research centres, the NSR method remained for many years a curiosity. It has, however, been reconsidered as a tool for quality assurance at the upcoming oncological hadrontherapy centres, exploiting the high energy proton and ion beams used for deep neoplasm treatment (see the end of this section).

Concurrently with the development of high-accuracy drift chambers, a detailed study of the processes of signal induction on electrodes led to the introduction of a method of localization based on the recording of the signals induced by the avalanches on cathodes (Charpak and Sauli, 1974). Unlike drift chambers, which require an external time reference to perform localization, cathode-readout MWPC can be self-triggered by the anodic signal, thus making the device suitable for detection and localization of short-range and neutral radiation. An example of application for radio-chromatography, the two-dimensional activity distribution in a sliced tritium-labelled rat brain is shown in Figure 1.14 (Dominik *et al.*, 1989).

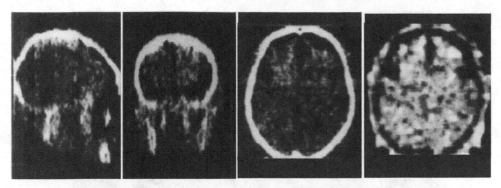

Figure 1.13   Nuclear scattering radiography: 3-D reconstruction of the density distribution on slices through a human head. The rightmost image corresponds to selected p-p elastic scatter events (Duchazeaubeneix *et al.*, 1980). By kind permission of Wolters Kluwer Health.

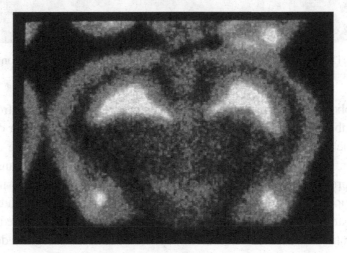

Figure 1.14   Radio-chromatography of a tritiated slice of a rat's brain (Dominik *et al.*, 1989). By kind permission of Elsevier.

Many variants of the two-dimensional MWPC have been developed for the needs of particle physics experimentation, and will be described in the next chapters, the most powerful being perhaps the time projection chamber developed by David Nygren and collaborators in the late 1970s (Nygren and Marx, 1978); for a review see for example Sauli (1992).

Introduced by Thomas Ypsilantis and Jacques Séguinot in the late 1970s (Seguinot and Ypsilantis, 1977), the Cherenkov ring imaging (RICH) technique is a particle identification method based on the detection and localization of UV photons emitted in a radiator by the Cherenkov effect with a gaseous counter.

Figure 1.15   Philippe Mangeot and Anna Peisert with one of the photon detectors of the E605 RICH at Fermilab. Picture by the author (1980).

Using a photosensitive gas filling, these detectors are prone to suffer from photon-
induced feedback problems when operated at the high gains needed for detection of
single photoelectrons, due to the copious emission of photons by the avalanches.
Developed in 1978 by Charpak and myself, the multi-step avalanche chamber (MSC)
solved the problem (Charpak and Sauli, 1978). The MSC combines in the same
device a region of high field between two semi-transparent meshes, and a standard
MWPC, separated by a low-field gap. A fraction of the electrons created in the first
multiplier transfers to the second and multiplies again, permitting the desired high
gain; with a proper choice of the photosensitive gas concentration, photons emitted
by the avalanches in the MWPC are absorbed before reaching the first amplifier and
do not induce the formation of secondary avalanches (Charpak *et al.*, 1979a).

Built by a CERN–CEN–Saclay collaboration, two UV-photon sensitive MSCs
mounted on a helium-filled radiator operated for several years as the Cherenkov
ring imaging particle identifier of the experiment E605 at Fermilab (Adams *et al.*,
1983); with a 3-D projective readout (the anodes and two sets of angled cathode
wires), the detector achieved ambiguity-free localization of multiple photoelectrons in a ring pattern. The picture in Figure 1.15 shows two members of the group
next to one of the UV-photons sensitive chambers; performances of the detectors
are discussed in Section 14.3.

The first generation of RICH detectors used vapours of triethyl amine (TEA)
added to the gas mixture as photosensitive agent; due to its ionization threshold in

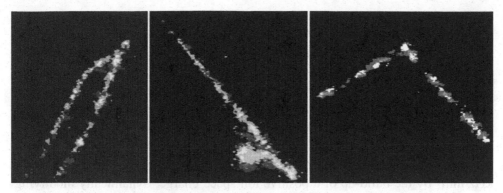

Figure 1.16   Cosmic rays recorded with the optical imaging chamber (Charpak *et al.*, 1987). By kind permission of Elsevier.

the far UV (7.5 eV), TEA requires the use of expensive fluoride windows to separate detector and radiator. A substantial improvement in the technology was made possible when David Anderson, who joined our CERN group in 1983, brought in samples of TMAE, a product with the amazingly low ionization threshold in the vapour phase of 5.4 eV; this permitted the use of quartz windows, and led to the construction of large acceptance RICH particle identifiers at CERN and elsewhere (see Section 14.4).

The research on low photo-ionization threshold vapours had an interesting spin-off: it was found that these compounds also act as efficient internal wavelength shifters, copiously emitting photons at wavelengths near to or in the visible range, easy to detect and image with optical means (see Section 15.1). In the imaging chambers, the ionized trails drift to an end-cap multiplier in conditions optimized to obtain a large scintillation yield; a solid-state camera records the projected images through the detector window (Charpak *et al.*, 1987). The detectors are continuously active, limited in rate capability only by the image acquisition hardware. Examples of cosmic ray activity recorded with an optical imaging chamber are shown in Figure 1.16. A chamber producing images visible to the naked eye, colloquially named 'Charpaktron', operated for some time in CERN's permanent exhibition Microcosm.[4]

Other applications of the optical imaging chambers include autoradiography of radioactive compounds (Dominik *et al.*, 1989) and detection of low-energy nuclear decays (Miernik *et al.*, 2007). As demonstrated in the last reference, the scintillation intensity is proportional to the ionization density, and the detector can be used for quantitative measurements of the energy loss as well as the interaction topology.

---

[4] Requiring several minutes of adaptation in the dark, the device was not a top hit with visitors, and was eventually discontinued.

Mass-produced in a variety of sizes and shapes, and integral components in many experimental setups, MWPCs have, however, some limitations intrinsic in their conception. The delicate stringing and the fragility of the thin anode wires affect the reliability; the production of various kinds of deposits on the electrodes in the charge multiplication processes results in a long-term deterioration of the detectors. More fundamentally, while localization accuracies of 50–100 µm can be achieved with a measurement of the drift time or of the cathode induced charge profiles, this goes at the expense of the multi-track resolution, which is around 10 mm at best. The rate capability of MWPCs is also limited to a few kHz/mm^2 by the build-up of a positive ion space charge, dynamically modifying the electric fields.

In 1988 at the Institute Laue-Langevin (ILL) in Grenoble, Anton Oed developed a new detector concept named micro-strip gas counter (MSGC), which promised to improve both the multi-track resolution and the rate capability by at least one order of magnitude (Oed, 1988). The MSGC structure consists of thin parallel metallic strips engraved on a glass substrate and alternately connected as anodes and cathodes. Strip widths are typically of 10 and 50 µm for anodes and cathodes respectively, at 100 µm distance; a drift electrode provides gas tightness and completes the detector. Thanks to the narrow pitch of the strips and to the fast collection of the majority of the positive ions by the closer cathode strips, the MSGC can operate efficiently at radiation fluxes above one MHz/mm^2, with localization accuracies and multi-track resolutions for fast particles around 50 µm and 500 µm, respectively (see Section 13.1).

The exceptional performances of the new device attracted considerable interest; I acted as spokesperson of the CERN-based international collaboration RD-28, approved in 1992 and aiming at the development of the MSGC technology.[5] Figure 1.17 shows one of the MSGC detectors, fully equipped with readout electronics, part of a setup built by the GDD group under my leadership, and used for systematic performance studies in high intensity beams (Barr *et al.*, 1998).

However, experience has shown that the detectors were rather fragile; because of the high electric field strength on the strips' edges, the avalanche charge occasionally grows large enough to exceed the so-called Raether limit and induce discharges (see Section 8.8). As an example, Figure 1.18 is a close view of a fresh, slightly pitted and seriously damaged MSGC plate after use. Despite a large effort dedicated to finding ways to improve the long-term reliability, with only few exceptions MSGC-based detectors were eventually discontinued for most applications.

---

[5] The research effort continued from 2008 to include other types of high-resolution detector, generically named micro-pattern gas detectors (MPGD) under the label RD-51.

Figure 1.17 A 10×10 cm^2 MSGC detector fully equipped with readout electronics (Barr *et al.*, 1998). By kind permission of Elsevier.

Figure 1.18 Close views of a MSGC plate before use, pitted by a moderate amount of micro-discharges and after a full discharge. Pictures by the author (Sauli, 1998).

The problems encountered with the microstrip detectors encouraged disappointed developers to seek for alternative devices. In 1997 I introduced a new concept, the gas electron multiplier (GEM) (Sauli, 1997), a development pursued by several groups and successfully used in many experiments (see Section 13.4). A GEM electrode is a polymer foil, copper-clad on both sides and with a high density of through-holes, typically 100 per square mm (Figure 1.19). On application of a high voltage gradient between the two sides, and in an appropriate gas environment, electrons released by ionization on one side of the electrode drift into the holes and multiply; most electrons in the avalanche exit the holes and transfer to a second element, another GEM foil or a printed circuit electrode with a pattern

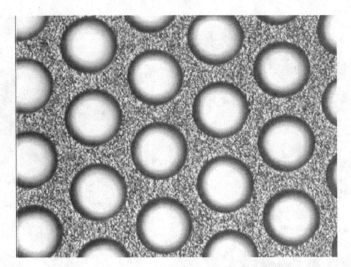

Figure 1.19   Close view of a GEM electrode; typical diameter and distance of the holes are 70 and 140 μm, respectively. Picture by the author (Sauli, 1997).

of strips or pads for charge collection and position readout. Not fortuitously, the structure bears a resemblance to the multi-step chamber described before, albeit at a miniaturized scale.

Offering performances comparable to those of the MSGC, the new device has distinctive advantages: a sturdy construction with a separation of the multiplication and signal pickup electrodes, thus minimizing the likelihood of damage due to discharges; the possibility of cascading several electrodes, in the so-called multi-GEM chamber, permits one to reach very high overall gains and a strong reduction of the positive ions backflow, a paramount feature in GEM-based time projection chambers (see Section 13.5). The readout electrode itself can be patterned at will, a common choice being two sets of perpendicular strips, typically at a few hundred microns pitch, to perform bi-dimensional localization of tracks (Bressan *et al.* (1999c)). An overall signal can be detected on the lower electrode of the last GEM in a cascade, thus providing an energy trigger for selection and recording of neutral radiation.

Manufactured originally with standard printed circuit tools, the early GEM foils had a surface of only a few cm^2; the technology has evolved both at CERN and in industry to satisfy the increasing demand for larger surfaces, reaching at the time of writing almost a square metre.[6] In Figure 1.20 I hold a 30×30 cm^2 GEM foil, one of a large production for the forward tracker of the COMPASS spectrometer at CERN that includes more than 20 medium-size triple-GEM

---

[6] The surface treatment group, led by Rui de Oliveira, has developed the technology used for the manufacturing of large size GEM electrodes at CERN.

Figure 1.20   The author with a large GEM foil used for the COMPASS tracker. Picture CERN (2001).

detectors (Altunbas *et al.*, 2002). Commissioned in 2002, the tracker was still in operation in 2012.

Owing to their reliability, fast response and high-accuracy localization properties, GEM-based detectors find numerous applications in particle physics, medical diagnostics, astrophysics and other fields (see Section 13.7).

Figure 1.21 is an example of X-ray absorption radiography of a small mammal recorded with a GEM detector (Bachmann *et al.*, 2001).

In an ideal continuance of the nuclear scattering radiography method, I contributed to the development of diagnostics instrumentation aimed at improving the quality assurance in hadrontherapy, the deep tumour treatment with high-energy ion beams. The proton range radiography (PRR) system relies on the measurement of the direction and residual range of particles at energies above full absorption in the patient, and provides a map of the integrated density in the body; in an alternative named interaction vertex imaging (IVI), the same instrument is placed at angles with the beam, and permits one to reconstruct the density of the interaction vertices. Both instruments use medium-size GEM detectors for the measurement of the beam or scattered particles direction (Amaldi *et al.*, 2011; Bucciantonio *et al.*, 2013).

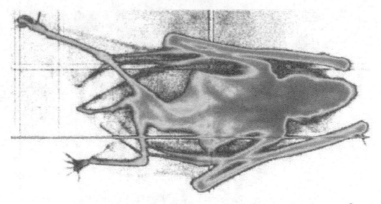

Figure 1.21   Soft X-ray radiography of a bat (image size 70×30 mm^2) recorded with a two-dimensional GEM chamber (Bachmann *et al.*, 2001). By kind permission of Elsevier.

1.3 Basic processes in gaseous counters

The physical processes governing the operation of gaseous counters are discussed extensively in the following chapters. As an introductory help to the novice reader, a short summary of the major events leading to the detection of ionizing radiation is presented here.

The process begins with the release in the gas of one or more ion–electron pairs, in a number and with a distribution that depend on the nature and energy of the radiation; the minimum energy loss required is of course the ionization potential of the concerned atom or molecule. Occasionally, the energy of the primary electrons is sufficient to further ionize the gas molecules, in a cascade of secondary interactions that stops when all the available energy loss is dissipated and the various yields reach thermal equilibrium. For charged particles, the distinction between primary and total ionization is paramount for understanding the detectors' performance. Table 1.1 provides typical values of the number of primary and total ion pairs produced in argon gas at normal conditions by various types of radiation. Needless to say, the larger the release, the more undemanding the requirements on the detector recording electronics; for example, α particles can easily be detected with ionization chambers having no gas gain.

Once released in the gas, electrons and ions may neutralize by mutual recombination or by collisions with the walls; subjected to an external electric field, they separate and migrate towards the electrodes of the counter. The field strength needed for separation depends on the primary ionization density and on the gas, but is typically of a few tens of V/cm in argon at NTP.

When separated, electrons and ions diffuse thermally in the gas volume, bouncing around as an effect of collisions with the molecules, with a global slow motion

Table 1.1 *Examples of ionization yields in argon at NTP of various kinds of radiation.*

Particle	Primary	Total
UV photon	1	1
1 keV X-ray	1	50
100 keV electron	1000 ip/cm	3000 ip/cm
1 GeV proton (minimum ionizing)	25 ip/cm	100 ip/cm
5 MeV α particle	~10^4	~3×10^4

Table 1.2 *Drift velocity w^+ (w^-) and transverse diffusion σ^+ (σ^-) of ions (electrons), for 1 cm drift in several gases (at NTP) and field values.*

Gas	Field (V cm^{-1})	w^+ (cm ms^{-1})	σ^+ (mm) (1 cm)	w^- (cm μs^{-1})	σ^- (mm) (1 cm)
Argon	100	1.7	0.22	0.24	2.8
	1000	17	0.07	0.45	0.58
CH_4	100	2.2	0.22	1.73	0.29
	1000	22	0.07	10.56	0.24
CO_2	100	1.1	0.22	0.08	0.22
	1000	11	0.07	0.73	0.08

in the direction of the field, named drift velocity. Due to their mass difference, the velocity for ions and electrons at a given field strength differs by several orders of magnitude; moreover, for electrons it strongly depends from the gas and field, as shown by the examples in Table 1.2.

The drift velocity of ions increases almost linearly with the field, up to very high values. On the contrary, for electrons, which can acquire energy from the external field between collisions, the dependence is more complex, often reaching a maximum at fields of a few hundred V/cm, and then saturating or decreasing, depending on the gas mixture. The diffusion has also a strong dependence on the field and gas; some values are given in the table, for 1 cm of drift.

By increasing the field strength, electrons acquire enough energy to induce inelastic processes in their collisions with the gas molecules; in argon at NTP the threshold for the appearance of ionizing collisions is around 10 kV/cm. As a result, new electron–ion pairs are formed, and the charge multiplication process continues in avalanche, with the fast electrons on the front and a tail of slower ions. The resulting voltage–current dependence has a characteristic exponential shape, as shown by the example in Figure 1.22 (Sharma and Sauli, 1992).

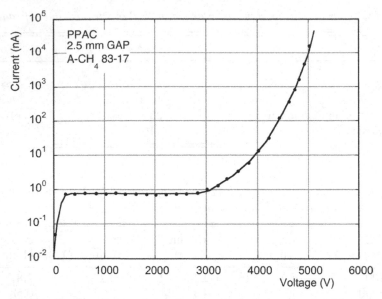

Figure 1.22   Characteristic voltage–current dependence measured in a gaseous counter. The gain, or multiplication factor, of the device is defined as the ratio of the current at a given voltage to the constant value before multiplication (Sharma and Sauli, 1992). By kind permission of Elsevier.

Within a few ns from the start of the multiplication process, all electrons in the avalanche have reached the anode; their collection contributes to the formation of the detected signal. The time taken by ions to reach the cathode, where they are neutralized, is tens to hundreds of μs, depending on the detector geometry and field strength. In their motion, particularly during the initial fast drift in the high field close to the anodes, ions induce charge signals in all electrodes; the (negative) charge induced by ions on the anode constitutes in fact the largest fraction of the detected signal.

Many other processes contribute to the response of a gas counter. In the presence of electro-negative pollutants, the most common being oxygen and water, electrons can be lost by capture. As the attachment cross sections depend on the electron energy, the capture probability is field-dependent; common practice shows that for efficient electron collection, fields above a few hundred V/cm may be needed.

In competition with the charge-amplifying ionizing collisions, atomic and molecular excitation can result in the emission of photons that can generate secondary electrons in the gas or at the electrodes, spreading the original avalanche. At very high gains, these processes can lead to the transition from the avalanche to a streamer, and in extreme cases induce a discharge. All these processes are discussed in detail in the next chapters.

1.4 Outline of the book

The technology of gaseous detectors has been in continuing and fast evolution, mostly thanks to the increasingly demanding requirements of particle physics experimentation. The phenomena describing the release, collection and multiplication of charges, basic in the operation of counters, have been studied extensively in the field referred to as gaseous electronics, and are discussed in numerous textbooks; the information on the newly developed devices is instead scattered in a large number of topical articles and conference proceedings. This book has been conceived as guidance to the field of gaseous radiation detectors, providing the essential data bibliography on their development and applications.

The first chapters describe the major phenomena providing detectable signals in gaseous counters, starting with the processes of energy loss for charged particles and for neutral radiation. The fate of the charges released in the gas is discussed then at increasing values of electric field, from the simple transport to the onset of inelastic collisions and charge multiplication. On each subject, simple approximate calculations based on classic theory are presented whenever possible, and the reader is referred to the more sophisticated and powerful computer programs developed to describe the various processes.

The following chapters then describe the evolution of detectors, from the simple parallel plate and single-wire counters to the more elaborated multi-wire and drift chambers and their siblings. The technologies aimed at realizing coarse very large area detectors and, conversely, smaller but very performing detectors are discussed in detail in the corresponding sections.

After the description of the many successful developments, the last chapter covers an unfortunate and still largely unsolved problem with all gaseous detectors, the deterioration of performances, or ageing, with the long-term exposure to radiation.

2

Electromagnetic interactions of charged particles with matter

2.1 Generalities on the energy loss process

The processes of transfer of part or all of its energy to a suitable medium, in which direct or indirect effects of the interaction can be recognized, mediate the detection of nuclear radiation. A variety of macroscopic mechanisms can be exploited for the conversion of the energy spent in the medium into a detectable signal: scintillation in fluorescent materials, chemical transformations intervening in photographic emulsions, condensation of droplets in saturated vapours or acoustic shock waves are just a few examples.

For charged particles, the largest fraction of energy dissipated in matter is due to electromagnetic interactions between the Coulomb fields of the projectile and of the molecules in the medium. Except for particles approaching the end of their range, where mechanical elastic collisions become relevant, the slowing down in matter is mainly due to multiple inelastic processes of excitation and ionization, whose probability is a function of the energy transfer involved. For fast electrons in condensed matter, Figure 2.1 gives an example of the dependence of the collision probability from the energy transfer. In the region between a few and a few tens of eV, the presence of atomic and molecular excitation levels with energy-dependent cross sections results in a rather complex structure: this is the region of distant collisions, since they involve a large impact parameter. Molecules can undergo radiationless rearrangements, dissociate or get excited or ionized, with the emission of photons or the appearance of free electron–ion pairs.

At increasing values of the energy transfer, the collision probability decreases exponentially with energy without particular structure up to the maximum cinematically allowed transfers, which depends on the projectile mass and energy. The outcome of these close collisions, involving increasingly small impact parameters, is the creation of excited species or the appearance of positive ions with the ejection of free electrons in the medium. Despite their low probability,

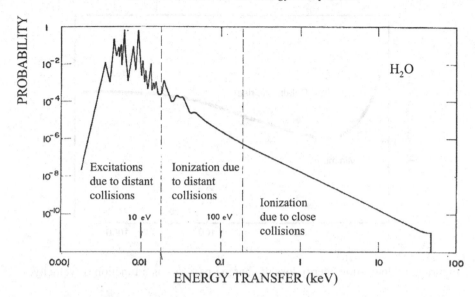

Figure 2.1   Collision probability of fast electrons in water as a function of energy transfer (Platzman, 1967). By kind permission of Elsevier.

large energy transfer yields (often named delta electrons) can further interact with the medium, and play a dominant role in determining the statistics of the energy loss process.

The number of electromagnetic collisions per unit length of material traversed, and therefore the resulting energy loss or stopping power, are a fast decreasing function of the particle velocity; on approaching the speed of light, the energy loss reaches a minimum and then slightly increases to a constant value, the so-named relativistic rise and Fermi plateau (Figure 2.2). For gases at moderate pressures, the increase can reach 40–50% above minimum, while it is reduced in condensed media.

Most of the considerations in the following sections refer to the ionization component of the energy loss; in gases, the yields of excitation processes, luminescence or scintillation photons, are usually too small in intensity to be exploitable for detection. This is not the case for heavily ionizing particles in high pressure and liquid rare gases, in which the primary scintillation is sufficiently intense to provide useful signals. Photon emission can also be enhanced by the presence of strong electric fields; the processes of primary and secondary photon emission are discussed in Chapter 5.

At very high particle energy, other mechanisms of electromagnetic interaction can occur: bremsstrahlung, coherent Cherenkov photon emission, and transition radiation. Except for electrons, for which bremsstrahlung is considerable even at low energy, these processes contribute little to the overall energy expenditure of

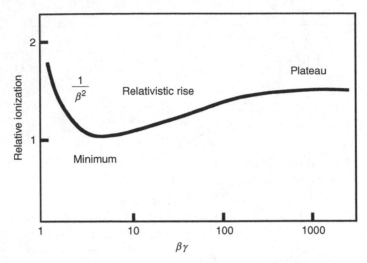

Figure 2.2   Ionization energy loss of charged particles as a function of velocity.

heavy charged particles; they can, however, be exploited for particle identification, through the detection and analysis of the angular distribution and energy spectra of emitted photons.

Electrons and photons created by the primary encounters can interact with the medium, releasing further excitations and ionizations; secondary mechanisms, particularly in composite materials, contribute to the overall photon or electron yield transferring part of the excitation energy into ionization or vice-versa. Of all outcomes of the energy loss processes, most gaseous detectors exploit the electrons created by the ionizing radiation; the presence of residual excited states, ions or photons is relevant only in that they may induce secondary phenomena, such as recombination, charge transfers and photoelectric effects. At low energies, the total specific ionization exceeds three or four times the primary, but this ratio decreases towards higher energies (Price, 1958).

Table 2.1 summarizes physical parameters useful to estimate the energy loss and ionization yields of fast charged particles in gases commonly used in proportional counters (Beringer, 2012). Data are provided at normal temperature and pressure (NTP, 20 °C and 1 atmosphere); appropriate scaling laws can be used for different conditions. The energy per ion pair W_I and the differential energy loss dE/dx refer to unit charge particles at the ionization minimum; they correspond to reasonable averages over existing data and should be considered approximate. The same comment applies to the number of primary and total ion pairs per unit length, N_P and N_T.

The number of primary ionizations, being an outcome of independent Coulomb interactions, follows a Poisson statistics:

Table 2.1 *Physical constants for various gases at NTP and approximate values of energy loss and ion-pair production (unit charge minimum ionizing particles).*

Gas	Density mg cm^{-3}	E_x eV	E_I eV	W_I eV	$dE/dx\|_{min}$ keV cm^{-1}	N_P cm^{-1}	N_T cm^{-1}
Ne	0.839	16.7	21.6	30	1.45	13	50
Ar	1.66	11.6	15.7	25	2.53	25	106
Xe	5.495	8.4	12.1	22	6.87	41	312
CH$_4$	0.667	8.8	12.6	30	1.61	37	54
C$_2$H$_6$	1.26	8.2	11.5	26	2.92	48	112
iC$_4$H$_{10}$	2.49	6.5	10.6	26	5.67	90	220
CO$_2$	1.84	7.0	13.8	34	3.35	35	100
CF$_4$	3.78	10.0	16.0	54	6.38	63	120

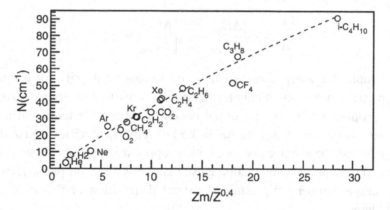

Figure 2.3   Primary ionizing collisions per cm as a function of atomic number of gases at NTP (Smirnov, 2005). By kind permission of Elsevier.

$$P_k^n = \frac{n^k}{k!} e^{-n},$$
(2.1)

where n and k are the average and actual numbers of pairs, respectively.

The theoretical detector efficiency, defined as the probability of having at least one interaction, is then:

$$\varepsilon = 1 - P_0^n = 1 - e^{-n}.$$
(2.2)

No simple expression exists for the number of primary ionizing encounters, and one has to resort to experimentally determined data or dedicated simulation programs. For fast singly charged particles, the specific primary ionization increases almost linearly with the average charge number of the medium, as shown in the compilation of Figure 2.3 (Smirnov, 2005).

The total number of ion pairs released in a medium in the absence of recombination or other secondary processes can be estimated from the expression:

$$N_T = \frac{\Delta E}{W_I},$$ (2.3)

where ΔE is the total energy loss in the material; the average energy per ion pair W_I varies between 20 and 40 eV for most gases (see Table 2.1) and depends little on the mass and energy of the ionizing particle.

In composite materials and for the gas mixtures used in proportional counters, a composition law based on the relative concentrations can be used with good approximation, neglecting interactions between excited species; the differential energy loss in a mixture of materials A, B, ... with relative mass concentrations p_A, p_B, ... is then given by:

$$\frac{\Delta E}{\Delta x} = p_A \left[\frac{\Delta E}{\Delta x}\right]_A + p_B \left[\frac{\Delta E}{\Delta x}\right]_B + \dots.$$ (2.4)

As an example, the average energy loss and ionization density for a relativistic charged particle in a gaseous counter filled with a mixture of argon–isobutane in the mass proportions 70–30, at normal conditions, from the table and using the appropriate composition laws, are $\Delta E = 3.5$ keV/cm, $N_P = 45$ ion pairs/cm, $N_T = 136$ ion pairs/cm; the average distance between primary ionizing collisions is about 220 μm, and each primary interaction cluster contains three ion pairs. These are of course average numbers; the actual statistical distribution of the yields will be discussed later.

2.2 The Bethe–Bloch energy loss expression

The energy loss processes due to multiple Coulomb interactions of charged particles have been subject of research since the original works of Rutherford on heavy particle scattering. In a semi-classical formulation, usually referred to as the Rutherford expression, the probability of a unit charge particle of velocity β to release an energy between ε and $\varepsilon + d\varepsilon$ in a layer of a material of thickness dx and density ρ can be written as:

$$\frac{d^2 N}{dx\, d\varepsilon} = K \frac{Z}{A} \frac{\rho}{\beta^2} \frac{1}{\varepsilon^2} \qquad K = \frac{4\pi N e^2}{mc^2}.$$ (2.5)

e and m are the charge and mass of the electron, Z, A and ρ the medium atomic number, mass and density, and N Avogadro's number; in the CGS system of units the rest mass of the electron $mc^2 = 0.511$ MeV and $K = 0.308$ MeV g^{-1} cm^2.

Expression (2.5) describes well the energy loss process of ions for intermediate velocities. Several corrections are, however, necessary both at low and very high velocities to obtain agreement with the experimental results (Fano, 1963; Northcliffe, 1963). In a general formulation the differential energy loss, or stopping power (Bethe–Bloch expression) is written as:

$$\frac{\Delta E}{\Delta x} = -\rho \frac{2KZ}{A\beta^2}\left[\ln \frac{2mc^2\beta^2}{I(1-\beta^2)} - \beta^2 - \frac{C}{Z} - \frac{\delta}{2}\right]. \tag{2.6}$$

The expression shows that the differential energy loss depends only on the particle's velocity β and not on its mass; the additional term C/Z represents the so-called inner shell corrections, that take into account a reduced ionization efficiency on the deepest electronic layers due to screening effects, and $\delta/2$ is a density effect correction arising from a collective interaction between the medium and the Coulomb field of the particle at highly relativistic velocities; its contribution is small for non-condensed media. It should be noted, however, that in thin absorbers electrons produced with high momentum transfer might escape from the layer, thus reducing the effective yield.

No simple analytical expression for the correction factors in expression (2.6) has been given; tables and compilations allow their estimate (Gray, 1963; Fano, 1963; Northcliffe, 1963). Alternatively, one can find tables and plots of energy loss for ions in both the low and intermediate (Williamson and Boujot, 1962; Ziegler, 1977) and the high energy regions (Trower, 1966). Web-based platforms permit one to compute stopping powers and ranges of charged particles in a wide range of materials and energies (Berger *et al.*, 2011). Expressing the material thickness in reduced units (length times density), the stopping power is around 2 MeV cm^{2}g^{-1} almost independently from the material, with the exception of very light materials, as shown in Figure 2.4 (Beringer, 2012).

Expressed as a function of momentum, the average energy loss depends on the mass of the particle; this can be exploited for particle identification, as discussed in the next section. An example is shown in Figure 2.5 for an argon–CO_2 mixture at atmospheric pressure (Allison *et al.*, 1974).

2.3 Energy loss statistics

The differential energy loss computed with the Bethe–Bloch expression or obtained from the described compilations represents only the average; event per event values fluctuate around the average, with a distribution that depends on the particle energy and the absorbing medium nature, thickness and conditions. The process is dominated by the statistics of emission of energetic delta electrons;

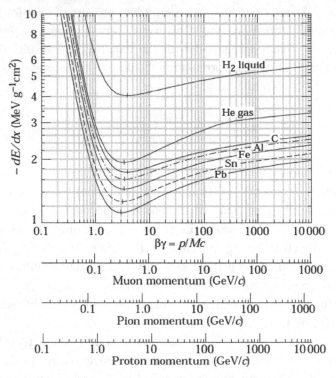

Figure 2.4   Differential energy loss as a function of velocity and momentum for singly charged particles in different materials (Beringer, 2012). By kind permission of the American Physical Society.

the simulation programs mentioned in the previous section can describe it in detail. However, it is instructive to use a simple formulation derived from Rutherford's approximation. Assuming that all energy in an interaction is imparted to a quasi-free electron, integration of expression (2.5) between ε_0 and ε_M (the maximum energy transfer) gives the probability of creating in a layer dx an electron of energy equal or larger than ε_0:

$$\frac{dN(\varepsilon > \varepsilon_0)}{dx} = W\left(\frac{1}{\varepsilon_0} - \frac{1}{\varepsilon_M}\right) \cong \frac{W}{\varepsilon_0}, \quad W = K\frac{Z}{A}\frac{1}{\beta^2}, \tag{2.7}$$

an approximation valid for $\varepsilon_M \gg \varepsilon_0$.

A comparison between the probability deduced from this expression and the result of a more sophisticated quantum-mechanical Monte Carlo calculation is shown in Figure 2.6 (Lapique and Piuz, 1980); the second shows clearly the contribution of the various electronic shells levels. For many practical purposes, however, the simpler formulation is often good enough.

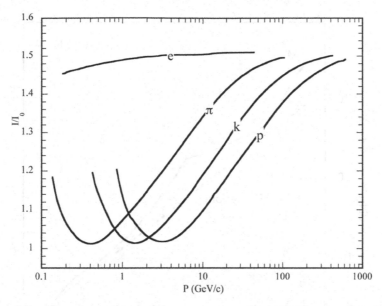

Figure 2.5   Mass dependence of the average specific ionization as a function of the particle momentum in the relativistic rise region (Allison *et al.*, 1974). By kind permission of Elsevier.

As an example, in one cm of argon at STP there is a ~5% probability of emission of an electron of energy equal to or larger than 2 keV; this has to be compared with the average energy loss of 2.4 keV/cm in the same conditions (see Table 2.1), meaning that in 5% of the events the observed energy loss is almost twice the average.

The production of energetic secondary (delta) electrons, with low probability but large ionization yields, determines the peculiar shape of the energy loss distribution; named the Landau expression, from the Russian physicist who studied the process in the forties, it can be written as:

$$f(\lambda) = \frac{1}{\sqrt{2\pi}} e^{-\frac{1}{2}(\lambda + e^{-\lambda})}, \tag{2.8}$$

where the energy variable λ represents the normalized deviation from the most probable energy loss $(\Delta E)_{MP}$:

$$\lambda = \frac{\Delta E - \Delta E_{MP}}{\xi}, \quad \xi = K \frac{Z}{A} \frac{\rho}{\beta^2} x.$$

For thin gas samples, the width of the energy loss distribution is close to the most probable value, with a characteristic asymmetric tail at large values. In Figure 2.7 (Igo *et al.*, 1952), values measured with a proportional counter are compared with

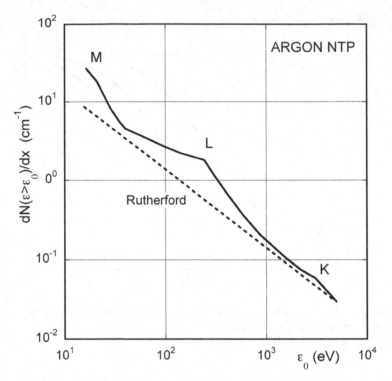

Figure 2.6   Number of electrons produced by ionization at an energy equal to or larger than ε_0; the full curve is a Monte Carlo calculation, the dashed line the prediction of Rutherford theory (Lapique and Piuz, 1980). By kind permission of Elsevier.

Landau's prediction, showing a good concordance; dedicated studies and refinements of the theory have since improved the agreement. In the figure, the dashed curve is the expected distribution for a Gaussian statistics, i.e. only determined by the fluctuations in the number of primary ionization encounters.

Figure 2.8 (Lehraus *et al.*, 1981) gives an example of energy loss distribution measured with a thin gaseous counter for particles of different mass (protons and electrons) and equal momentum. As expected, the most probable values differ by about 30%, but the distributions largely overlap due to the Landau energy loss statistics. To achieve particle identification, multi-sampling devices are used to measure many independent segments of the same tracks and combine them with appropriate statistical analysis; a truncated mean algorithm on 64 measured samples results in the distributions shown in Figure 2.9, with a good separation between protons and pions and, to a lesser extent, electrons (Lehraus *et al.*, 1981).

Motivated by the development of multi-wire chamber arrays for relativistic particle identification, systematic measurements have been made to find gas

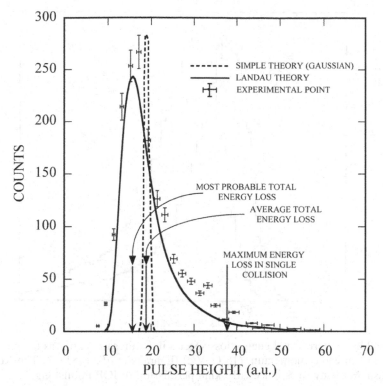

Figure 2.7   Comparison of experimental data (points with error bars) and Landau theory calculations of the energy loss in a thin sample of gas. The dashed curve represents the Gaussian expectation for the same average energy loss (Igo *et al.*, 1952). By kind permission of the American Physical Society.

mixtures with the best resolution. A compilation of results is given in Figure 2.10, providing the relative resolution (FWHM/most probable energy loss) in a wide range of gases and as a function of sample thickness (Lehraus *et al.*, 1982). As shown also in Figure 2.11, from the same reference, the best resolution is obtained with hydrocarbons; unfortunately, in light molecular gases the relativistic rise is smaller than for the noble gases, balancing the improvements due to resolution. In consideration also of their flammability and tendency to create deposits under irradiation (see Chapter 16), the use of hydrocarbons gas fillings has been basically abandoned, except in small percentages.

A model for the calculation of ionization losses of relativistic particles in thin absorbers, with comparison to experimental results, is given in Grishin *et al.* (1991).

Using an improved photo-absorption ionization model, Smirnov (2005) has computed the distribution of the energy loss of fast particles in a range of conditions and compared the results with experimental measurements; the program

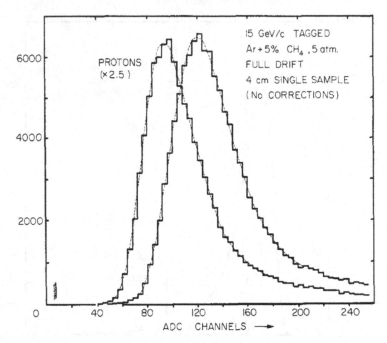

Figure 2.8   Experimental energy loss spectra in a thin sample of gas for protons and pions of equal momentum (15 GeV/c) (Lehraus *et al.*, 1981). © The Royal Swedish Academy of Sciences. By kind permission of IOP Publishing.

HEED, available on-line, allows one to compute the energy loss processes in a wide range of conditions (Smirnov, 2012). The agreement is excellent, as shown in Figure 2.12 and Figure 2.13, providing the energy loss for pions and electrons, in the region of the relativistic rise, as a function of the particle velocity and for several model calculations.

The probability of a primary ionization center consisting of several secondary ion–electron pairs, or cluster size, can be computed with the programs mentioned above, or directly measured; it is a fast decreasing function of the number of charges in the cluster, and depends little on the medium. Figure 2.14 shows measured values for argon and methane (Fischle *et al.*, 1991).

Attempts have been devised to exploit the primary ionization information, a method named cluster counting, that would allow one to improve considerably the particle identification resolution (Walenta, 1981). The experimental problem is to preserve the structure of the primary clusters during their drift to a collecting electrode, since the cloud is quickly smeared by diffusion. In the time expansion chamber (TEC), described in Section 9.5, this is partly achieved with a suitable optimization of the gas mixture having low drift velocities at moderate electric fields. The primary clusters distribution for relativistic particles,

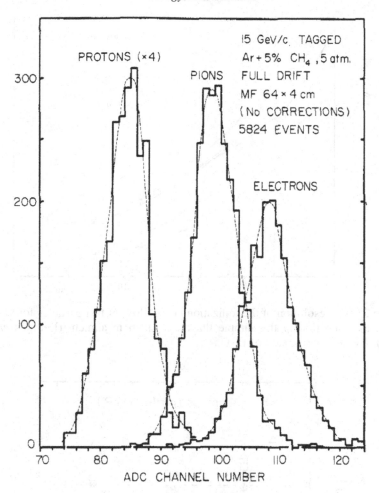

Figure 2.9 Particle identification resolution obtained from statistical analysis of energy loss measured on 64 track samples (Lehraus *et al.*, 1981). © The Royal Swedish Academy of Sciences. By kind permission of IOP Publishing.

measured with a TEC prototype, exhibits a Poisson-like shape, as against a Landau distribution that would result when recording the total ionization loss (Walenta, 1979).

By operating a drift chamber with helium-containing gas fillings to increase the distance between clusters, some efforts to improve on particle identification resolution at high energies have met a moderate success (Cerrito *et al.*, 1999).

Model calculations suggested, however, that the primary ionization increase in the relativistic rise region is only a fraction of the total, as shown in Figure 2.15, demonstrating that an important fraction of the relativistic rise is produced by the

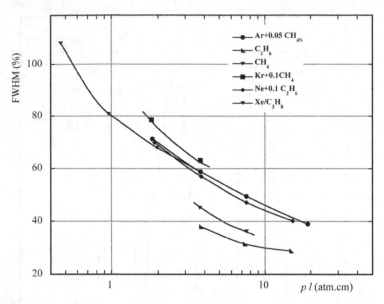

Figure 2.10   Resolution of the ionization energy loss of fast particles for several gases, as a function of the sample thickness given in atm.cm (Lehraus *et al.*, 1982). By kind permission of Elsevier.

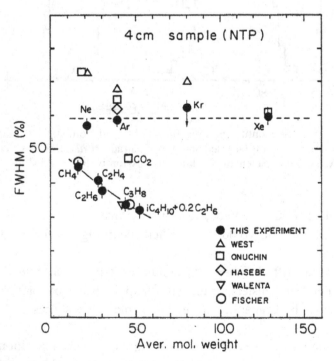

Figure 2.11   Relative ionization energy loss resolution in several gases in 4 cm gas samples (Lehraus *et al.*, 1982). By kind permission of Elsevier.

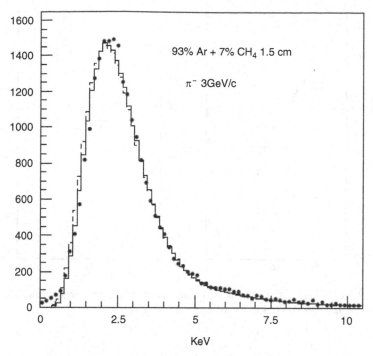

Figure 2.12   Comparison between measured and computed energy loss distributions (Smirnov, 2005). By kind permission of Elsevier.

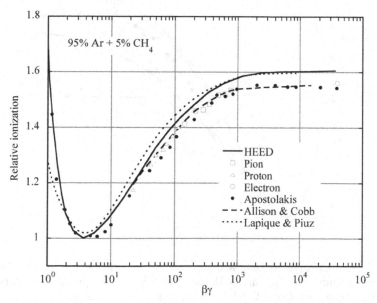

Figure 2.13   Relative most probable ionization loss measured and computed with several models (Smirnov, 2005). By kind permission of Elsevier.

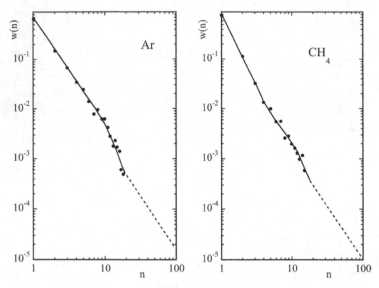

Figure 2.14   Cluster size probability for fast particles in argon and methane (Fischle *et al.*, 1991). By kind permission of Elsevier.

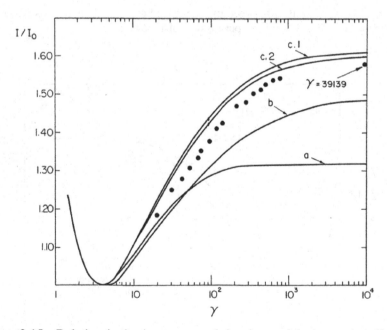

Figure 2.15   Relative ionization computed for the total losses under slightly different assumptions (c1 and c2), for the number of primary clusters (a) and the mean number of released electrons (b). The dots represent an experimental measurement. (Lapique and Piuz, 1980). By kind permission of Elsevier.

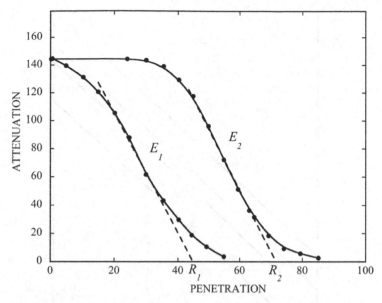

Figure 2.16   Definition of the practical electron range from attenuation curves.

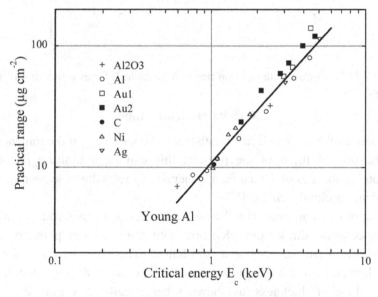

Figure 2.17   Practical electron range as a function of energy in several materials (Kanter, 1961). By kind permission of the American Physical Society.

secondary delta electrons produced by primary encounters; this reduces the possible advantages of the improved statistics and brings the expected resolution of the cluster counting technique close to the one of a simpler total ionization loss measurement (Lapique and Piuz, 1980).

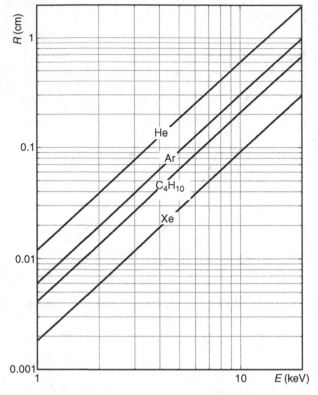

Figure 2.18   Approximate electron range in gases at NTP as a function of their energy.

2.4 Delta electron range

A consequence of the delta electron statistics is the smearing of the ionization trails outside the line of flight of the particle; this can be a limiting factor in the localization capabilities of detectors, in general only recording the average position of the ionization clouds (Sauli, 1978).

Due to multiple scattering with the gas molecules, slow electrons do not follow straight trajectories; the average distance from the emission point, or practical range, is shorter than the integrated path length. The practical range for a given energy is defined from the extrapolation to the abscissa of the attenuation curve at increasing absorber thickness, as shown schematically in Figure 2.16 for two mono-energetic beams of electrons. Expressed in reduced units, the practical range for slow electrons in light materials is almost independent of the element, as seen in the compilation of Figure 2.17 (Kanter, 1961).

For slow electrons (between 1 and 40 keV), a good approximation of the electrons' practical range in light elements is given by:

$$R = 10.0\,E^{1.7}, \tag{2.9}$$

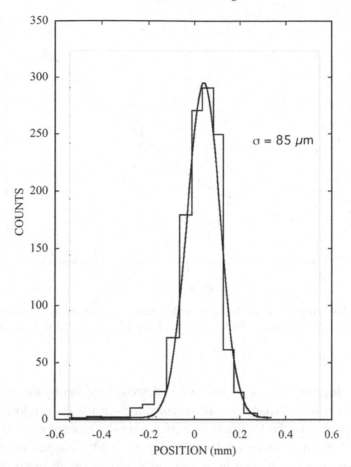

Figure 2.19   Localization accuracy in drift chambers, showing larger deviations at short drift times due to ionization produced by long-range delta electrons (Breskin *et al.*, 1974b). By kind permission of Elsevier.

with R in $\mu g\ cm^{-2}$ and E in keV (Kobetich and Katz, 1968). Figure 2.18 gives the electron range in several gases at NTP computed from the expression; in argon, a 2 keV electron has a practical range of 180 μm. As indicated in the previous section, an electron of at least this energy is produced in 5% of the events; for those tracks, a corresponding systematic shift of the measured position is expected, contributing to the tails in the distributions, compared with typical localization error of 50–100 μm achieved with high accuracy drift chambers (Breskin *et al.*, 1974b). An example is shown in Figure 2.19, providing the measured drift time distribution in the detector for tracks perpendicular to the drift direction; the asymmetric tail on the left side is due to the earlier arrival of the charge released by long-range delta electrons, and corresponds to about 5% of the events.

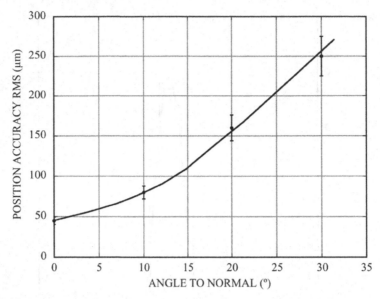

Figure 2.20   Angular dependence of the localization accuracy measured with the cathode induced charge method (Charpak *et al.*, 1979b). By kind permission of Elsevier.

A similar dispersive effect is observed in detectors exploiting the measurement of the cathode-induced charge profile; asymmetries in the energy loss result in a strong dependence of the localization accuracy on the incidence angle of tracks, see Section 8.9; an example is shown in Figure 2.20 (Charpak *et al.*, 1979b). The two dispersive effects add up in devices exploiting both the drift time and the induced charge measurements, as the time projection chambers (Chapter 10).

Use of heavier gases or higher pressures helps reduce the range of electrons of a given energy, but may be compensated by the increase in their number.

3

Interaction of photons and neutrons with matter

3.1 Photon absorption and emission in gases

Photons interact with matter through several electromagnetic processes, with a probability determined by their energy as well as by the density and atomic number of the medium. Unlike charged particles, which release energy all along their trajectory in a trail of ionizing collisions, photons interact with matter in single encounters, with or without the creation of secondary particles. Gases are basically transparent in the visible and near ultra-violet regions; absorption begins at shorter wavelengths, in the far and vacuum ultra-violet, when the photon energy exceeds the threshold for inelastic interactions with atoms or molecules. In atomic gases, the first ionization potential defines a sharp threshold for absorption, accompanied by the emission of a photoelectron; for molecules, the presence of mechanical excitation states can lead to radiationless absorption below the photo-ionization threshold.

From the onset of absorption and up to energies of a few tens of keV, the dominant interaction process is the photoelectric effect, resulting in the release of one or more free electrons and lower energy (fluorescence) photons. Gases become gradually more transparent at photon energies exceeding a few tens of keV, and are therefore not used directly as detection media; the region of sensitivity can be extended with the use of solid converters, optimized to allow the electrons released by the interactions to enter the gaseous device and be detected. In this chapter, the absorption and photoelectric processes are discussed in detail, while the processes occurring at higher energies (Compton scattering, pair production) are only briefly mentioned.

Equally important for understanding the gas counters operation are the processes of photon emission in gas discharges, followed by absorption in the gas itself or on the counter's electrodes, often resulting in the emission of secondary electrons or photons; this subject is covered in Chapter 5.

3.2 Photon absorption: definitions and units

The absorption of a mono-energetic beam of photons by a uniform layer of material of thickness x is described by an exponential law:

$$I = I_0\, e^{-\mu\rho x} = I_0\, e^{-\alpha x}, \qquad (3.1)$$

where I_0, I are the photon fluxes entering and leaving the layer, μ the mass absorption coefficient (in cm^2/g), ρ the density of the material (g/cm^3) and x is in cm; $\alpha = \mu\rho$ (in cm^{-1}) is the linear absorption coefficient, and represents the probability of interaction per unit length of absorber; its inverse, $\lambda = \alpha^{-1}$, is the absorption length.

Introducing in the expression the reduced material thickness $\chi = \rho x$ (in g/cm^2):

$$I = I_0\, e^{-\mu\chi}. \qquad (3.2)$$

The fraction of photons removed from the beam (the theoretical detection efficiency, if all interactions result in a visible signal) is then:

$$\varepsilon = \frac{I_0 - I}{I_0} = 1 - e^{-\mu\chi}. \qquad (3.3)$$

The linear absorption coefficient relates to the absorption cross section (in cm^2) through the expression:

$$\alpha = N\sigma, \qquad (3.4)$$

where N is the number of atoms or molecules per unit volume:

$$N = N_A \rho / A. \qquad (3.5)$$

In the expression, $N_A = 6.0247 \times 10^{23}$ molecules/g mole is the Avogadro number and A the atomic weight (in g/mole). For an ideal gas at STP (0 °C, 1 atm), $N = 2.687 \times 10^{19}$ molecules/cm^3; expressing the cross section in megabarn (1 Mb $= 10^{-18}$ cm^2):

$$\alpha_{STP}(cm^{-1}) = 26.87\, \sigma(Mb). \qquad (3.6)$$

The linear absorption coefficient in different conditions can be computed using a density-dependent scaling law:

$$\alpha = \frac{\rho_{STP}}{\rho}\alpha_{STP}, \qquad \rho = P\frac{273.14}{T}\rho_{STP}, \qquad (3.7)$$

with the pressure P in atmospheres and the absolute temperature T in °K.

For molecules, under the assumption that the constituent elements maintain their characteristics (chemical reactions can modify the optical properties at long wavelengths), the mass absorption coefficient can be deduced from the weighed sum of the individual cross sections:

$$\mu = \frac{N_A}{M} \sum n_i \sigma_i, \tag{3.8}$$

where n_i is the number of atoms of type i and $M = \Sigma n_i A_i$ the molecular weight.

The absorption coefficient for mixtures can be obtained as the sum of the values of components weighted by their mass fraction:

$$\alpha = \sum \mu_i \rho_i. \tag{3.9}$$

The absorption cross sections and derived quantities depend on photon wavelength. Values for elements and molecules, in a wide range of energies, are scattered in a myriad of articles, and summarized in numerous textbooks and compilations, from the ultra-violet (Marr, 1967; Robin, 1974; Berkowitz, 2002) to the X- and γ-ray energies (McMaster, 1969; White-Grodstein, 1957). Recent compilations are also available on-line (Berger *et al.*, 1998; Henke *et al.*, 1993; Thompson, 2004). A variety of units have been used in the literature to describe absorption processes, generally with the purpose of providing values not too dependent on atomic number and density; some are summarized here (units are given in the GGS system):

the absorption length $\lambda = 1/\alpha$ in cm;
the mass absorption coefficient μ (often improperly called cross section) in cm^2/g;
the absorption cross section σ, in cm^2, barn (1 barn $= 10^{-28} m^2$) or Megabarn
   (1Mb $= 10^{-18}$ cm^2). In view of expression (3.4), the cross section is often
   named 'per atom' or 'per molecule';
the molar extinction coefficient η, in litres/mole cm, defined through the
   expression:

$$\eta = \frac{M}{x} \log_{10} \frac{I}{I_0}, \tag{3.10}$$

where M is the molar volume (litres/g mole) and x the thickness of the absorber (in cm). The extinction coefficient relates to the absorption coefficient through the expression:

$$\eta = M\alpha \log_{10} e. \tag{3.11}$$

For an ideal gas at STP, $M = 22.42$ litres/g mole; combining the previous expressions:

$$\sigma(\text{Mb}) = 3.826 \times 10^{-3} \, \eta. \tag{3.12}$$

The absorption cross sections, and therefore all other quantities, are a function of the photon wavelength λ (or of its energy E). Throughout this section we have preferred to use the photon energy as variable; the corresponding wavelength can be derived from:

$$\lambda(\text{nm}) = 1240/E(\text{eV}). \tag{3.13}$$

3.3 Photon absorption processes: generalities

As an example, Figure 3.1 shows, in a wide range, the energy dependence of the absorption cross section of photons in tungsten, redrawn from data in Thompson (2004) between 10 and 1000 eV, and from White-Grodstein (1957) for higher energies.

For low to intermediate energies above the material's ionization threshold photoelectric absorption dominates, with distinctive jumps when the photon energy exceeds the various electronic shell levels. The result of the interaction is the emission of a photoelectron with a characteristic energy equal to the difference between the initial and the shell energy, accompanied by a cascade of lower energy photons or electrons resulting from the return of the atom to the ground state.

Towards higher energies, the absorption process is gradually dominated by Compton scattering, the diffusion of the incoming photon by a quasi-free electron in the atom; in coherent (or Raleigh) scattering the atom is left in the ground state, while for incoherent Compton scattering the outcome is the appearance of a lower energy photon and of an electron, emitted in a wide angular range with an energy sharing determined by conservation laws.

When the photon energy exceeds twice the mass of the electron (1.02 MeV), the process of pair production in the nuclear field of the material opens up and

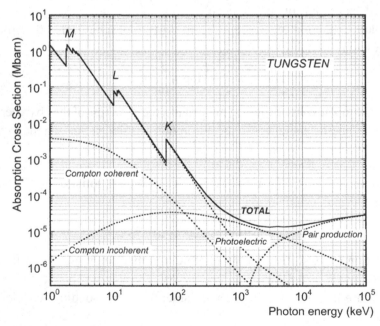

Figure 3.1   Photon absorption cross section for tungsten. Data from Thompson (2004) and White-Grodstein (1957).

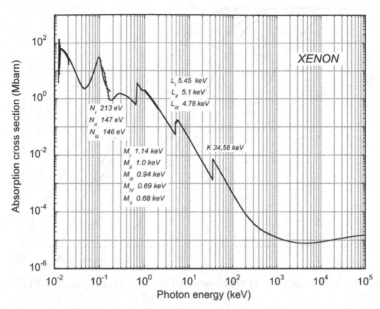

Figure 3.2   Photon absorption cross section for xenon. Data from McMaster (1969); Henke *et al.* (1993); Massey *et al.* (1969).

becomes gradually dominant, resulting in the disappearance of the photon and the creation of an electron–positron pair; appropriate energy and momentum conservation laws determine the correlations between angles and energies of the pair.

For gaseous detectors, only the domain extending from the near ultra-violet to the hard X-rays is of practical concern. For atomic gases, in this region and above the ionization potential, the dominant process is photo-ionization; Figure 3.2 provides the absorption cross section for xenon between 10 eV and 1 MeV, compiled from different sources (McMaster, 1969; Henke *et al.*, 1993; Massey *et al.*, 1969). A slight mismatch is seen between the sets of data coming from different sources.

Xenon is transparent to photons up to the first ionization potential (12.1 eV); above, the cross section reaches a maximum and decreases with upward jumps when crossing the energy levels of the various electronic shells. The region between 10 and 100 eV corresponds to the near, far and vacuum ultra-violet domain, while the X-ray region begins above 100 eV.

Knowledge of the absorption cross section for a given material permits one to estimate, using expressions (3.1) to (3.3), both the losses in the detector windows and the theoretical maximum efficiency of photon detection as a function of wavelength. As an example, Figure 3.3 shows the computed detection efficiency of a 1 cm thick counter filled with several gases at STP. Modulated by the onset of the absorption edges, it decreases from 100% at low photon energy to a few per cent

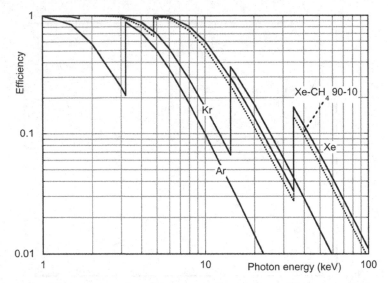

Figure 3.3   X-ray conversion efficiency in 1 cm of several gases at STP.

at shorter wavelengths. Addition to noble gases of organic quenchers, due to their smaller cross sections, results in a reduction of efficiency corresponding to an equivalent decrease of the effective mass of the main component, as shown for the Xe-CH_4 mixture.

At increasing energies, the fast decrease in conversion efficiency of gases can be partly compensated by the use of heavier molecules, thicker volumes or higher pressures. Alternatively, solid converters can be placed in front of the active volume, detecting the products of the interaction in the gas. For tungsten at 511 keV, the photon energy relevant in positron emission tomography (PET) imaging, the linear absorption coefficient is about 3 cm^{-1}, almost equally shared between photoelectric and Compton (Figure 3.1). From expression (3.2), one can infer that in a 100 μm tungsten foil about 3% of the photons interact; this is the upper limit for efficiency, assuming that all interactions result in the release of detectable ionization in the gas. As the thickness of the foil increases, so does the efficiency of conversion, but the electron ejected by photoelectric or Compton effect has a fading chance of emerging into the gas and being detected. To achieve a high conversion efficiency and at the same time a good charge extraction probability, multi-layer stacks of thin converters with open gas channels designed to collect the ionization charge have been developed for the so-called heavy drift chamber, see Section 3.8 (Jeavons *et al.*, 1975).

The occurrence of a specific absorption process affects the response of detectors. When the conversion is photoelectric, the emitted electron carries a well-defined fraction of the incident photon energy, thus permitting its wavelength to be deduced. In Compton scattering, on the other hand, the ejected electron energy depends on its emission angle, generally unknown; at best, a detector can localize the primary interaction point, but has no energy resolution. Further interactions within the detecting medium of the scattered photon complicate the issue.

3.4 Photon absorption in gases: from the visible to the near ultra-violet domain

Gases are transparent below their absorption threshold, but owing to the fast increase and large value of the cross section they quickly become opaque above; for many practical purposes, the knowledge of the absorption curve for a known thickness, such as those shown in Figure 3.4 (Lu and McDonald, 1994), is sufficient to determine their role in the counter action.

In atomic gases, absorption occurs above a well-defined threshold corresponding to the lowest ionization potential (the energy of the valence electron), and always results in the emission of a photoelectron. The process of photo-absorption is associated with the jump of an orbital electron to a higher energy state, followed by internal rearrangements bringing the atom back to the ground state with the emission of lower energy photons or electrons. When the energy of the absorbed photon exceeds the ionization potential of the atom or molecule, the outcome of the interaction can be the ejection in the medium of a free electron; the atom or

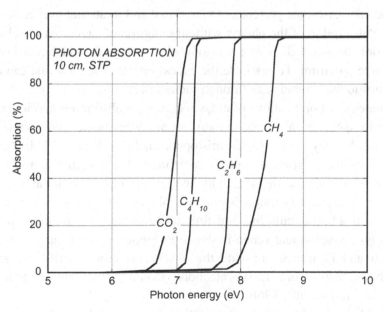

Figure 3.4 Photon absorption curves for several gases (Lu and McDonald, 1994). By kind permission of Elsevier.

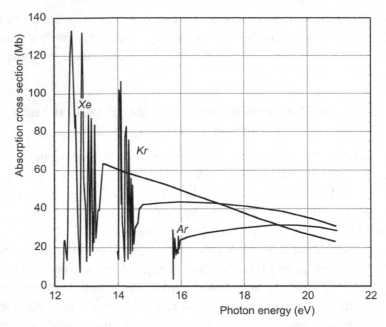

Figure 3.5   Absorption cross sections for argon, krypton and xenon. Data from Huffman *et al.* (1955).

molecule is left in an excited state, and returns to the ground state through a single transition or a cascade of processes.

In molecular gases, the presence of vibrational and rotational excited states may lead to the absorption of the photon without emission of radiation; the absorption cross section begins well before the ionization potential, and has a broad and structure-free spectrum. To dissipate the excess energy, the molecule can dissociate or return to the ground state through radiationless processes.

Comprehensive compilations of cross sections for absorption and ionization in atomic and molecular gases can be found in many textbooks and review articles (Marr, 1967; Massey *et al.*, 1969; Christophorou, 1971; Robin, 1974; Berkowitz, 2002). Values from different sources sometimes disagree; this may be due to different methods of measurement, or to the presence of trace pollutants in the gas.

For atomic gases, absorption begins sharply at the ionization potential, and continues with a broad multi-line structure until reaching a continuum; the behaviour for argon, krypton and xenon is shown in Figure 3.5 (Huffman *et al.*, 1955). Figure 3.6 and Figure 3.7 provide the absorption cross section for saturated hydrocarbons (Au *et al.*, 1993), alcohols (Robin, 1974) and dimethyl ether (DME) (Calvert and Pitts, 1966).

Figure 3.8 is the absorption cross section for water (Watanabe and Zelikoff, 1953), oxygen (Friedman, 1960), ozone (Tanaka *et al.*, 1953); Figure 3.9 for

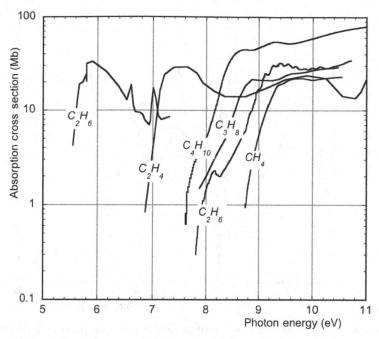

Figure 3.6   Absorption cross sections for hydrocarbons. Data from Au *et al.* (1993). By kind permission of Elsevier.

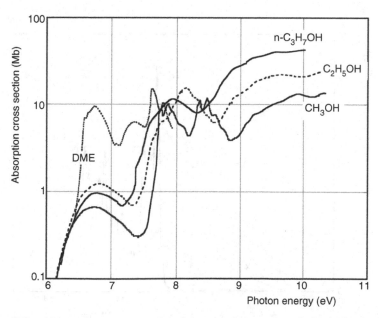

Figure 3.7   Absorption cross section for several vapours and alcohols. Author's compilation from data in different sources: dimethyl ether (Calvert and Pitts, 1966) and alcohols (Au *et al.*, 1993).

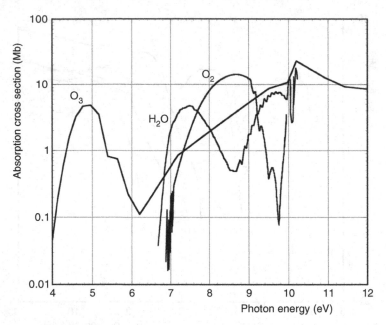

Figure 3.8   Absorption cross section for several gases. Compilation from various sources: water (Watanabe and Zelikoff, 1953), oxygen (Friedman, 1960) and ozone (Tanaka *et al.*, 1953). By kind permission of the American Optical Society.

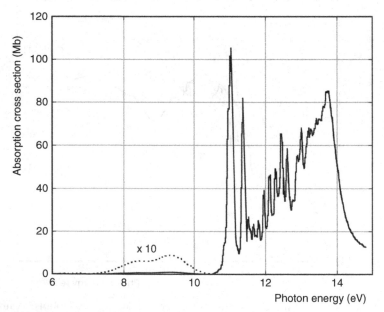

Figure 3.9   Absorption cross section for carbon dioxide (Chan *et al.*, 1993a). By kind permission of Elsevier.

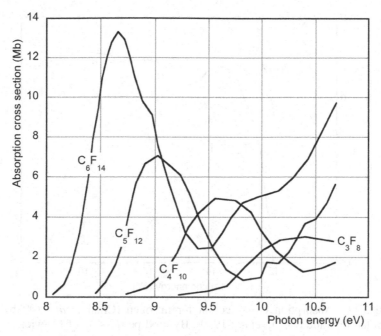

Figure 3.10 Absorption cross sections for fluorinated carbon compounds (Bélanger *et al.*, 1969). By kind permission of Elsevier.

carbon dioxide (Chan *et al.*, 1993a). Of particular relevance are the values for water and oxygen, the most common pollutants in a detector; ozone is produced in the avalanches and can affect the performances at very high rates due to its electronegativity. Figure 3.10 shows the absorption cross section in gaseous fluorinated carbon compounds (Bélanger *et al.*, 1969); freons are used as additives in detectors as well as radiators in Cherenkov ring imaging.

Figure 3.11 provides the absorption cross section for carbon tetrafluoride (CF_4) (Zhang *et al.*, 1989) and nitrogen (Chan *et al.*, 1993b).

By mixing species, one can cover a wide region of photon absorption, extending well into the near ultra-violet, and design a gas filling effectively quenching photon-mediated propagation processes; such mixtures are currently used to achieve high stable gains in proportional counters, and more recently in spark and resistive plate counters (see Chapter 12).

3.5 Photo-ionization: near and vacuum ultra-violet

Gaseous counters have only sporadically been used for detection of low energy photons, due to the limited wavelength coverage between the ionization potential and the cut-off of available detector windows. The field has been revitalized by the

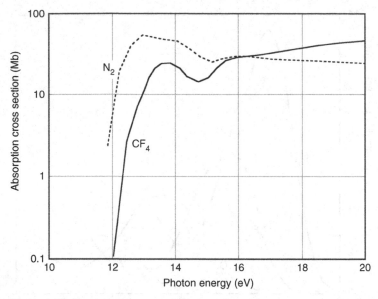

Figure 3.11   Absorption cross section for nitrogen (Chan *et al.*, 1993b) and carbon tetrafluoride (Zhang *et al.*, 1989). By kind permission of Elsevier.

development of particle identification methods based on the imaging of the pattern generated in suitable radiators by the Cherenkov effect (see ring imaging Cherenkov counters, Chapter 14); this has led to an intensive search of vapours with low ionization potential, which added to a main gas constituent extend the spectral response and therefore the detection efficiency. Absorption and photo-ionization cross sections of vapours used for UV photon detection are shown in Figure 3.12 for triethyl-amine, $(C_2H_5)_3N$ (TEA) and Figure 3.13 for tetrakis dimethyl amino ethylene $C_2[(CH_3)_2N]_4$ (TMAE) (Holroyd *et al.*, 1987).[1]

For completeness, although not used in a vapour phase, Figure 3.14 gives the quantum efficiency of caesium iodide, employed as thin-layer internal photo-cathode in several detectors, in particular for Cherenkov ring imaging (RICH) devices (Séguinot *et al.*, 1990).

Not only can photons from external sources be converted into charge and detected, but the development and spread of the signals in gaseous counters can be largely dominated also by photon emission and absorption processes. At high electric fields, photons in the ultra-violet region are copiously emitted by the inelastic collisions of electrons with atoms and molecules; depending on their wavelength, they can be absorbed or reconvert into secondary electrons, both in the gas and on the detector electrodes, spreading the original charge and often

[1] Cross section values and detailed shapes of the curves depend on the source.

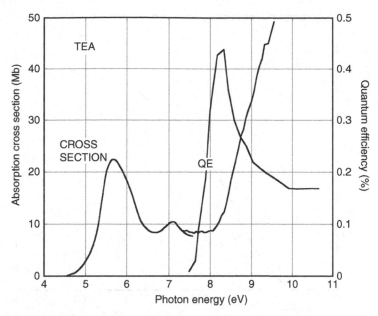

Figure 3.12   Absorption cross section and photo-ionization efficiency for triethyl amine (TEA) (Holroyd *et al.*, 1987). By kind permission of Elsevier.

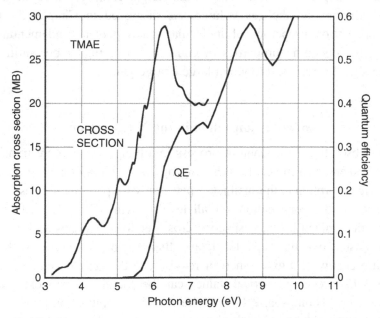

Figure 3.13   Absorption cross sections and photo-ionization efficiency for TMAE (Holroyd *et al.*, 1987). By kind permission of Elsevier.

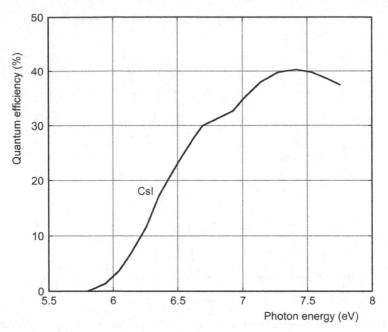

Figure 3.14   Quantum efficiency of caesium iodide (Séguinot *et al.*, 1990). By kind permission of Elsevier.

leading to discharge. An extreme case is the Geiger counter, where the discharge propagates through all the gas volume. As discussed in Chapter 5, the non-radiative absorption properties of molecular gases play a fundamental role in controlling the operating properties of counters, in particular permitting one to reduce the signal spread and to attain large charge gains.

3.6 Photo-ionization in the X-ray region

The X-ray absorption is a quantum process involving one or more transitions in the electron shells of a molecule. Denoting by E_j the binding energy of a shell j, photoelectric absorption in the shell takes place only for photon energies $E_0 > E_j$; at a given energy, the contributions of all levels having $E_j < E_0$ add up with a probability determined by the respective cross sections. The absorption reaches a maximum just above the shell level, and then rapidly decreases with energy. The binding energy of a given shell increases with the atomic number, as shown in Figure 3.15; precise numerical values can be found in textbooks and tables (McMaster 1969; Thompson, 2004). Figure 3.16 is a compilation photo-ionization cross section for noble gases in the X-ray energy range from one to one hundred keV; Figure 3.17 shows the cross section for some molecular gases used in a

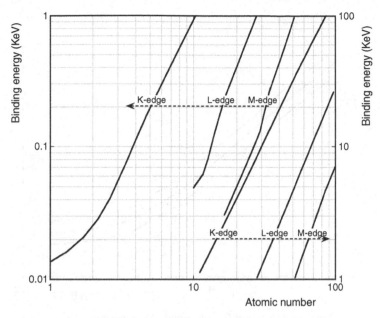

Figure 3.15   Binding energy as a function of atomic number.

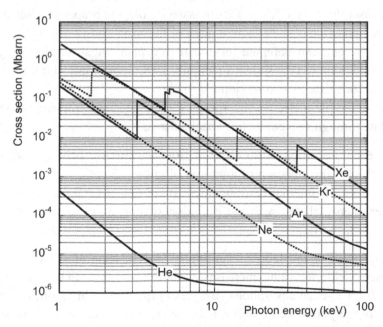

Figure 3.16   Photon total cross section for noble gases.

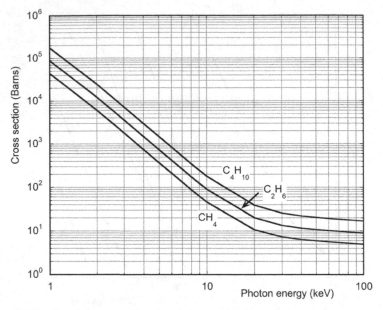

Figure 3.17   Total cross section for saturated hydrocarbons.

proportional counter, computed using the data for carbon and hydrogen in the quoted tables and expression (3.8). As already mentioned, in the mixtures of a noble gas and a hydrocarbon quencher used in gaseous counters, the contribution of the additives to photon absorption in the X-ray domain is negligible, but proper account should be taken of the reduction of the effective mass of the main constituent.

From (3.6), one can compute the mean free path for absorption in gases at standard conditions (STP), shown in Figure 3.18; appropriate composition rules can be used for mixtures (expression (3.9)).

Following the photoelectric absorption, the excited molecule can return to the ground state through two competing mechanisms:

fluorescence: transition of an electron from an energy shell $E_i < E_j$ into the j-shell, with the emission of a photon of energy E_j - E_i;

Auger effect: an internal rearrangement involving several electrons from the lower energy shells, with the emission of one or more electrons of total energy close to E_j.

The fraction of de-excitations resulting in the emission of a photon is named fluorescence yield; its value increases with the atomic number, as shown in Figure 3.19 for K-edge absorptions (Broyles *et al.*, 1953). In argon, for example, about 8% of the photoelectric absorptions are accompanied by the emission of a photon, while in 92% of the events, together with the photoelectron, one or more

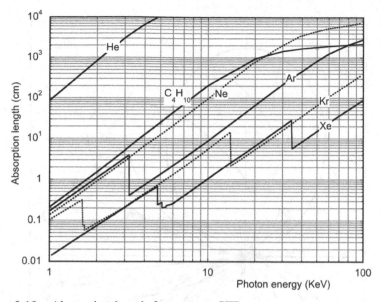

Figure 3.18   Absorption length for gases at STP.

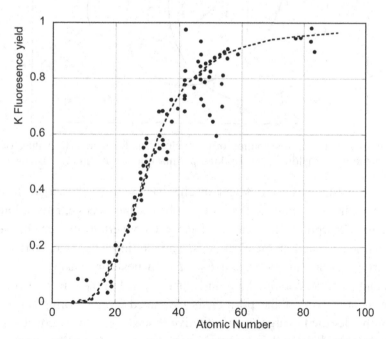

Figure 3.19   K-edge fluorescence yield as a function of atomic number (Broyles *et al.*, 1953). By kind permission of the American Physical Society.

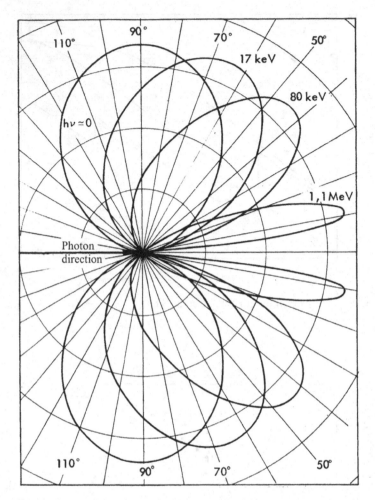

Figure 3.20   Angular distribution of photoelectrons for increasing values of the photon energy (Amaldi, 1971). By kind permission of the author.

electrons are produced by the Auger mechanism, with total energy close to E_k. Fluorescence photons, emitted at energy just below the K-edge, have a long mean free path for absorption; they can be locally reconverted, or flee the detection volume and be absorbed on the electrodes. This produces the characteristic escape peak of argon, at an energy around $E_0 - E_k$. In position-sensitive detectors the emission and reconversion of a long-range photon within the sensitive volume can introduce a large error if the position is estimated by measuring the centre of gravity of the detected charge. A quantitative discussion on this effect for xenon-filled counters can be found for example in Bateman *et al.* (1976).

Depending on its energy, the primary photoelectron is emitted in a preferential direction, as shown in Figure 3.20 (Amaldi, 1971); up to a few tens of keV, the direction of emission is centred around the perpendicular to the incoming photon direction, while at increasing energy the emission points towards the forward direction. As discussed in Chapter 2, however, multiple scattering on the gas molecules quickly randomises the motion of the heavily ionising photo-electron, and spreads the released charge around the interaction point. The range of electrons in gases was also discussed in the previous chapter.

Similarly to the case of charged particles (Section 2.1), the number N of electron–ion pairs released in a gas by a converted X-ray can be estimated by dividing the total energy loss (the photon energy in case of a complete absorption) by a phenomenological quantity W_i, energy per ion pair, whose value depends little on the gas, energy and nature of the primary ionizing particle;

$$N = \frac{E_x}{W_i}. \tag{3.14}$$

However, while for charged particles the statistics of the energy loss is dominated by the Landau fluctuations, due to rare but very energetic encounters, for X-rays the constraint imposed by a maximum energy loss that cannot exceed the photon's total available energy modifies the statistics over a simple $\sqrt{N}$ dispersion. The process, first described in the early sixties (Fano, 1963) results in a reduction of the fluctuations, and is expressed by a gas-dependent quantity called the Fano factor; the statistical fluctuation in the number of ion pairs is then written as:

$$\sigma_N = \sqrt{FN} \tag{3.15}$$

with $F \leq 1$. Theoretical values of the Fano factor have been computed for noble gases and their mixtures, and measured for a wide range of other gases (De Lima *et al.*, 1982; Dias *et al.*, 1991; Doke *et al.*, 1992; Pansky *et al.*, 1993; do Carmo *et al.*, 2008); its value is particularly low in Penning mixtures, where a fraction of the energy loss spent in excitations can reconvert into ionization. As the values of W_i and F are correlated, they are usually estimated together. Table 3.1 is a

Table 3.1 *Average energy per ion pair and Fano factor for some gases (Sipilä, 1976).*

Gas	W_I (eV)	F (theory)	F (exp.)
Ne	36.2	0.17	
Ar	26.2	0.17	
Xe	21.5		≤ 0.17
Ne+0.5% Ar	25.3	0.05	
Ar+0.5% C_2H_2	20.3	0.075	0.09
Ar+0.8% CH_4	26.0	0.17	0.19

representative summary (Sipilä, 1976); values can differ sometimes substantially depending on sources.

As will be discussed in Section 7.5, the value of F sets a statistical lower limit for the achievable energy resolution of counters, and has therefore extensively been studied in the development of high resolution X-ray detectors, and in particular the gas scintillation proportional counter (Policarpo, 1977).

3.7 Compton scattering and pair production

For photon energy above the highest electron shell level, Compton scattering becomes the dominant process. Quantum-mechanical theory describes the differential cross section and angular distributions of the interaction, see for example Evans (1958) and Davisson and Evans, (1952). The incident photon with energy E_0 is scattered by a quasi-free electron at an angle θ and energy E_1, while the electron emerges with an energy $E_0 - E_1$ at an angle ϕ, related by the expressions:

$$\frac{1}{E_1} - \frac{1}{E_0} = \frac{1 - \cos\theta}{mc^2}, \tag{3.16}$$

$$\cot\phi = \left(1 + \frac{E_0}{mc^2}\right)\tan\frac{\theta}{2}. \tag{3.17}$$

Except for very thick and dense detectors, it is unlikely for the scattered photon to be absorbed again in the same counter; the energy deposit depends therefore on the (generally unknown) angle of scattering, and no energy resolution is achievable in the Compton region.

The direction of the photon can be determined with a method named double Compton scattering. With reference to Figure 3.21 (left), the incident photon interacts in a first position-sensitive detector, recording the energy and coordinates of the recoil electron; the scattered photon interacts again in a following detector, providing energy and coordinates of the second Compton electron. Combining the previous expressions for the double scatter event, one can deduce the angle φ between the primary and scattered photons:

$$\varphi = \sin^{-1}\left(1 - \frac{m_0c^2}{E_2} + \frac{m_0c^2}{E_1 + E_2}\right); \tag{3.18}$$

the source has to lie then on a circle subtending an angle φ from the line joining the two scattering points. For a point source, the intersection of two or more circles determines its position (Figure 3.21 right).

The double Compton scattering method is exploited in astrophysics to determine the location of gamma sources, and in medicine for single photon emission

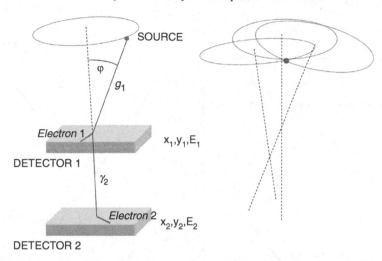

Figure 3.21   Principle of the double Compton scattering method.

tomography (SPECT) imaging; normally implemented with high density solid state detectors to achieve good efficiency, it is used also with liquid noble gas detectors (Section 15.2).

Electron–positron pair production can take place by an interaction in the nuclear field of materials at photon energies above the threshold energy of 1.02 MeV, corresponding to twice the electron mass; at high incident energies, the electrons and positrons are emitted mostly in the forward direction.

Both for Compton scattering and pair production, the probability of interaction in a gas counter is generally too small to be exploited for detection, but can take place on the counter walls or on suitably added converters. Electromagnetic shower counters are either constructed sandwiching chambers with conversion plates, or by using high-density drift chambers with suitable conversion electrodes in the gas volume itself.

3.8 Use of converters for hard photons detection

Due to the fast decreasing cross section, gaseous devices are not useful for detection of photons above a few tens of keV. Various methods of enhancing the efficiency at higher energies have been devised, making use of internal thin converters, particularly in view of applications in portal imaging and positron emission tomography. The conversion can be achieved with a stack of insulated metallic grids at suitable potentials, as in the high-density drift chamber (Jeavons *et al.*, 1975), with stacks of multi-wire proportional chambers having high-Z cathodes (Bateman *et al.*, 1980), sets of resistive plate chambers (Blanco *et al.*, 2003) or multi-electrode structures using gas electron multiplier (GEM) detectors (Iacobaeus *et al.*, 2000).

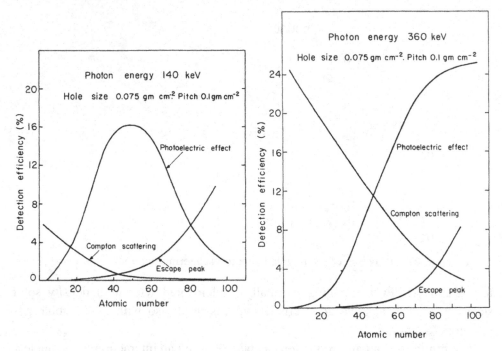

Figure 3.22   Detection efficiency of the heavy drift chamber as a function of the converter's atomic number, computed for 140 and 500 keV photons (Jeavons and Cate, 1976). © IEEE Transactions on Nuclear Sciences.

Depending on the converter material and photon energy, the detectable yield is an electron created by photoelectric effect or Compton scattering. While the conversion efficiency increases with the electrode thickness, the probability of the electron reaching the gas and being detected decreases; for each material and photon energy there is therefore an optimum thickness of the converter above which the efficiency remains constant.

Due to the higher ejected electron energy and the absence of scattered photons, which can reconvert in the detector, best performances are obtained using a material with a high photoelectric to Compton ratio. Figure 3.22 (Jeavons and Cate, 1976) shows the computed detection efficiency of a high-density drift chamber as a function of atomic number, for 140 and 500 keV, resulting from the K-shell photoelectric effect and Compton scattering. The curve labelled 'escape peak' corresponds to the background due to the de-excitation of the K-shell from a first photoelectric interaction. For the highest photon energy, a good compromise is tungsten ($Z = 74$), used as converter in the high-density drift chambers developed for positron emission tomography applications (Jeavons, 1978).

Stacks of multi-wire proportional chambers (MWPCs) with thin (125 μm) lead cathodes have been use for a similar purpose; Figure 3.23 gives the measured

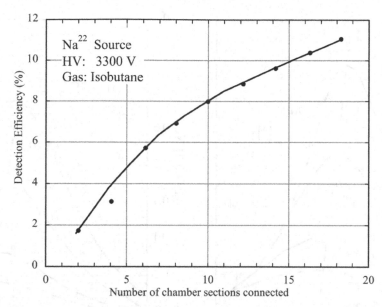

Figure 3.23   Detection efficiency for 511 keV photons as a function of the number of modules. Each module is a MWPC with 125-µm thick lead-coated cathodes (Bateman *et al.*, 1980). By kind permission of Elsevier.

detection efficiency for 511 keV photons as a function of the number of detector modules (Bateman *et al.*, 1980).

The intrinsically simpler design of resistive plate counters (RPCs), widely used in particle physics, has been adapted to the detection of hard photons. Figure 3.24 shows the computed efficiency of stacks of RPCs with lead and glass converters as a function of the converter thickness and number (Blanco *et al.*, 2009). While lead provides the highest efficiency, use of high-resistivity glass as converters and electrodes allows one to reduce the required number of readout channels, combining several detection elements in a single multi-gap resistive plate structure, an approach pursued by several groups, see Chapter 12.

The photon conversion efficiency of GEM detectors (Chapter 13) can be enhanced coating the active electrodes with a high-Z metal; Figure 3.25 shows computed and measured values of detection efficiency of a single detector with the electrodes coated with 3 µm of gold, as a function of photon energy (Koike *et al.*, 2011); multiple GEM structures can be used to increase the efficiency.

Metallic electrodes used in gaseous counters have a non-vanishing probability of ejecting a photoelectron when the incident photon energy exceeds their work function, typically between 4 and 5 eV (Fowler, 1931). Although in general not exploitable for detection, due to the very low efficiency, this process is often invoked to explain various kinds of internal photon-mediated secondary processes; sensitivity to external light sources as fluorescent lamps has also been observed.

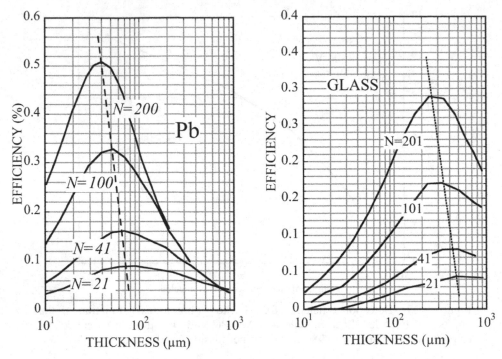

Figure 3.24   Detection efficiency for 511 keV γ rays as a function of thickness and number of conversion layers of lead and glass in a multi-gap resistive plate chamber (Blanco *et al.*, 2009). By kind permission of Elsevier.

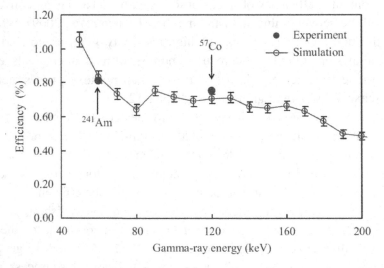

Figure 3.25   Computed and measured detection efficiency of a gold-coated GEM electrode as a function of photon energy (Koike *et al.*, 2011). By kind permission of Elsevier.

3.9 Transparency of windows

To be detected, and unless they are emitted within the counter itself, photons have to enter the gaseous counters through a suitable window, where they can be partly converted; absorption losses depend on the window composition and thickness and on the energy of the photons. Figure 3.26 shows the transparency of several materials commonly used as windows for photon detectors in the visible and near ultra-violet, compiled from various sources. The region of sensitivity is the convolution of the window transparency and the quantum efficiency of the detecting medium, either in the gas phase or as internal photocathode. Examples of spectral response of several photosensitive compounds used in Cherenkov ring imaging detectors were given in Section 3.5; the high ionization threshold of TEA (7.5 eV) requires the use of expensive fluoride windows, while with TMAE or CsI a UV-quality quartz window permits the photon detection at lower energies.

The thin polyimide windows often used for gas containment because of their good mechanical strength have a rather low transparency cut-off, preventing most of the more energetic photons from entering the detector. Figure 3.27 shows the absorption length of mylar,[2] with a cut-off around 4 eV, and aclar,[3] a fluoropolymer transparent at shorter wavelengths often added as second layer to reduce water absorption caused by the hydroscopic polyimides.

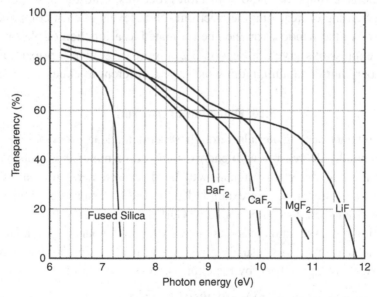

Figure 3.26 Optical transparency of fluoride and fused silica windows.

---

[2] DuPont polyethylene terephthalate   [3] Polychlorotrifluoroethylene

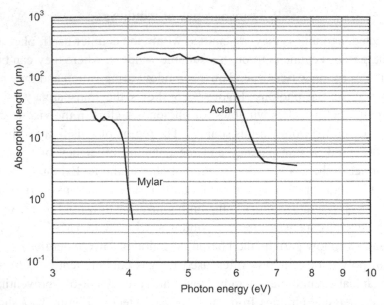

Figure 3.27   Absorption length of thin polymer windows in the near UV.

Figure 3.28 provides the computed absorption length of materials that can be used as detector windows in the soft X-ray domain, above 100 eV; the curves have the characteristic shape of photo-ionization processes, with the absorption length sharply decreasing on reaching the electronic shell levels of each component. The transparency of the window can then be estimated as a function of photon energy, using the expressions given in Section 3.2. Figure 3.29 gives an example, computed for beryllium and aluminium windows of different thickness.

3.10 Detection of neutrons

Neutrons interact with materials through several nuclear processes, with cross sections generally decreasing with energy. Only some major mechanisms leading to the creation of detectable signals will be discussed here; for an exhaustive coverage of the subject, see for example Knoll (1989). Extensive tables and plots of neutron cross sections can be found in Garber and Kinsey (1976).

Interaction processes and cross sections vary largely depending on materials and energy. The neutron energy, usually given in eV, is also expressed as wavelength in Å; the two units are related by the expression (Plank's law):

$$\lambda(\text{Å}) = 0.28601/\sqrt{E(\text{eV})}.   \tag{3.19}$$

For high and intermediate-energy neutrons, above a few MeV, the major mechanism of interaction is elastic scattering, particularly efficient with low-atomic number materials, like hydrogenated compounds; while slowing down the

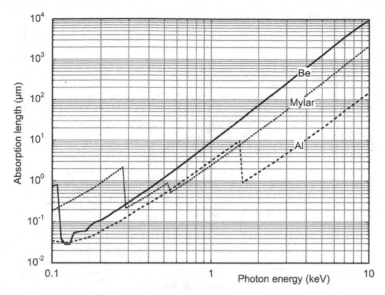

Figure 3.28   Absorption length of window materials for soft X-rays.

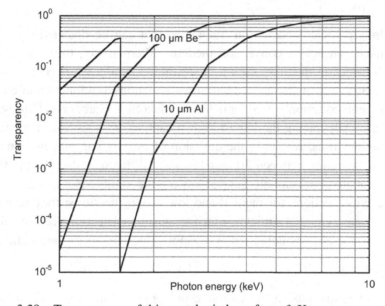

Figure 3.29   Transparency of thin metal windows for soft X-rays.

neutrons, in a process called moderation, elastic scattering does not provide a mean of detection. At energies around 1 MeV, inelastic scattering may occur, leaving the nucleus in an excited state that returns to the ground state through a radiative decay; this is the process exploited in activation devices. Around and below thermal energies, 0.025 eV, other nuclear interaction processes dominate; radiative capture and nuclear reactions with the emission of charged particles, protons,

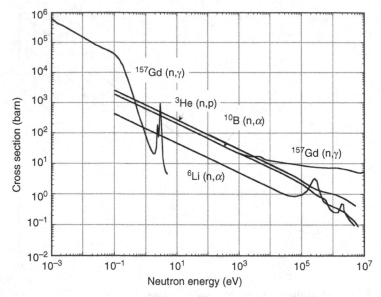

Figure 3.30   Neutron cross sections for several isotopes. Data from Garber and
Kinsey (1976).

tritons, alpha particles and fission fragments, that can be detected directly. The
detector itself can act as conversion medium; alternatively, thin converter foils
surround a detection medium, such as a scintillator or a gaseous counter.

Figure 3.30 is a compilation of neutron cross sections as a function of energy for
several elements commonly used for detectors (data from Garber and Kinsey, 1976).
Gadolinium, while having the largest known capture cross sections for epithermal
and thermal neutrons, results in the emission of prompt gammas and conversion
electrons in the range between 30 and 180 keV, difficult to disentangle from the
Compton scattered background. Several low atomic number isotopes have particularly large cross sections for thermal and low-energy neutrons; they result in the
emission of heavy charged particles, which are easier to detect, can be used in the gas
phase and are therefore widely used in detectors, through one of the reactions:

$$\tfrac{3}{2}\text{He} + n \rightarrow \tfrac{3}{1}\text{H} + p,\ \tfrac{10}{5}\text{B} + n \rightarrow \tfrac{7}{3}\text{Li} + \alpha \text{ and } \tfrac{6}{3}\text{Li} + n \rightarrow \tfrac{3}{1}\text{H} + \alpha.$$

The first process results in the emission in opposite directions of two ionizing
prongs (triton and proton), with energies of 191 and 573 keV respectively, easily
detected, and widely exploited in gaseous proportional counters. With a cross
section of around 5000 barns at thermal energies, at atmospheric pressures the
mean interaction length is around 10 cm (see expression (3.6)); higher efficiencies
can be obtained by increasing the counter's gas pressure. An additional advantage
of helium-filled counters is the reduced sensitivity to gamma rays, making their use
suitable in mixed radiation fields.

A systematic study of performances of ^{3}He-filled proportional counters can be found in Mills *et al.* (1962) and Ravazzani *et al.* (2006). A by-product of the nuclear power industry, the helium isotope has high costs and limited availability, making its use problematic for large detection volumes.

^{10}B-enriched boron trifluoride (BF$_3$) is used as gas filling in proportional counters (Fowler, 1973); the higher energy of the reaction products (840 keV for the lithium ion and 1.47 MeV for the α particle) makes their detection easier, but the efficiency is affected by the enrichment factor (natural boron contains around 20% ^{10}B isotopes). A comparison of performances of He and BF$_3$-filled counters can be found in Lintereur *et al.* (2011).

Detectors are designed to collect the ionization released by the interaction products; they can be positional, but do not provide information on the neutron energy. If the ionized trails have sufficient space extension, as in gaseous counters, detection and imaging of the yields can be exploited to provide information on the interaction vertex (see for example Fraga *et al.*, 2002; Miernik *et al.*, 2007). A long trail, however, has the drawback of being absorbed for part of the events on the counter's walls, resulting in a shift of the energy loss spectrum towards smaller amplitudes.

Alternatively to the use of absorbers in the gas phase, the neutron efficiency of conventional gaseous counters can be enhanced by the use of internal solid converters, deposited on the cathodes or on suitably segmented additional electrodes or baffles (Dighe *et al.*, 2003). Aside from boron, other converters used are thin layers of lithium, through the reaction $^6_3\mathrm{Li} + n \rightarrow {}^3_1\mathrm{H} + \alpha$, and ^{157}Gd, releasing low energy conversion electrons. As for photons, the detection efficiency is a convolution of the neutron reaction cross section and the probability of the reaction yield leaving the converter into the counter's gas; for boron and lithium, typical layers thickness of a few microns are optimal. As the two charged prongs in each event are emitted back to back, only one contributes to detection.

A systematic study of efficiency and optimization for thermal neutron counters with thin-film ^{10}B, ^{6}Li and ^{6}LiF coatings is given in McGregor *et al.* (2003); although the work aims at estimating the performances of solid state counters, most of the consideration apply to gaseous detectors. From the quoted reference, Figure 3.31 and Figure 3.32 show the computed residual energy of different ions as a function of the layer thickness, for neutrons perpendicular to a ^{10}B and ^{6}Li layer, and Figure 3.33 shows the estimated detection efficiency for both front and back irradiation. The calculation assumes full detector efficiency as far as a non-zero residual energy is lost in the active device. For thermal neutrons of two wavelengths, Figure 3.34 gives the calculated triton escape probability, forward, backward and total from ^{6}Li foils as a function of thickness (Dagendorf *et al.*, 1994).

Figure 3.35 (Klein and Schmidt, 2011) is an example of measured residual energy spectra measured for neutron irradiation in a gaseous counter

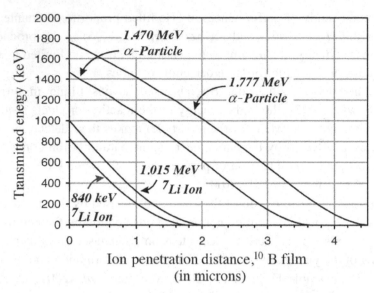

Figure 3.31   Residual energy for different reaction yields on ^{10}B as a function of film thickness (McGregor *et al.*, 2003). By kind permission of Elsevier.

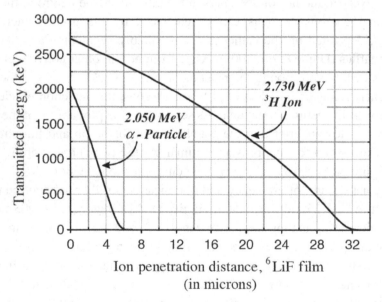

Figure 3.32   Residual energy for different reaction yields on ^{6}Li as a function of film thickness (McGregor *et al.*, 2003). By kind permission of Elsevier.

with 400 and 2300 nm-thick ^{10}B converters. For the thinner layer, the peaks correspond to the maximum energy of the prongs, tailing down to smaller values due to the energy loss in the converter; the degradation is more pronounced for the thicker layer.

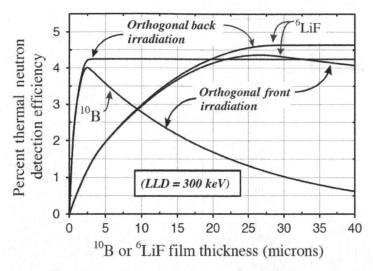

Figure 3.33 Thermal neutron detection efficiency computed as a function of conversion layer thickness, for perpendicular front and back irradiation (McGregor *et al.*, 2003). By kind permission of Elsevier.

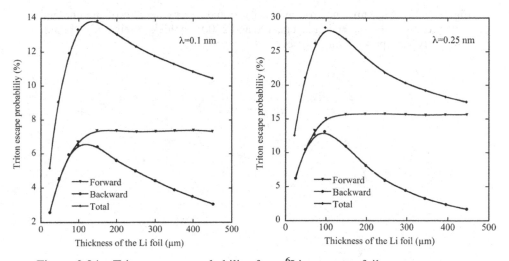

Figure 3.34 Triton escape probability from ^{6}Li converter foils at two neutron wavelengths (Dagendorf *et al.*, 1994). By kind permission of Elsevier.

The development of micro-pattern gas detectors (Sauli and Sharma, 1999; Titov, 2007), with their excellent localization and energy resolution, has enhanced the research on position-sensitive neutron detectors (see Chapter 13). In devices based on the gas electron multiplier, the multi-electrode structure can easily be optimized for neutron detection coating the electrodes with thin converter layers; Figure 3.36 (Klein and Schmidt, 2011) provides the expected detection efficiency of a

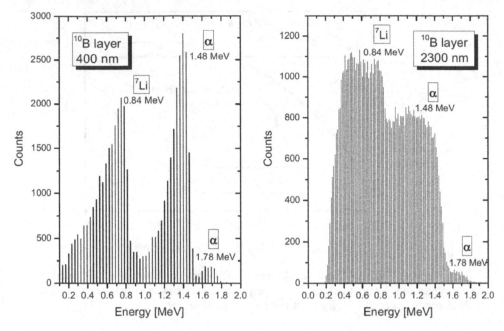

Figure 3.35   Measured neutron-induced residual energy spectra of the charged yields for two ^{10}B conversion layers, 400 and 2300 nm thick (Klein and Schmidt, 2011). By kind permission of Elsevier.

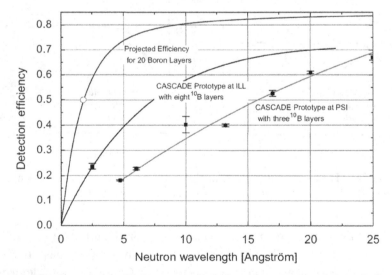

Figure 3.36   Computed efficiency as a function of neutron wavelength detector for multiple ^{10}B converter layers. Points with error bars are experimental measurements (Klein and Schmidt, 2011). By kind permission of Elsevier.

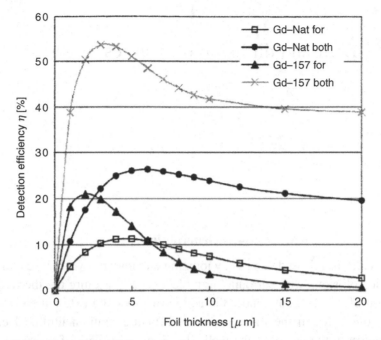

Figure 3.37   Detection efficiency for thermal neutrons as a function of thickness of natural Gd and ^{157}Gd. Values are given for all conversion electrons and for the forward fraction only (Masaoka *et al.*, 2003). By kind permission of Elsevier.

multi-GEM detector as a function of neutron wavelength for several choices in the number of layers; points with error bars are the measurements with three ^{10}B coatings, about 1.2 μm thick each. With a similar structure, using thin-foil natural Gd and ^{157}Gd Masaoka *et al.* (2003) have computed the detection efficiency for thermal neutrons as a function of converter thickness both for forward and total conversion electrons (Figure 3.37).

It should be mentioned that the copious neutron background usually met in high-energy physics experiments, unwillingly converted by the detector materials, has deleterious effects on the instrumentation: damages to materials and electronics and discharges in gaseous counters; this issue will be discussed in Chapter 16.

4

Drift and diffusion of charges in gases

4.1 Generalities

Ions and electrons released in a gas by ionizing encounters swiftly lose their energy in multiple collisions with surrounding molecules and acquire the thermal energy distribution of the medium. Under the action of moderate external electric fields, charges move through the medium while diffusing, until neutralized either by recombination in the gas or at the walls. For ions, a process of charge transfer is possible with a molecule of its own gas or of another species having lower ionization potential. Electrons, wandering in the gas and colliding with the molecules, can be neutralized by a positive ion, attach to a molecule having electron affinity, or get absorbed at the walls of the containment vessel.

Most quantities to be discussed here (mobility, drift velocity, diffusion, ...) depend on the gas density, a function itself of temperature, and are therefore inversely proportional to pressure. The appropriate invariant for the field-dependent variables is the ratio E/P, expressed in classic works in units of V/torr, or E/N where N is the Loschmidt constant (number of molecules per unit volume). In modern detector practice it is customary, however, to present data as a function of field at normal temperature and pressure (NTP: 20 °C, one atmosphere). For different conditions, appropriate density-dependent scaling rules should be used (see Section 3.2).

4.2 Experimental methods

The drift and diffusion of charges under the effect of electric and magnetic fields has been a classic subject of research for decades in the field named gaseous electronics. Major outcomes of these studies are a detailed knowledge of molecular collision processes and of the electron–molecule cross sections; methods and results are discussed in numerous textbooks (Loeb, 1961; Brown, 1959; Hasted, 1964; Massey *et al.*, 1969; Huxley and Crompton, 1974).

Drift properties of ions and electrons are studied with devices named drift tubes, which have designs that can vary from the very sophisticated systems used for fundamental gaseous electronics studies to the simpler structures introduced in the development of modern gaseous detectors (see Section 4.7). For ions, due to many possible processes of charge transfer and chemical reactions between excited species, and the consequent sensitivity to very small amounts of pollutants, the systems used for fundamental studies are built with clean materials and are thoroughly outgassed using vacuum technologies. A detailed description of drift tubes and experimental methods can be found for example in McDaniel and Mason (1973).

In its basic design, a drift tube generates charges (ions or electrons) at a known time at one end of a region with a uniform electric field, and permits one to measure the distribution of arrival time of the charges at a collecting electrode, wire or plate, after the drift; the measurement can be done in a current mode, or with detection of the individual pulses. For ions, the source is usually an electron-impact ion beam, admitted through a control grid to the main section of the tube by microsecond-wide gating pulses, much shorter than the total drift time to be measured, which is typically from several hundred microseconds to milliseconds. The distance between source and detector can be varied; in some systems, the ion beam enters the tube through several ports, at different distances form the collecting anode. The tube's end can be coupled to a mass spectrometer, permitting a positive identification of the ion species. Figure 4.1 is a schematic cross section of

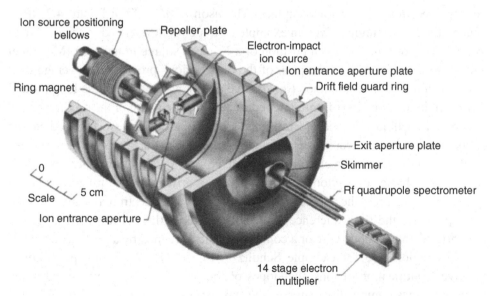

Figure 4.1   A drift tube with ion source, drift volume and mass spectrometer for ion identification (Thomson *et al.*, 1973). By kind permission of The American Institute of Physics.

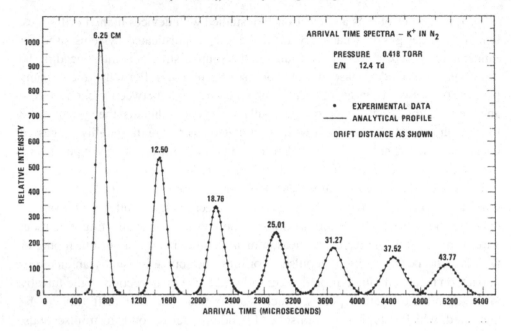

Figure 4.2   Spectra of arrival time of ions in a drift tube for several source positions (Thomson *et al.*, 1973). By kind permission of The American Institute of Physics.

one such devices, showing the movable ion source, the drift volume and the end detector spectrometer for ion sampling (Thomson *et al.*, 1973); Figure 4.2, from data by the same authors, gives an example of ion current recorded in the drift tube as a function of time for seven position of a K^+ ions source in nitrogen (McDaniel and Mason, 1973). A fit to the peak of the distributions provides the average drift time per unit length of migration (the drift velocity), and the width the longitudinal diffusion, in the direction of drift, after correcting for the width of the original beam.

Simpler methods to measure the ions drift properties have been used in the course of modern detectors development, restricting the study to the species produced in the gas by the electron–molecule collisions during avalanche multiplication. In its basic conception, the instrument is a drift chamber structure exposed to an ionizing source; the anodic signal, produced by the electron avalanche on the wire, provides the time reference, and the ions' drift time is recorded detecting their arrival on a cathode wire or a collection plate, screened by a grid, at the end of the drift region (see for example Schultz *et al.*, 1977). While not providing a positive identification of ions, the slopes of the mobility curves, measured while varying the gas composition, suggest the type of dominant ion (see Section 4.4).

Drift tubes used for the measurement of electron drift properties are similar in design, but require a time reference that can be obtained with short electron bursts

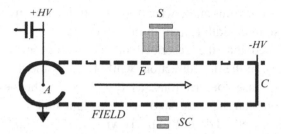

Figure 4.3 Schematics of a drift chamber. Electrons produced by a collimated charged particle source S drift to the anode A where they are detected after avalanche multiplication. Field electrodes at suitable potentials create a uniform electric field; a pair of scintillation counters SC in coincidence provides the time reference.

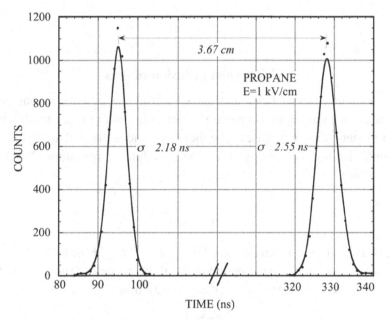

Figure 4.4 Example of electron drift time spectra for two positions of a collimated source (Jean-Marie *et al.*, 1979). By kind permission of Elsevier.

controlled by electrical shutters, or generated by a UV or laser pulse hitting an internal electrode (Huxley and Crompton, 1974; Christophorou *et al.*, 1966). In more recent setups making use of a collimated charged particle beam or radioactive electron sources, the time reference is given by an external set of scintillator counters in coincidence, as shown schematically in Figure 4.3.

Figure 4.4 is an example of arrival time spectra for electrons released by a collimated source at two distances from the anode (Jean-Marie *et al.*, 1979); the time difference between the peaks provides the drift velocity, while the width of the distribution is a convolution between the longitudinal diffusion of the drifting electrons and the source width (1 mm in this example).

It should be noted that the described devices record only the longitudinal diffusion of charges, in the drift direction; while for ions the diffusion is symmetric, this is not the case for electrons, as discussed in the next sections. The transverse diffusion can be measured using segmented collection electrodes sharing the charge, a method described by Townsend (1947) and used extensively, for example in the development of time projection and micro-pattern chambers (Chapters 10 and 13).

Drift properties in a magnetic field are measured by inserting the detectors into a magnet, if needed adjusting the voltage applied to the field shaping electrodes to compensate for the Lorentz angle; an example is described in Breskin *et al.* (1974b).

4.3 Thermal diffusion of ions

In the absence of external fields and inelastic collision processes, ions and electrons released in a gas behave like neutral molecules, with properties described by the classic kinetic theory of gases. The theory provides the probability of an atom or molecule having an energy ε at the absolute temperature T (Maxwell–Boltzmann law):

$$F(\varepsilon) = 2\sqrt{\frac{\varepsilon}{\pi(kT)^3}} \ e^{-\frac{\varepsilon}{kT}}, \tag{4.1}$$

where k, Boltzmann's constant, equals 1.38×10^{-16} erg/°K or $8.617 \ 10^{-5}$ eV/°K; the distribution does not depend on the mass of the particles. The average thermal energy is obtained by integration over the distribution:

$$\bar{\varepsilon}_T = kT$$

at normal conditions, $\bar{\varepsilon}_T = 0.025$ eV (Figure 4.5).

The corresponding distribution of velocity v for a particle of mass m is:

$$f(v) = 4\pi \left(\frac{m}{2\pi kT}\right)^{\frac{3}{2}} v^2 e^{-\frac{mv^2}{2kT}}; \tag{4.2}$$

by integration, one gets the average value of the velocity:

$$\bar{v} = \int_0^\infty v \ f(v) \mathrm{d}v = \sqrt{\frac{8kT}{\pi m}}; \tag{4.3}$$

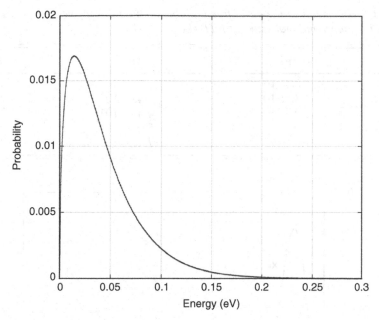

Figure 4.5   Energy distribution of molecules at normal conditions.

the most probable value v_{MP} is:

$$v_{MP} = \sqrt{\frac{2kT}{m}}. \tag{4.4}$$

Figure 4.6 is the computed distribution of the velocity probability for atoms of noble gases at NTP; one can see that they are largely supersonic.

A localized distribution of molecules or ions diffuses symmetrically by multiple collisions following a Gaussian law:

$$\frac{dN}{N} = \frac{1}{\sqrt{4\pi Dt}} e^{-\frac{x^2}{4Dt}} dx, \tag{4.5}$$

where dN/N is the fraction of particles found in the element dx at a distance x from the origin and after a time t; D denotes a diffusion coefficient. The root mean square of the distribution, or standard deviation, is given for linear and volume diffusion, respectively, by:

$$\sigma_x = \sqrt{2Dt} \text{ and } \sigma_V = \sqrt{6Dt}. \tag{4.6}$$

Classic values of the diffusion coefficient D, mean free path between collisions λ and average velocity v of atoms and molecules in their own gas are given in Table 4.1 (Loeb, 1961).

Expression (4.5) can be used to estimate the time-dependent dilution in a gas volume of a given species, for example ions released in a given position or

Table 4.1 *Classic values of the mean free path between collisions, velocity and diffusion coefficient for atoms and molecules at NTP (Loeb, 1961).*

Gas	λ(cm)	v(cm/s)	D(cm^2/s)
H_2	1.8×10^{-5}	2×10^5	0.34
He	2.8×10^{-5}	1.4×10^5	0.26
Ar	1.0×10^{-5}	4.4×10^4	0.04
O_2	1.0×10^{-5}	5.0×10^4	0.06
H_2O	1.0×10^{-5}	7.1×10^4	0.02

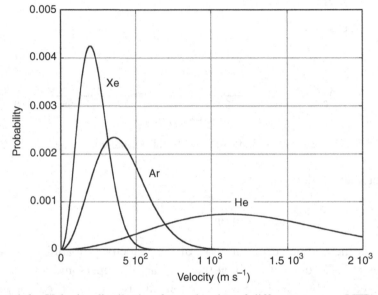

Figure 4.6   Velocity distribution for molecules of different masses (NTP).

pollutants penetrating from a hole in a counter. Figure 4.7 is an example of space distributions of oxygen ions in air, at normal conditions, after different time intervals, computed from the previous expression; after a few seconds, the thermal diffusion spreads a foreign species through the whole counter volume. This implies also that the detector volume can be easily contaminated by a leak; even a small overpressure, resulting in a gas outflow, can hardly compete with the penetration of the very fast foreign molecules into the volume.

4.4 Ion mobility and diffusion in an electric field

When an electric field is applied to the gas volume, a net movement of ions along the field direction is observed. The average velocity of this slow motion (not to be confused with the instant ion velocity v) is named the drift velocity w^+, and is

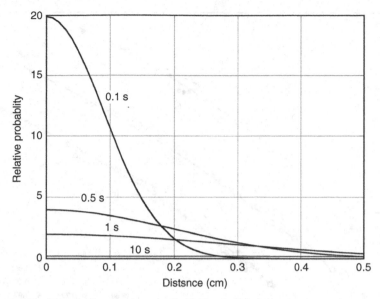

Figure 4.7   Space distribution of initially localized oxygen molecules after increasing time intervals (NTP).

linearly proportional to the electric field up to very high values of E, Figure 4.8 (McDaniel and Mason, 1973). It is therefore customary to define a quantity μ, the ion mobility, as:

$$\mu = \frac{w^+}{E}.$$

$$(4.7)$$

The value of the mobility is specific to each ion moving in a given gas, and depends on pressure and temperature through the expression:

$$\mu(P, T) = \frac{T}{T_0}\frac{P_0}{P}\mu(P_0, T_0).$$

$$(4.8)$$

A constant mobility is the direct consequence of the fact that, up to very high fields, the average energy of ions is almost unmodified; as will be seen later, this is not the case for the electrons.

A classic argument allows one to obtain the following relationship between mobility and diffusion coefficient (Nernst–Townsend formula[1]):

$$\frac{D}{\mu} = \frac{kT}{e},$$

$$(4.9)$$

where D is the diffusion coefficient, and e the electron charge.

---

[1] Often improperly named the Einstein formula, from his later works on Brownian motion.

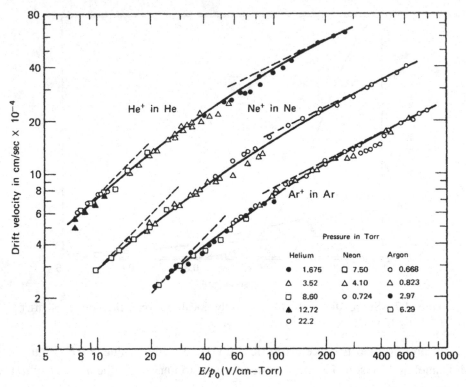

Figure 4.8   Drift velocity of several ions in their own gas as a function of field (McDaniel and Mason, 1973). Reproduced with kind permission of John Wiley & Sons, Inc.

Ions migrating for a time t over a length x diffuse with a probability distribution given by (4.5) and with a linear standard deviation along the drift direction obtained by combining the previous expressions:

$$\sigma_x = \sqrt{\frac{2kT}{e}\frac{x}{E}};\qquad(4.10)$$

the space diffusion therefore does not depend on the type of ion and pressure, but only on the field (Figure 4.9).

The mobility of an ion in a different gas follows in good approximation a simple dependence on the mass ratio (Langevin's law):

$$\mu_I = \sqrt{\left(1 + \frac{M_M}{M_I}\right)},\qquad(4.11)$$

where M_M and M_I are the molecular weights of the support gas and of the migrating ions.

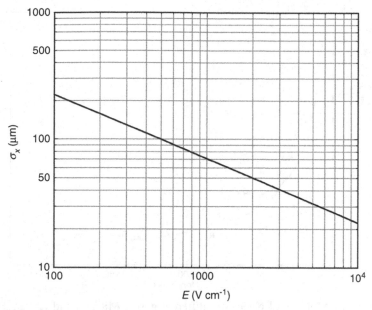

Figure 4.9   RMS of diffusion of ions in the drift direction as a function of field (NTP).

Figure 4.10 is a classic measurement of the mobility of nitrogen ions in various gases, and illustrates the inverse square root dependence on mass (Mitchell and Ridler, 1934). A noteable exception is the mobility of the N_2 ion itself, due to a process of charge transfer between the ion and its molecule, energetically possible for ions in their own gas (all other molecules in the plot have ionization potentials larger than that of nitrogen).

In a mixture of gases $G_1, G_2, \ldots, G_N$ the mobility μ_i of the ion G_i^+ is given by the relation (Blanc's law):

$$\frac{1}{\mu_i} = \sum_{j=1}^{n} \frac{p_j}{\mu_{ij}}, \qquad (4.12)$$

where p_j is the volume concentration of gas j in the mixture, and μ_{ij} the mobility of ion G_i^+ in the gas G_j.

In gas mixtures, a very effective process of collisional charge transfer can take place, quickly removing all ions except those with the lowest ionization potential. Depending on the nature of the ions and on the difference in ionization potentials (small differences increase the charge transfer probability), it takes between 100 and 1000 collisions for an ion to transfer its charge to a molecule having a lower potential. Since the mean free path for collisions under normal conditions is around 10^{-5} cm (see Table 4.1), after a drift length between $10^{-3}/p$ and $10^{-2}/p$ centimetres,

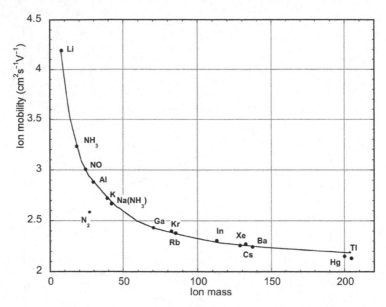

Figure 4.10   Mobility of N_2 ions in different gases (Mitchell and Ridler, 1934).

where p is the percentage of the molecules with lowest ionization potential, the charge transfer mechanism will have left only one species of ions migrating.

Experimental values of mobility for ions in various gases, including their own, are given at normal conditions in Table 4.2, from various sources. In most cases, the exact nature of the drifting ions has not been identified directly; therefore, the values indicated might correspond to an average over several species. With the exception of light ions, most values are rather similar; as will be discussed later, this has the practical consequence that the ions' clearing time in detectors is almost independent of the gas used.

Figure 4.11 and Figure 4.12 provide examples of the measured dependence of the inverse mobility of ions in binary mixtures of Ar-CO_2 (Schultz *et al.*, 1977), Ar-C_3H_8 and CF_4-C_3H_8 (Yamashita *et al.*, 1992) as a function of the gas density or fraction of the molecular component; the values follow Blanc's law well, under the assumption that the drifting ion is the molecule with the lowest ionization potential.

As can be inferred from expression (4.12), for a given ion drifting in a mixture, the inverse mobility depends linearly on the mixture's specific weight; lines of equal slope therefore represent the migration of the same kind of ion. Figure 4.13 shows experimental values of ion mobility in mixtures of argon–isobutane and argon–isobutane–methylal as a function of their density (Schultz *et al.*, 1977).

Although the ion species were not identified in this measurement, the converging slopes of the curves can be interpreted as curve F providing the mobility of

Table 4.2 *Mobility of ions, from various sources. Data without a reference are average values from compilations in McDaniel and Mason, 1973).*

Gas	Ion	μ (cm^2 V^{-1}s^{-1})
H$_2$	Self	13.0
He	Self	10.2
Ar	Self	1.7
Ar	CH$_4$	1.87 (Schultz *et al.*, 1977); 2.07 (Yamashita *et al.*, 1992)
Ar	C$_2$H$_6$	2.06 (Yamashita *et al.*, 1992)
Ar	C$_3$H$_8$	2.08 (Yamashita *et al.*, 1992)
Ar	i-C$_4$H$_{10}$	1.56 (Schultz *et al.*, 1977); 2.15 (Yamashita *et al.*, 1992)
Ar	CO$_2$	1.72
Ar	(OCH$_3$)$_2$CH$_2$	1.51
CH$_4$	Self	2.22 (Yamashita *et al.*, 1992)
C$_2$H$_6$	Self	1.23 (Yamashita *et al.*, 1992)
C$_3$H$_8$	Self	0.793 (Yamashita *et al.*, 1992)
i-C$_4$H$_{10}$	Self	0.612 (Yamashita *et al.*, 1992)
i-C$_4$H$_{10}$	(OCH$_3$)$_2$CH$_2$	0.55 (Schultz *et al.*, 1977)
(OCH$_3$)$_2$CH$_2$ (Methylal)	Self	0.26 (Schultz *et al.*, 1977)
O$_2$	Self	2.2
CO$_2$	Self	1.09
H$_2$O	Self	0.7
CF$_4$	Self	0.96 (Yamashita *et al.*, 1992)
CF$_4$	CH$_4$	1.06 (Yamashita *et al.*, 1992)
CF$_4$	C$_2$H$_6$	1.04 (Yamashita *et al.*, 1992)
CF$_4$	C$_3$H$_8$	1.04 (Yamashita *et al.*, 1992)

isobutane ions in argon–isobutane mixtures, curve A the mobility of methylal in argon–methylal and curves B, C, D, E the mobility of methylal ions in argon–isobutane–methylal mixtures. In the range of electric fields considered (a few hundred to a thousand V/cm) and for 1 cm of drift, if more than 3–4% methylal is added to the mixture, the exchange mechanism is fully efficient and the measurement is consistent with the assumption that only those ions are left migrating. The process is exploited in gaseous detectors to effectively remove organic ions produced in the avalanches, having the tendency to form polymers, transferring the charge to a non-polymerizing species, such as alcohols, methylal and others (see Chapter 16).

4.5 Classic theory of electron drift and diffusion

Electrons released in a gas by ionization quickly reach thermal equilibrium with the surrounding molecules; in the absence of external fields, their energy

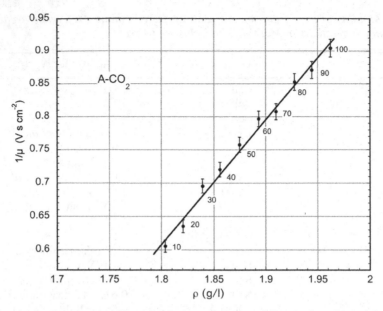

Figure 4.11   Inverse mobility of CO_2 ions in argon–CO_2 as a function of density of the mixture (Schultz *et al.*, 1977). By kind permission of EDP Science Journals.

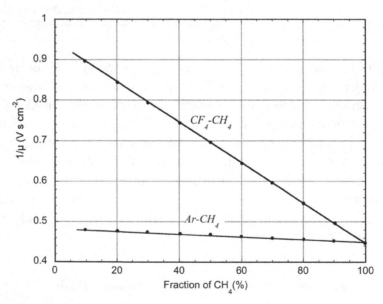

Figure 4.12   Inverse ion mobility in argon–methane and carbon tetrafluoride–methane mixtures (Yamashita *et al.*, 1992). By kind permission of Elsevier.

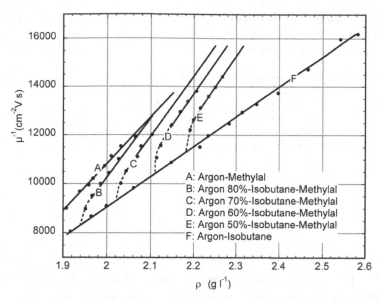

Figure 4.13   Inverse mobility in argon–isobutane–methylal mixtures, as a function of density (Schultz *et al.*, 1977). By kind permission of EDP Science Journals.

distribution follows the same law as for ions, see expression (4.1). However, due to the lower mass, their thermal velocity is several orders of magnitude higher, as one can deduce from (4.3), with an average value at room temperature of about 10^7 cm/s. Thermal diffusion occurs with a correspondingly larger value of the diffusion coefficient.

Free electrons can be neutralized by an ion, absorbed in the walls, or attach to a molecule having electron affinity or being electro-negative; the probability of attachment per collision, h, negligible for noble gases, has finite values for gases having incomplete outer electronic shells, as can be seen in Table 4.3 (Fulbright, 1958; Loeb, 1961; Townsend, 1947). The table also shows the average attachment time $t_h = 1/hN$, where N is the number of collisions per unit time. In oxygen, for example, the average time needed for a thermal electron to be attached is about 200 ns.

Adding a percentage p of an electro-negative molecule to a main gas mixture, the probability of attachment is hp. As discussed later, the attachment coefficient is a strong function of the electron energy, hence of the field.

When an electric field is present, the electron swarm moves in the direction opposite to the field vector; a simple theory of mobility can be formulated along the same lines as for positive ions. It was found very early, however, that except for very low fields, the mobility of electrons is not constant: because of their small mass, electrons can substantially increase their energy between collisions with the

Table 4.3 *Electron attachment probability, frequency of collisions and average attachment time for several gases at NTP.*

Gas	h	$N(s^{-1})$	$t_h(s)$
CO_2	6.2×10^{-9}	2.2×10^{11}	0.71×10^{-3}
O_2	2.5×10^{-5}	2.1×10^{11}	1.9×10^{-7}
H_2O	2.5×10^{-5}	2.8×10^{11}	1.4×10^{-7}
Cl	4.8×10^{-4}	4.5×10^{11}	4.7×10^{-9}

gas molecules. In a simple formulation (Townsend, 1947), the electron drift velocity can be written as:

$$w^- = k\frac{eE}{m}\tau, \tag{4.13}$$

where τ is the mean time between collisions. The value of the constant k, between 0.75 and 1, depends on assumptions about the energy distribution of electrons, see for example Palladino and Sadoulet (1974). While convenient for qualitative consider-ations, Townsend's expression is not very useful in practice, since the values of w^- and τ depend on the gas and field.

During the drift in electric fields, and as a result of multiple collisions with molecules, electrons diffuse, spreading the initially localized charge cloud. The extent of diffusion depends on the gas, but also strongly on E, due to the increase of the electron energy. To take this into account, expression (4.9) can be modified by introducing a phenomenological quantity ε_k, named characteristic energy:

$$\frac{D}{\mu} = \frac{\varepsilon_k}{e}; \tag{4.14}$$

for thermal electrons, $\varepsilon_k = kT$, and the expression reduces to the previous one. The linear space diffusion over the distance x can then be written as:

$$\sigma_x = \sqrt{\frac{2}{e}\frac{\varepsilon_k}{E}\frac{x}{E}}. \tag{4.15}$$

Expression (4.10) can be rewritten to show explicitly the dependence on the reduced field E/P:

$$\sigma_x = \sqrt{\frac{2\varepsilon_k}{e}}\sqrt{\frac{P}{E}}\sqrt{\frac{x}{P}}. \tag{4.16}$$

4.6 Electron drift in magnetic fields

The presence of an external magnetic field modifies the drift properties of the swarm of electrons. The Lorentz force exerted on moving charges alters the linear segment of motion between two collisions into circular trajectories, and can affect the energy distribution; the net effect is a reduction of the drift velocity, and a drift of the swarm along a direction at an angle with the electric field lines.

The simple theory provided by expression (4.13) permits computation of the effect on velocity and diffusion for the two particular cases of perpendicular and parallel fields:

$$\vec{E} \perp \vec{B} \qquad \tan \theta_B = \omega \tau$$

$$w_B = \frac{E}{B} \frac{\omega \tau}{\sqrt{1 + \omega^2 \tau^2}}, \qquad (4.17)$$

$$\vec{E} // \vec{B} : w_B = w_0,$$

$$\sigma_L = \sigma_0,$$

$$\sigma_T = \frac{\sigma_0}{\sqrt{1 + \omega^2 \tau^2}}, \qquad (4.18)$$

where θ_B is the angle between the drifting swarm and the electric field in the plane perpendicular to $\vec{B}$, $\omega = EB/m$ is the Larmor frequency and τ is the mean collision time. For perpendicular fields, the swarm drifts at an angle to the electric field, with a reduced drift velocity; for parallel fields in contrast the magnetic drift velocity and the longitudinal diffusion are unaffected, while the transverse diffusion, perpendicular to the fields, is reduced by a factor that depends on the product $\omega \tau$. This effect is exploited to substantially improve the space resolution in drift and time projection chambers.

In the general case of arbitrary directions of electric and magnetic fields, the so-called friction force theory provides the following expression:

$$\vec{w} = \frac{e}{m} \frac{\tau}{1 + \omega^2 \tau^2} \left[\vec{E} + \omega\tau \frac{\vec{E} \times \vec{B}}{B} + \omega^2\tau^2 \frac{\vec{B} \, (\vec{E} \cdot \vec{B})}{B^2} \right]. \qquad (4.19)$$

4.7 Electron drift velocity and diffusion: experimental

The experimental measurements of electron drift and diffusion properties, accompanied by the development of electron transport theories, have been a major subject of research from the 1920s through to the 1950s. Due to the extreme sensitivity of the measurements to trace pollutants, particularly in pure noble gases, results were often controversial. Addition to rare gases of controlled amounts of molecular additives, as needed in gaseous counters to guarantee

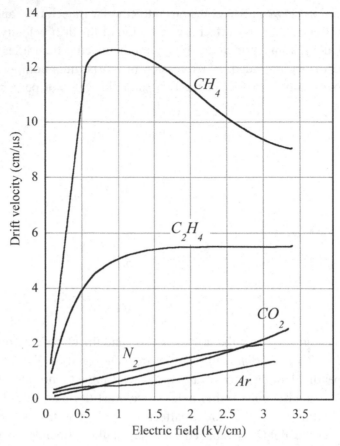

Figure 4.14   Electron drift velocity as a function of field in pure gases at NTP
(Sauli, 1977). By kind permission of CERN.

stable and high gain operation, generated abundant phenomenological literature
on the subject. Many compilations exist, from classic works (Brown, 1959; Loeb,
1961; Christophorou, 1971) to detector-oriented data collections (Peisert and
Sauli, 1984). With the development of dedicated electron transport simulation
programs, described in Section 4.10, drift properties of electrons in most gases
and mixtures commonly used in detectors can be computed with high accuracy in
a wide range of electric fields, including the region of avalanche multiplication.
Only selected representative examples of experimental measurements will be
presented here.

The compilation in Figure 4.14 for several pure gases illustrates the wide
span of values and shape of the field-dependence of the drift velocity (Sauli,
1977). The addition of even very small fractions of one gas to another, which
modify the average energy, can dramatically change the drift properties;

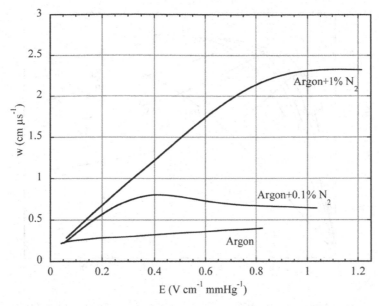

Figure 4.15   Effect on drift velocity of small nitrogen addition (Colli and Facchini, 1952). By kind permission of the American Institute of Physics.

as mentioned, it has a particularly strong effect for noble gases, as illustrated in Figure 4.15 (Colli and Facchini, 1952).

In the course of development of the drift chambers, electrons drift and diffusion properties have been extensively measured as a function of field in many mixtures of noble gases and hydrocarbons. Figure 4.16 is a collection of measurements in argon–isobutane; the curves are the results of early calculations making use of the transport theory, discussed in Section 4.10 (Schultz and Gresser, 1978). The mixture containing around 30% of i-C_4H_{10} exhibits a constant, or saturated, drift velocity at moderate values of field, and was therefore selected for the early operation of high-accuracy drift chambers (Breskin *et al.*, 1974b). Figure 4.17 is another example for argon–methane mixtures at NTP (Jean-Marie *et al.*, 1979). The same authors measured the effect on drift velocity of the addition of small quantities of nitrogen, a frequent pollutant in detectors having small leaks (Figure 4.18).

A good knowledge of the dependence of drift velocity on the gas composition and ambient conditions is needed for obtaining stable operation, particularly in high-accuracy drift chambers. This has been for long an experimental issue, until the availability of accurate calculation programs based on the electron transport theory. For 'cold' gases, in which the electrons remain thermal up to high field values, the expected relative change in drift velocity is simply related to the change in the gas density and therefore to the absolute temperature, $\Delta w/w = \Delta T/T$.

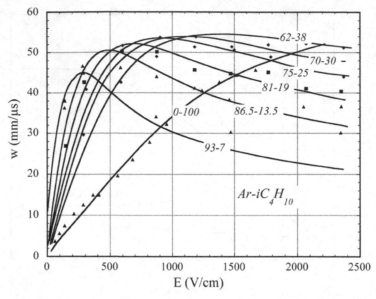

Figure 4.16 Electron drift velocity in argon–isobutane mixtures at NTP (Breskin *et al.*, 1974b). By kind permission of Elsevier.

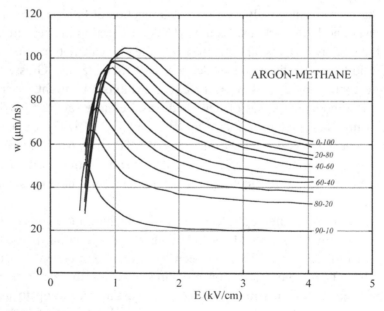

Figure 4.17 Electron drift velocity in argon–methane mixtures at NTP (Jean-Marie *et al.*, 1979). By kind permission of Elsevier.

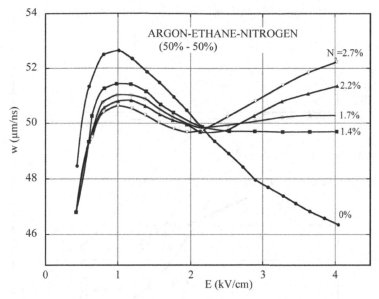

Figure 4.18   Effect on drift velocity of small nitrogen additions to an argon–ethane mixture (Jean-Marie *et al.*, 1979). By kind permission of Elsevier.

When electrons 'warm up' under the effect of the electric field, the variation can be larger or smaller, and even negative, depending on the detailed dependence of the electron–molecule cross sections on the electron energy. Figure 4.19 is an example of the estimated relative variation of drift velocity as a function of field for several gas mixtures at NTP per degree of temperature increase; the point with error bars is a measurement (Schultz and Gresser, 1978) and corresponds to the operating conditions selected for the high-accuracy drift chambers (Breskin *et al.*, 1974b). A fundamental parameter for the long-term operation of drift detectors, the temperature dependence of drift velocity has been measured for many other gases; Figure 4.20 is an example of drift time variation as a function of temperature, measured in dimethyl ether (DME) with a time expansion chamber for 34 mm of drift at a field of 164 V/cm, compared with calculations (Hu *et al.*, 2006).

In Figure 4.21 the computed result is shown for a standard argon–methane 90–10 gas mixture; the point with error bars is an experimental measurement (Peisert and Sauli, 1984). Note that, as it could be expected, both the variations due to the field and to temperature changes are minimized in the region of saturated drift velocity.

In the framework of the development of micro-pattern detectors, electron drift velocities in several gas mixtures have been measured up to very high fields, above 10 kV/cm; Figure 4.22 is an example for a mixture of helium–isobutane 80–20 at atmospheric pressure (Colas *et al.*, 2001).

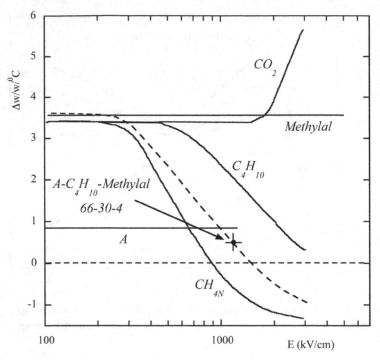

Figure 4.19   Computed variation of drift velocity as a function of field for 1 °C temperature increase (Schultz and Gresser, 1978). By kind permission of Elsevier.

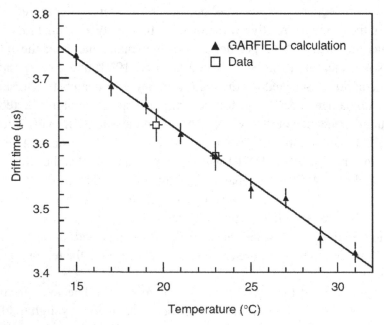

Figure 4.20   Temperature dependence of drift time for dimethyl ether at a field of 164 V/cm (Hu *et al.*, 2006). By kind permission of Elsevier.

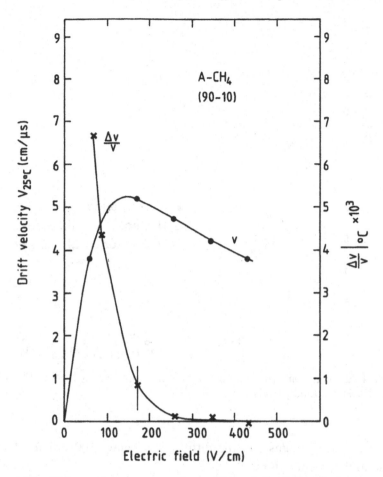

Figure 4.21 Value and temperature dependence of the drift velocity in argon–methane (Peisert and Sauli, 1984). By kind permission of CERN.

Figure 4.23 is a compilation of values of the electron linear diffusion as a function of field for several gases at NTP and 1 cm of drift (Palladino and Sadoulet, 1975). Diffusion is very large in pure noble gases, while it approaches the thermal limit in molecular gases; intermediate values are obtained for mixtures. The choice of the best gas mixture for a specific detector is often a compromise between the values of drift velocity and diffusion, and will be discussed in the following chapters.

The time expansion chamber (TEC), described in Chapter 9, relies on the use of a gas mixture having the conflicting requirements of a low diffusion, achieved at high fields, and a low drift velocity. Figure 4.24 combines measured and computed values for the mixture carbon dioxide–isobutane used in a high-resolution drift detector at DESY (Commichau *et al.*, 1985); at a field of 1 kV/cm, the drift

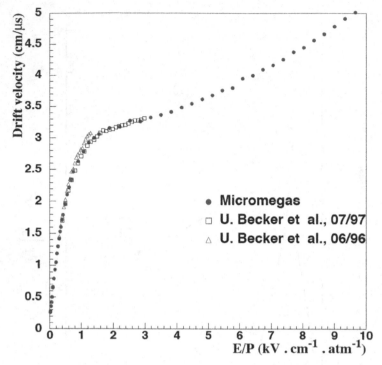

Figure 4.22 Electron drift velocity in helium–isobutane at high fields (Colas *et al.*, 2001). By kind permission of Elsevier.

velocity is around 7 μm/ns and the diffusion is below 100 μm for 1 cm drift, very close to the thermal limit.

An interesting gas in this respect is dimethyl ether, $(CH_3)_2O$ (DME); its drift and diffusion properties are compared with other gases in Figure 4.25 and Figure 4.26 (Villa, 1983). Considered very promising in the early developments of micro-pattern gas detectors, particularly because of its non-ageing properties, it has the drawback of being flammable and chemically aggressive for some materials.

Carbon tetrafluoride also has fast drift velocity and low diffusion, comparable to methane; CF_4-based gas mixtures for use in high-rate detectors have been studied extensively, Figure 4.27 (Christophorou *et al.*, 1979). Their main advantages for use in large volume detectors in particle physics are non-flammability and low sensitivity to neutrons; also, they do not form polymers in the avalanches, and even have etching properties capable of removing existing deposits on electrodes, as discussed in Chapter 16.

The classic theory assumes symmetric electron diffusion, described by a single coefficient D. This appeared not to be the case for some gases at high fields for which the longitudinal diffusion coefficient in the direction of drift, D_L, can be smaller that

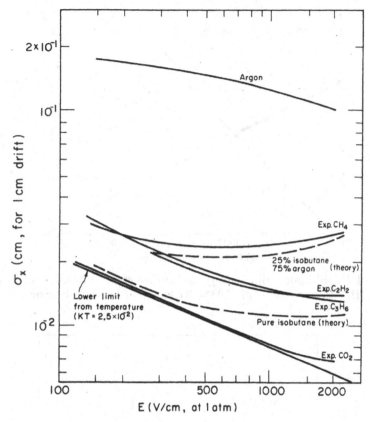

Figure 4.23   Electron diffusion (rms) for 1 cm drift in several gases (Palladino and Sadoulet, 1975). By kind permission of Elsevier.

the transverse coefficient D_T. Figure 4.28 shows the measured values of D_L/μ and D_T/μ as a function of field in argon, compared to the theoretical predictions (Lowke and Parker, 1969). In carbon dioxide, on the contrary, the diffusion remains thermal up to very high values of field, with practically no difference between longitudinal and transverse diffusion, as shown in Figure 4.29 (Christophorou, 1971). The difference can be explained by considering that the increase of electron energy between collisions is affected by the direction of the electrons in respect to the field, particularly in gases where the mean free path is long (Parker and Lowke, 1969). The reduction of longitudinal diffusion in some gas mixtures helps in improving the drift time resolution; Figure 4.30 is a measurement in an argon–methane–isobutane mixture made with the JADE detector (Drumm *et al.*, 1980).

For some gases, a violation of the *E/P* invariance of the drift velocity and of the inverse square root dependence on pressure of the diffusion are observed, due to a process of temporary electron capture and release from the molecules that results in slowing down of the drift at increasing pressures. Examples are given in

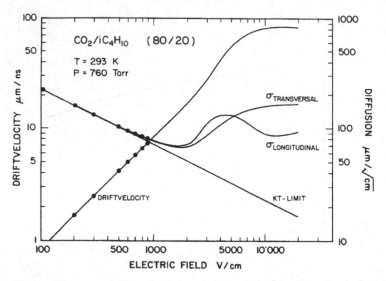

Figure 4.24 Electron drift velocity and diffusion as a function of field in CO_2/iC_4H_{10} (Commichau *et al.*, 1985). By kind permission of Elsevier.

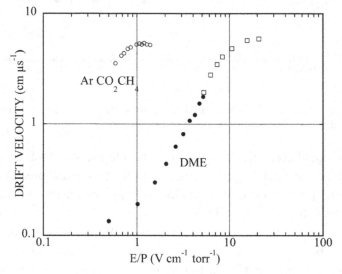

Figure 4.25 Drift velocity for 1 cm drift in DME, compared to other gases (Villa, 1983). By kind permission of Elsevier.

Figure 4.31 and Figure 4.32 for carbon dioxide (Bobkov *et al.*, 1984); a similar observation has been reported for dimethyl ether, limiting the improvement in localization accuracy that can be obtained with increasing pressures.

Figure 4.33 and Figure 4.34 are examples of measured drift velocity and Lorentz angles in a widely used drift chamber gas mixture (argon–isobutane–methylal),

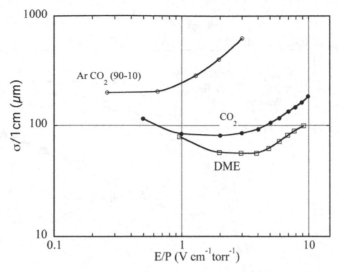

Figure 4.26   Diffusion for 1 cm drift of DME compared to other gases (Villa, 1983). By kind permission of Elsevier.

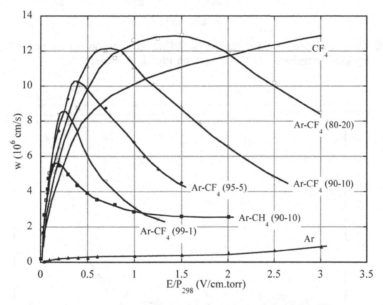

Figure 4.27   Electron drift velocity as a function of field in CF_4, pure and in mixtures with argon (Christophorou *et al.*, 1979). By kind permission of Elsevier.

for perpendicular electric and magnetic fields (Breskin *et al.*, 1974b). The drift velocity, reduced at low electric fields, tends to reach the same saturation value for all values of magnetic field; the angle of drift follows an almost linear dependence on B for large electric fields.

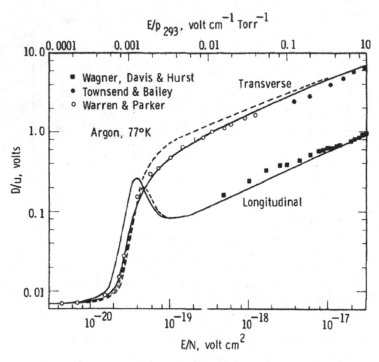

Figure 4.28   Longitudinal and transverse diffusion in argon (Lowke and Parker, 1969). By kind permission of the American Physical Society.

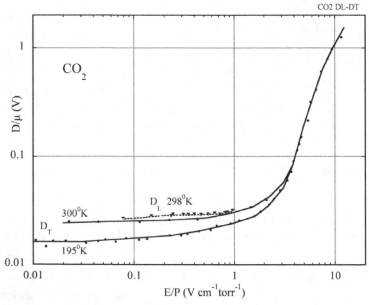

Figure 4.29   Longitudinal (D_L) and transverse (D_T) diffusion of electrons in carbon dioxide. Redrawn from data of Warren and Parker (1962) and Wagner *et al.* (1967).

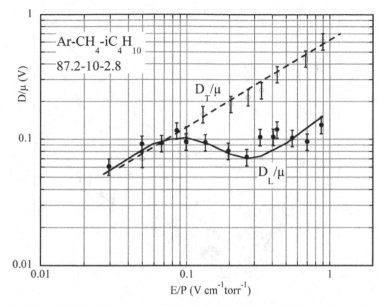

Figure 4.30 Longitudinal and transverse diffusion coefficients measured in a drift chamber (Drumm *et al.*, 1980). By kind permission of Elsevier.

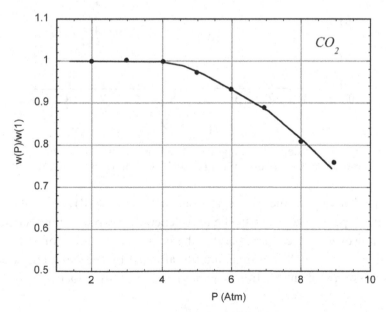

Figure 4.31 Pressure dependence of drift velocity in carbon dioxide at equal reduced field (Bobkov *et al.*, 1984). By kind permission of Elsevier.

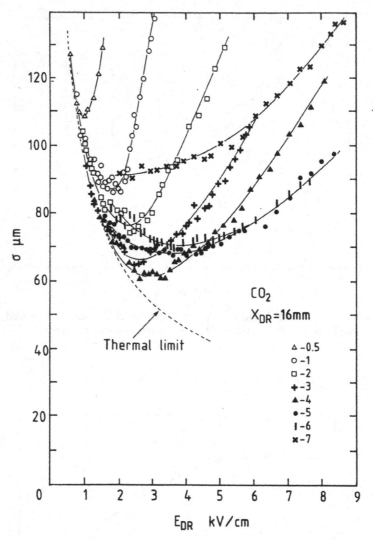

Figure 4.32   Dependence of longitudinal diffusion on the electric field for several pressures in CO_2 (Bobkov *et al.*, 1984). By kind permission of Elsevier.

While qualitatively instructive, expressions (4.17) and (4.18) are of little practical use since the dependence of τ on the fields is generally unknown. For moderate field strengths, however, one can assume that the energy distribution of electrons, and therefore the average collision time, are not affected by the field, and use Townsend's expression to deduce τ from w_0, the measured drift velocity for $B = 0$:

$$\tau \approx \tau_0 = \frac{m}{eE} w_0; \tag{4.20}$$

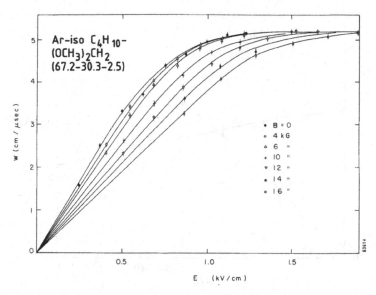

Figure 4.33   Drift velocity as a function of electric field and several values of perpendicular magnetic field (Breskin *et al.*, 1974b). By kind permission of Elsevier.

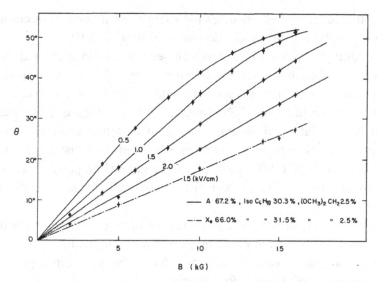

Figure 4.34   Drift angle as a function of electric and magnetic fields (Breskin *et al.*, 1974b). By kind permission of Elsevier.

magnetic drift angles and velocities can then be estimated using the quoted expressions. An example of comparison between the predictions of this simple model and measurements, for perpendicular fields, is shown in Figure 4.35 (Breskin *et al.*, 1974b); the agreement is reasonably good. At higher fields, the

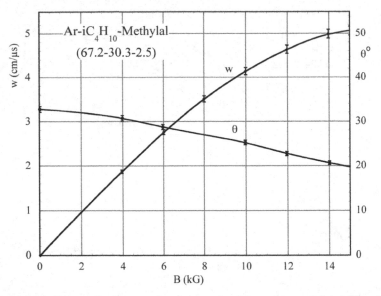

Figure 4.35 Comparison between measured values of magnetic drift velocity and angle with the predictions of a simplified model (Breskin *et al.*, 1974b). By kind permission of Elsevier.

presence of the magnetic field modifies the energy distribution of electrons and a more rigorous analysis is necessary, described in Section 4.10.

Magnetic drift velocities and Lorentz angles have been measured for many gases, in view of their relevance for detectors operated in high magnetic fields. Results for carbon tetrafluoride–isobutane and argon–carbon dioxide in equal percentage are given in Figure 4.36 (Ogren, 1995) and Figure 4.37 (Bittl *et al.*, 1997), and for a mixture of several gases with dimethyl ether in Figure 4.38 (Angelini *et al.*, 1994). Other measurements in a range of fast gas mixtures are reported in Kiselev *et al.* (1995) and confirm the general property of 'cold' gases, where electrons remain close to thermal at high fields, to have small magnetic angles.

With the availability of computer programs permitting one to calculate drift and diffusion properties in a wide range of gases and their mixtures the interest in systematic measurements of drift properties has decreased, although an experimental verification is often needed for special conditions.

4.8 Electron capture

Drifting in the gas under the effect of an electric field, electrons can be captured by molecules with electronic affinity creating negative ions. The major capture process, resonance dissociative attachment, has been abundantly described in the

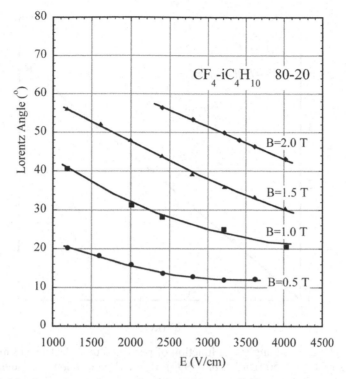

Figure 4.36   Lorentz angle as a function of perpendicular electric and magnetic fields in a CF_4-iC_4H_{10} 80–20 (Ogren, 1995). By kind permission of Elsevier.

literature, see for example Christophorou (1971); modern theories of electron transport include the capture cross sections in the estimate of drift properties.

In the classic theory (Brown, 1959), the capture process can be described by defining an attachment coefficient h, the probability of attachment per collision. The number of electron–molecule collisions per unit drift length is given by $(w\tau)^{-1}$, where w and τ are the drift velocity and mean collision time. Introducing Townsend's expression (4.13), the number of collisions with attachment per unit drift length is:

$$h\frac{1}{w\,\tau} = h\frac{e}{m}\frac{E}{w^2},$$

the electron loss during drift is then described by the differential equation:

$$dn = -\frac{e}{m}\frac{E}{w^2}\,dx;$$

an integration provides the number of surviving electrons after a drift length s:

$$n = n_0\,e^{-h\frac{e\,E_s}{mw^2}} = n_0 e^{-\frac{s}{\lambda}}, \tag{4.21}$$

where λ is a phenomenological mean capture length.

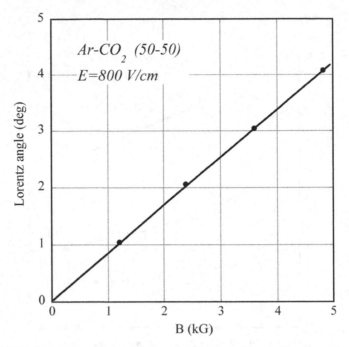

Figure 4.37   Lorentz angle as a function of perpendicular electric and magnetic fields in Ar-CO$_2$ 50–50 (Bittl *et al.*, 1997). By kind permission of Elsevier.

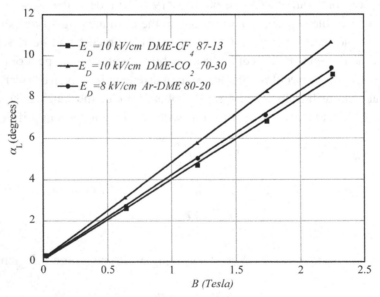

Figure 4.38   Lorentz angle in DME mixtures (Angelini *et al.*, 1994). By kind permission of Elsevier.

With the addition to a main gas of electro-negative molecules in a percentage p, and assuming that neither the drift velocity nor the attachment coefficient are modified, the previous expression can be written as:

$$n = n_0 e^{-ps/\lambda}. \tag{4.22}$$

For a detector uniformly exposed to radiation, with a drift cell size S, and assuming that the capture length is not affected by field variations within the cell, the average collected charge, normalized to the initial ionization charge, can be obtained by integrating the previous expression over the full drift path length S:

$$\frac{Q_S}{Q_0} = \int_0^S e^{-ps/\lambda} ds = \frac{\lambda}{P}\left(1 - e^{-ps/\lambda}\right). \tag{4.23}$$

As the detected charge fraction depends on the drift length within the cell, the energy resolution will of course be correspondingly affected.

Values of attachment coefficients for several gases, in the absence of an electric field, were given in Table 4.3. The cross section for electron capture varies considerably, however, with the electron energy, hence the field, and has been extensively studied experimentally, particularly for thermal electron energies; classic examples are given for oxygen in Figure 4.39 (Bloch and Bradbury, 1935) and for water in Figure 4.40 (Bradbury and Tatel, 1934). In both cases, the attachment probability is large for low fields and electron energies close to thermal, but decreases at increasing fields; for water, it becomes negligible at fields of a few $Vcm^{-1}torr^{-1}$.

Figure 4.41 is a collection of values of attachment rate as a function of electron energy for chlorinated vapours (Christodoulides and Christophorou, 1971); while usually not intentionally present as constituents in the main gas mixture, these compounds are often used as cleaning agents during the detector construction, and special care should be taken to remove these residual contaminations.

Modern compilations and measurements of the electron–molecule cross sections, including capture, and their relevance in the detectors' performances will be discussed in Section 4.10.

As discussed in the next section, due to the field dependence of electron–molecules cross sections, their energy distribution at a given value of field depends on the gas or gas mixture; as a consequence, the capture probability can differ considerably in different mixtures, for equal amounts of electro-negative contaminants. This is seen in Figure 4.42, where the experimentally measured fractions of electrons surviving after a 20 cm drift in a detector, at a drift field of 200 V/cm, are given as a function of oxygen content for two gas mixtures. For equal oxygen

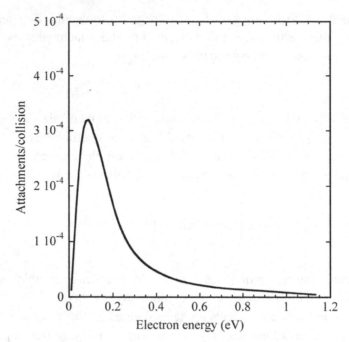

Figure 4.39 Electron attachment coefficient for oxygen (Bloch and Bradbury, 1935).

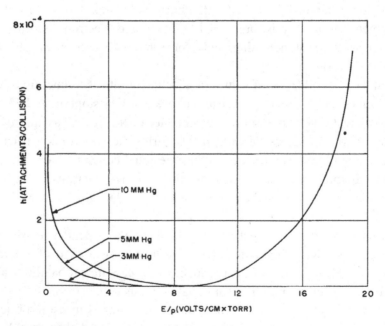

Figure 4.40 Electron capture probability in water vapours as a function of reduced field (Bradbury and Tatel, 1934).

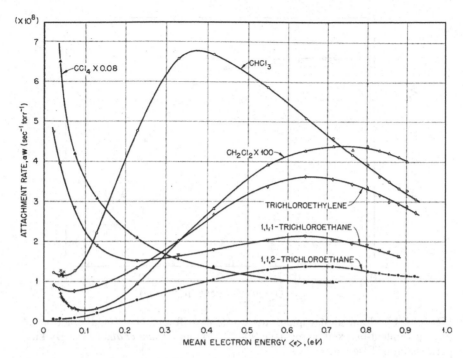

Figure 4.41 Attachment rates of chlorinated vapours (Christodoulides and Christophorou, 1971). By kind permission of the American Institute of Physics.

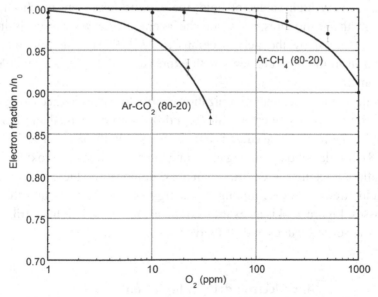

Figure 4.42 Fraction of electrons surviving after 20 cm drift in two gas mixtures, as a function of oxygen content ($E = 200$ V/cm). Compilation from Ar-CO$_2$ data of Price *et al.* (1982) and Ar-CH$_4$ data of Lehraus *et al.* (1984).

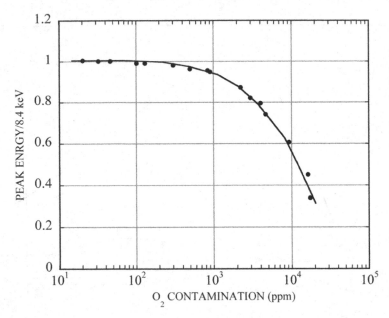

Figure 4.43 Ionization capture losses in a drift chamber as a function of oxygen content in a xenon–methane 90–10 mixture (Chiba *et al.*, 1988). By kind permission of Elsevier.

content, in argon–carbon dioxide (Price *et al.*, 1982), where electrons remain thermal up to very high values of field, the capture losses are much larger than for argon–methane mixtures, in which the average electron energy is increased well above thermal by the field (Lehraus *et al.*, 1984). The full lines are an exponential fit to the measurements with expression (4.22), using suitable values of the parameters.

The results of measurements of ionization energy loss in a charged particles beam uniformly illuminating a detector with 1 cm drift in a xenon–methane gas mixture are shown in Figure 4.43; the curve is a fit to the data with $\lambda p = 7.8 \times 10^{-3}$ (Chiba *et al.*, 1988). Similar studies, aiming at finding the effect of residual oxygen and SF_6 contaminations in xenon-filled drift chambers, demonstrate the crucial role of the electric field strength in determining the energy of the electrons and the ensuing capture losses. Figure 4.44 is an example of measured pulse height distributions for soft X-rays at several values of drift field (Andronic *et al.*, 2003).

4.9 Electron drift in liquid noble gases

Drift velocity and diffusion of electrons in liquefied noble gases have been studied years ago. The major difficulty both for measurements and exploitation in detectors

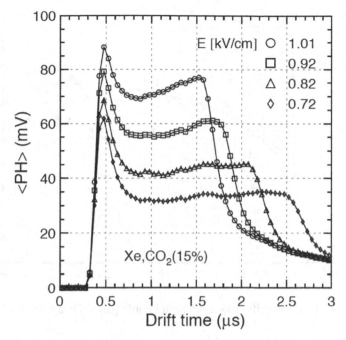

Figure 4.44   Average pulse height as a function of drift time for several values of drift field (Andronic *et al.*, 2003). By kind permission of Elsevier.

lies in the extreme sensitivity of the electron drift to the gas purity and the presence of electro-negative contaminations, often expresses in terms of electron lifetimes; this explains discrepancies between measurements. Figure 4.45, an eye-fit compilation from data from Miller *et al.* (1968), provides the electron drift velocity in liquid argon, krypton and xenon; at comparable values of field, the drift velocities are about ten times lower than in the corresponding gas phase.

The electron drift in liquid argon and xenon has been studied systematically in the course of development of the cryogenic and dual-phase detectors, described in Section 15.2. Measured with an LAr-TPC prototype, Figure 4.46 shows the field dependence of the velocity at two temperature values; measurements are compared with an empirical expression (Walkowiak, 2000). Figure 4.47 is the field dependence of electron drift velocity for xenon (Aprile *et al.*, 1991); experimental values of electron diffusion in liquid argon and xenon as a function of the reduced field E/N are given in Figure 4.48 (Doke, 1982).

For applications requiring charge multiplication in the gas phase, small additions of methane permit one to reach higher gains and faster drift velocity, with only a small reduction of the scintillation yield; electron drift properties in a double-phase xenon cell with small methane concentrations are described in (Lightfoot *et al.* (2005).

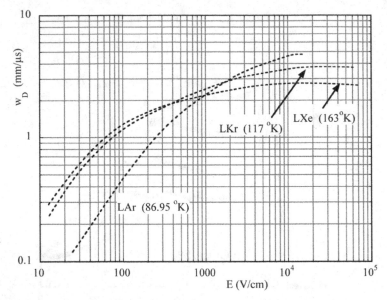

Figure 4.45 Electron drift velocity in liquid noble gases. Data from Miller *et al.* (1968).

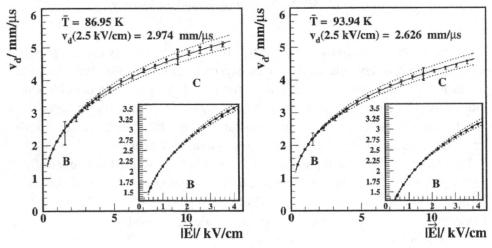

Figure 4.46 Electron drift velocity in liquid argon at two values of temperature (Walkowiak, 2000). By kind permission of Elsevier.

4.10 Transport theory

A rigorous theory of electron drift in gases has been developed by many authors, based on seminal works on Boltzmann transport theory describing the behaviour of free electrons in a gas under the influence of an electric field (Morse *et al.*, 1935). The algorithm can be outlined as follows: at each value of the field, the

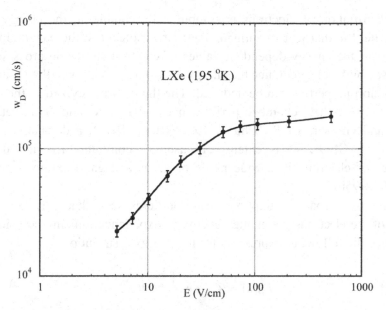

Figure 4.47   Electron drift velocity in liquid xenon (data from Aprile *et al.*, 1991). By kind permission of Elsevier.

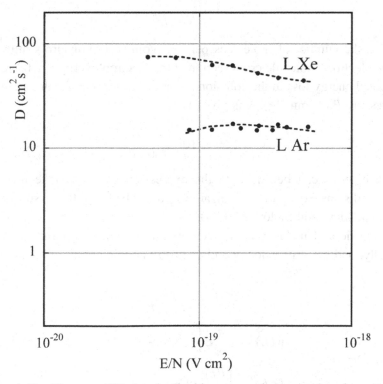

Figure 4.48   Electron diffusion in liquid argon and xenon (data from Doke, 1982). By kind permission of Elsevier.

energy distribution of electrons is computed by equalizing the energy gained from the field to that lost in collisions with the molecules; the calculation takes into account the energy-dependent values of electron–molecule cross sections, both elastic and inelastic. Once the energy distribution is known, the various drift and diffusion properties can be deduced. The theory was revived in the eighties in the study of drift chamber performances (Palladino and Sadoulet, 1975; Schultz and Gresser, 1978; Schultz, 1976; Biagi, 1999); a dedicated software program, MAGBOLTZ, is extensively used to compute drift and diffusion properties of electrons in a wide range of gases and gas mixtures (Biagi and Veenhof, 1995b).

Under rather broad assumptions, and for fields such that only a negligible fraction of the electrons get enough energy to experience ionizing collisions, one can deduce the following expression for the energy distribution:

$$F_0(\varepsilon) = C\sqrt{\varepsilon}\, e^{-\int \frac{3\varepsilon\, \Lambda(\varepsilon)\, d\varepsilon}{[e\, E\, l(\varepsilon)]^2}}, \tag{4.24}$$

with the mean free path between collisions given by:

$$l(\varepsilon) = \frac{1}{N\sigma(\varepsilon)}, \tag{4.25}$$

where N is the number of molecules per unit volume (Avogadro's number) and $\sigma(\varepsilon)$ is the electron–molecule cross section at the electron energy ε; $\Lambda(\varepsilon)$ represents the fractional energy loss in the collisions. For an ideal gas at absolute temperature T and pressure P (in mmHg), N is given by:

$$N = 2.69 \times 10^{19} \frac{P}{760} \frac{273}{T}. \tag{4.26}$$

Appropriate terms can be added to the expression (4.24) to take into account inelastic collisions (excitation, ionization), described by the respective cross sections (Palladino and Sadoulet, 1974).

If the elastic and inelastic cross sections are known, $F_0(\varepsilon)$ can be computed numerically, and the drift velocity and diffusion coefficient are obtained from the expressions:

$$w(E) = \frac{2}{3}\frac{e}{m}E\int \varepsilon l(\varepsilon)\, \frac{\partial \frac{F_0(\varepsilon)}{v}}{\partial \varepsilon}\, d\varepsilon,$$

$$D(E) = \int \frac{l(\varepsilon)}{3}\, vF_0(\varepsilon)d\varepsilon, \tag{4.27}$$

where $v = \sqrt{2\varepsilon/m}$ is the instant velocity of electrons of energy ε. Simple composition rules hold for gas mixtures, with obvious meanings:

$$\sigma(\varepsilon) = \sum p_i \sigma_i(\varepsilon) \ \text{ and } \ \sigma(\varepsilon)\Lambda(\varepsilon) = \sum p_i \sigma_i(\varepsilon)\Lambda_i(\varepsilon).$$

It is customary to define a characteristic energy ε_k as follows:

$$\varepsilon_k = \frac{e \, E \, D(E)}{w(E)}; \tag{4.28}$$

a comparison with expression (4.9) shows that the characteristic energy replaces the factor kT, and represents the average 'heating' of the electron swarm by the field. Further refinements in the theory and computational algorithms permit one to take into account the non-symmetric diffusion and the effect of the magnetic field on drifting electron swarms (Biagi, 1999).

The electron–molecule collision cross sections depend very strongly on E for most gases; for argon they go through maxima and minima (the Ramsauer effect). This is a consequence of the fact that the electron wavelengths approach those of the electron shells of the atom or molecule, and complex quantum-mechanical processes take place. At increasing values of E, the energy distribution therefore changes from the original Maxwellian shape, and the average energy can exceed the thermal value by orders of magnitude; eventually, cross sections reach a minimum, resulting in a saturation and even decrease of the drift velocity at increasing field.

The program MAGBOLTZ includes extensive compilations of the electron–molecule cross sections for a wide list of gases (Biagi and Veenhof, 1995a); detailed and updated values for the electron–molecule cross sections can be found on the open-access web site LXCAT. Figure 4.49 to Figure 4.54, simplified plots from the numerical data provided by LXCAT from the Morgan compilation (Montgomery and Montgomery, 1941) give examples for several atomic and molecular gases. For noble gases, the cross section is elastic until the electrons reach the first excitation and ionization energy, above 10 eV; on the contrary, for molecular gases inelastic channels, vibrational and rotational cross sections open up at energies above 0.1 eV. For electro-negative species, the attachment cross section is also given; in oxygen, a peak between 0.06 and 0.1 eV explains the large electron capture probability near thermal energies. Carbon tetrafluoride, on the contrary, has a peak around 8 eV, causing electron losses at high field values.

High values of the cross sections reduce the electron diffusion and increase the drift velocity; a large inelasticity implies that high fields are required to raise the electron energy. One can also understand why the addition of very small

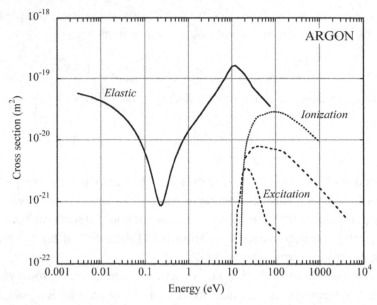

Figure 4.49 Electron–molecule cross section for argon (data from LXCAT).

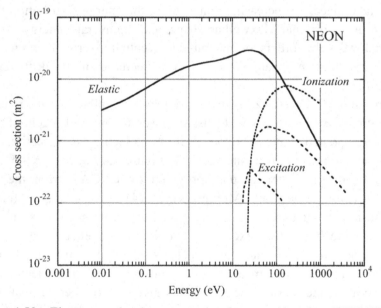

Figure 4.50 Electron–molecule cross section for neon (data from LXCAT).

amounts of a molecular gas to noble gases has the observed large effect on drift properties: for argon, at an energy close to the Ramsauer minimum, a 1% addition of carbon dioxide results in equal contributions to the cross sections of the two species.

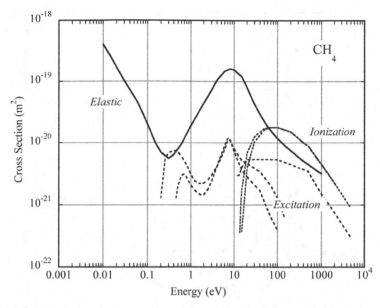

Figure 4.51   Electron–molecule cross section for methane (data from LXCAT).

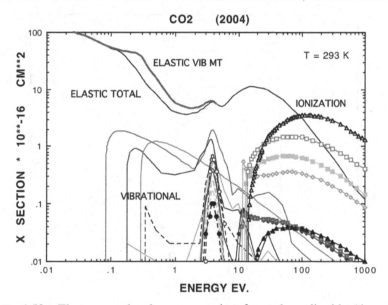

Figure 4.52   Electron–molecule cross section for carbon dioxide (data from LXCAT).

The predictive accuracy of the program has been verified in many cases when accurate data are available; Figure 4.55 and Figure 4.56 (Biagi, 1999) give representative examples.

Figure 4.57 shows the computed electron energy distribution at moderate fields (100 V cm^{-1}) in pure argon and in a 70–30 mixture of argon and carbon dioxide at

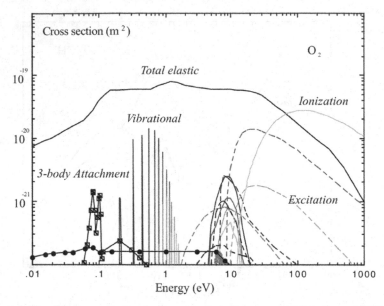

Figure 4.53   Electron–molecule cross section for oxygen (data from LXCAT).

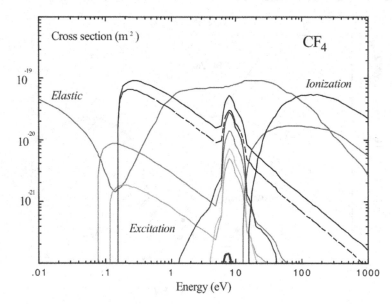

Figure 4.54   Electron–molecule cross section for carbon tetrafluoride (data from LXCAT).

NTP; the already mentioned 'cooling' effect of the addition of a molecular gas is apparent, with an average electron energy close to the thermal value for the mixture, while it reaches several eV in pure argon.

The electron energy is increased towards larger values with the application of higher electric fields. An example of computed distributions for an

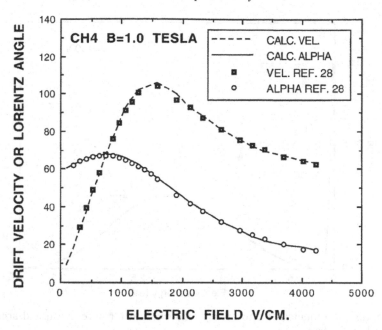

Figure 4.55   Computed and measured electron drift velocity in methane (Biagi, 1999). By kind permission of Elsevier.

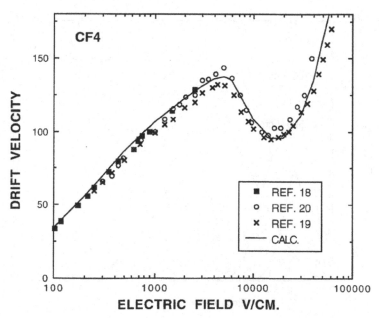

Figure 4.56   Computed and measured electron drift velocity in carbon tetrafluoride (Biagi, 1999). By kind permission of Elsevier.

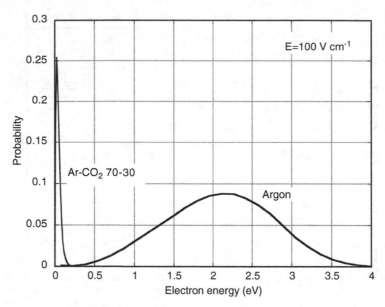

Figure 4.57   Computed electron energy distribution for pure argon and argon–CO_2 at low fields.

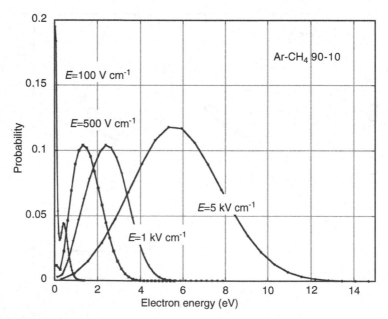

Figure 4.58   Computed electron energy distribution as a function of field in argon–CH_4.

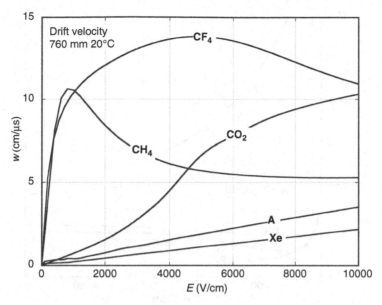

Figure 4.59   Computed drift velocity as a function of field for pure gases at NTP.

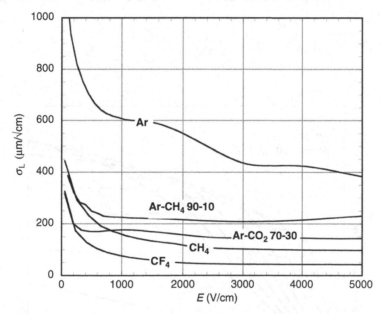

Figure 4.60   Computed longitudinal diffusion for 1 cm drift in several gases at NTP.

argon–methane 90–10 mixture at NTP is given in Figure 4.58; for a field of $5\,\text{kV}\,\text{cm}^{-1}$ the upper tail of the distributions exceeds the inelastic excitation and ionization levels of methane (8.8 and 12.6 eV, respectively), causing the onset of inelastic processes.

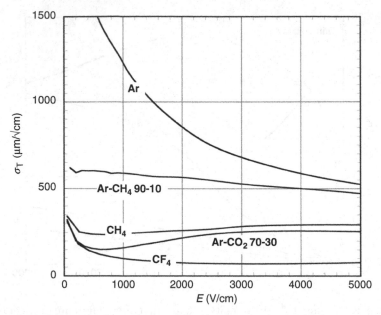

Figure 4.61   Computed transverse diffusion for 1 cm of drift in several gases at NTP.

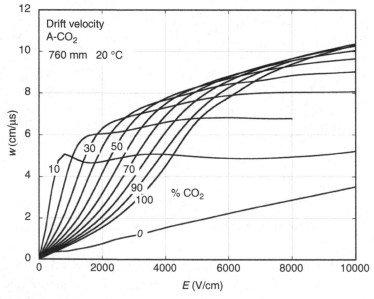

Figure 4.62   Computed drift velocity as a function of field for argon–carbon dioxide mixtures at NTP.

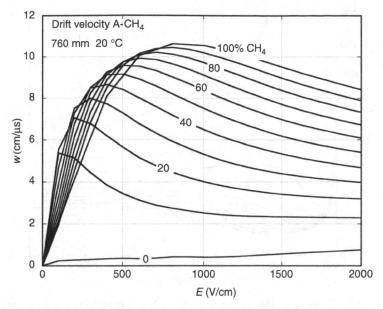

Figure 4.63 Computed drift velocity as a function of field for argon–methane mixtures at NTP.

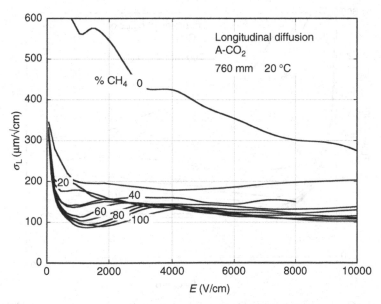

Figure 4.64 Standard deviation of the longitudinal diffusion for 1 cm drift, as a function of field, for argon–carbon dioxide mixtures at NTP.

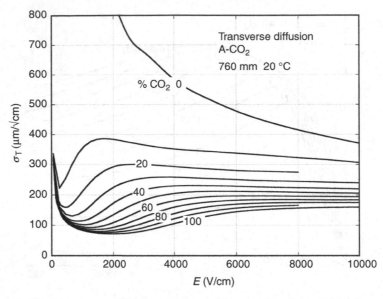

Figure 4.65   Transverse diffusion for 1 cm drift as a function of field for argon–carbon dioxide mixtures at NTP.

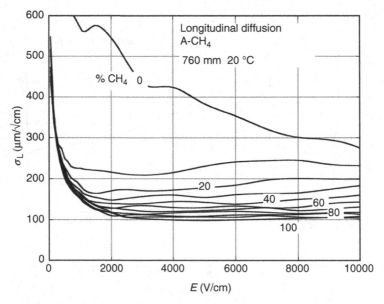

Figure 4.66   Longitudinal diffusion for 1 cm drift as a function of field for argon–methane mixtures at NTP.

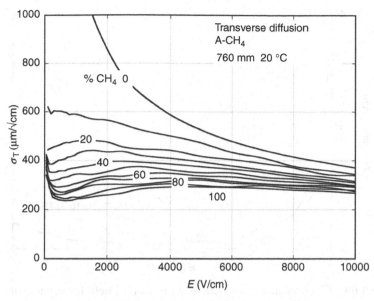

Figure 4.67   Transverse diffusion for 1 cm drift as a function of field for argon–methane mixtures at NTP.

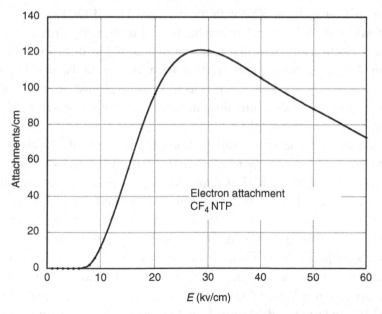

Figure 4.68   Electron attachments/cm of drift in carbon tetrafluoride as a function of field.

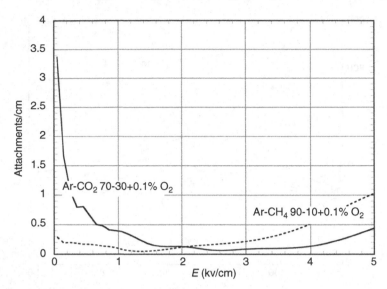

Figure 4.69   Electron attachments/cm as a function of field for equal addition of oxygen to argon–carbon dioxide and argon–methane mixtures.

Examples of drift velocity and diffusion for a range of gases and mixtures commonly used in detectors are given in Figure 4.59 to Figure 4.67; all data have been computed for gases at NTP with the program MAGBOLTZ.

The presence of an external magnetic field modifies the drift and diffusion properties of electrons, in particular reducing the transverse diffusion, as discussed in Section 4.6. This basic property, that permits one to largely improve the localization accuracy on long drift detectors, is exploited in time projection chambers; plots of drift and diffusion properties for several cases used in these devices are given in Section 10.5.

As it includes the attachment cross sections, the program allows also computation of the electron losses in pure gases and mixtures. Figure 4.68 provides the capture probability, expressed in attachments/cm, for pure carbon tetrafluoride at NTP, as a function of field: the large value at around 30 kV/cm reflects the corresponding peak in the cross section for electron energies approaching 10 eV (Figure 4.54). As at these high fields electrons already experience charge multiplication, losses are partly compensated by the increase of the overall charge. Experimental data on electron attachment in CF_4 are given in Anderson *et al.* (1992).

As another example Figure 4.69 shows the capture losses for equal additions of oxygen (0.1%) to argon–carbon dioxide and argon–methane mixtures; the very large difference in capture probability at low field in the first mixture is apparent, as already shown by the experimental results presented in Section 4.8.

5

Collisional excitations and charge multiplication in uniform fields

5.1 Inelastic electron–molecule collisions

When the electric field is increased above a few kV/cm, more and more electrons can acquire enough energy between two collisions to produce inelastic phenomena, excitations of various kinds and ionizations. A molecule can have several characteristic modes of excitation, increasing in number and complexity for polyatomic molecules; typical values for the lowest excitation levels are between ten and twenty electron volts. In each collision, an electron can spend an amount of energy equal to or larger than the excitation energy; the atom or molecule, left in an excited state, returns to the ground state through single or multiple transitions.

Noble gas atoms de-excite through photon emission, while weakly bound polyatomic molecules, as for example the hydrocarbons widely used in proportional counters as a quencher, have rotational and vibrational radiationless transitions. Addition of an organic vapour to noble gases therefore allows the dissipation of a good fraction of energy without the creation of photons or ions, an important factor to permit high gain and stable operation of proportional counters.

At higher fields, the probability of ionizing collisions increases and exceeds the decreasing probability of excitations; the outcome of an ionizing collision is the creation of an electron–ion pair, while the primary electron continues its motion in the gas. When the mean free path for ionizing collisions is small compared to the gas layer thickness, the two electrons soon acquire sufficient energy from the field to ionize further; this leads to a rapid growth of an electron–ion avalanche, and constitutes the basic mechanism of signal amplification in gas proportional counters.

Examples of absolute cross sections for electron–molecule collisions as a function of electron energy were given in the previous chapter. For a given value of the field, the statistical electron energy distribution spans a range of values of the cross sections, determining the relative probability of the various processes; at high

Table 5.1 *Major processes in electron and ion–molecule collisions; A and B designate two atoms (molecules), A*, B* their excited states and A^+, A^-, B^+ the corresponding ions.*

Process	Initial	Final
Excitation	$A+e$	A^*+e
Ionization	$A+e$	A^++e+e
De-excitation	A^*+e	$A+e$
Photo-excitation	$A+h\upsilon$	A^*
Photo-ionization	$A+h\upsilon$	A^++e
Photo-emission	A^*	$A+h\upsilon$
Electron capture	$A+e$	A^-
Radiative recombination	A^++e	$A+h\upsilon$
Excimers formation	A^*+A+A	A_2^*+A
Radiative excimer dissociation	A_2^*	$A+A+h\upsilon$
Collisional de-excitation	A^*+B	$A+B^*$
Charge exchange	A^++B	$A+B^+$
Penning effect	A^*+B	$A+B^++e$

fields, such as those encountered in gaseous detectors, the contribution of elastic processes becomes negligible, while excitations and ionizations tend to share the spent energy almost equally.

A large number of processes can follow the inelastic interaction of electrons and molecules; more than 20 are described in Meek and Cragg (1953). Table 5.1 summarizes those having a major impact on the operation of gaseous detectors in a pure gas A and mixtures of two gases A and B. Depending on the species, the probability of a process can be very different, and not all processes are energetically permitted. The process of molecular dissociation, direct or going through a phase of pre-dissociation, which result in breaking the molecule into two or more fragments, competes with the other de-excitation mechanisms; radiationless internal conversions can also occur, relaxing the molecule into the lowest excited state of a given multiplicity. Even a superficial description of these mechanisms would be far beyond the scope of this book; the reader is referred to the textbooks quoted above and in particular Christophorou (1971) for a more detailed analysis.

5.2 Excitations and photon emission

Electronic excitation processes have been extensively studied, both theoretically and experimentally, for rare gases and their mixtures. Originally motivated by the research on gaseous discharges and by the related field of proportional counters operation, this branch of applied research has received more interest in connection with the development of excimer gas lasers, see for example Rhodes (1979). The mechanisms leading to the creation and decay of excited states in pure noble gases

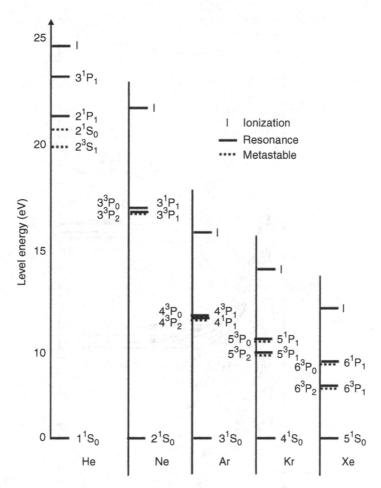

Figure 5.1 First electronic excitation and ionization levels for the noble gases (compiled from different sources).

and their mixtures are rather well understood, and quantitative calculations are possible using the appropriate formulations. This is to a much lesser extent the case for molecular gases; even for the simplest ones, such as the hydrocarbons often used in proportional counters, the study is often reduced to a purely phenomenological description of the energy exchange processes involved.

Figure 5.1, from different sources, shows schematically the lowest excitation and ionization levels for noble gases; numerical values of the corresponding energies were given in Table 2.1. Except for helium, rare gases have a very similar electronic structure, with the lowest excited states corresponding to one electron raised from the ground state $^1S_0$ to a $^1P_1$ singlet or to one of the $^3P_2$, $^3P_1$ or $^3P_0$ triplets.

Excited by electron impact, isolated noble gas atoms return to the ground state with the emission of a photon with an energy corresponding to the excitation level.

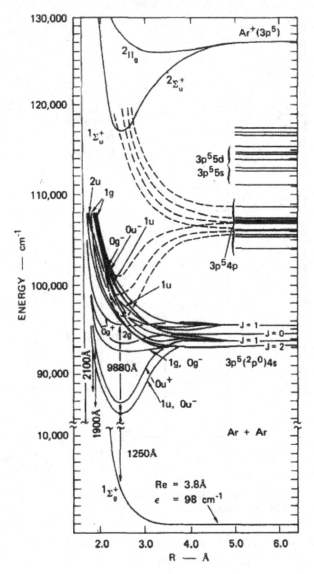

Figure 5.2   Argon excimer potential curves (Lorents, 1976). By kind permission of Elsevier.

However, at increasing pressures, collision of an excited atom with a neutral one can result in temporary formation of bound states, short-lived molecules named dimers for which lower energy transitions can occur. The energy levels of the argon dimer are shown in the Debye plot, Figure 5.2 (Lorents, 1976); the atomic levels correspond to the asymptotic values of the curves at large inter-atom separations. The atomic radiative de-excitation energy, 11.6 eV, corresponds to

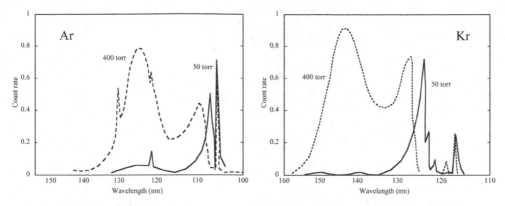

Figure 5.3   Fluorescence emission of argon and krypton excimers at different pressures (Hurst and Klots, 1976). By kind permission of John Wiley & Sons.

the emission of a photon with a corresponding wavelength (1050 Å), while the excimer emission is peaked at 1250 Å, as indicated in the figure. The relative fraction of dimeric and atomic emission depends on pressure, with the longer wavelength emission dominating above a few hundred torr, as shown in Figure 5.3 for argon and xenon (Hurst and Klots, 1976).

The creation of excited states by electron or particle impact and their successive evolution into radiative levels of the excimers is a very complex process; the timing properties of the emitted radiation depend both on the formation time, i.e. the time it takes for the excited states to form the diatomic molecules, and on the subsequent decay time to the ground state. For a review of the rare gases luminescence processes see for example Salete Leite (1980).

The photon emission increases rapidly with the applied electric field due to the increasing number of electrons with energy above the excitation level, and to the onset of charge multiplication. All gases emit photons when excited by electron impacts; however, the yields vary in a wide range, and depend on the wavelength sensitivity range covered by the detection method. As an example, Figure 5.4 shows the relative luminescence yields of a proportional counter measured in several gases as a function of the anode voltage (Keirim-Markus, 1972); the scintillation was measured with a standard bialkali photomultiplier through a quartz window and does not include the vacuum ultra-violet (VUV) region.

Secondary photon emission from gases under electron impact has been extensively studied experimentally during the development of the scintillation proportional counters; the process also plays a major role in the photon-induced discharge mechanisms that will be discussed later. Good summaries can be found in numerous review articles (Teyssier *et al.*, 1963; Thiess and Miley, 1975; Policarpo, 1977; Policarpo, 1981). In particular, the second reference includes summary tables of

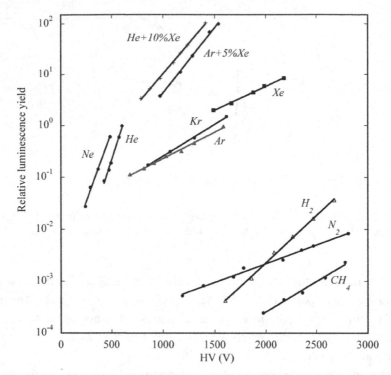

Figure 5.4   Normalized scintillation light yield as a function of field for different gases (Keirim-Markus, 1972). By kind permission of Springer Science+Business Media.

gas mixtures and pressures for which data are available. It is rather unfortunate that most of the measurements have been realized using a thin wire proportional counter, i.e. in a very inhomogeneous electric field; moreover, often only relative emission intensities are given, making difficult any quantitative estimate of the photon yield. Figure 5.5 (Suzuki and Kubota, 1979) shows the secondary scintillation spectra of argon, krypton and xenon measured at atmospheric pressure in a single wire proportional counter at a voltage below threshold for avalanche multiplication. Only the second continuum emission is observed, as noted above, with peaks centred at 128 nm (9.7 eV), 148 nm (8.4 eV) and 170 nm (7.3 eV) for A, Kr and Xe respectively; no emission is observed at longer wavelengths, between 2000 and 6000 Å. It is interesting to compare the emission spectra obtained at low fields with those in gas discharges, see the insets in the figure; Figure 5.6 shows the emission spectrum of krypton for a discharge at 400 torr and for proportional scintillation at 560 torr (Suzuki and Kubota, 1979). In the discharge, that involves the appearance of higher excited states, energetic transitions are possible, leading to emission at visible wavelengths.

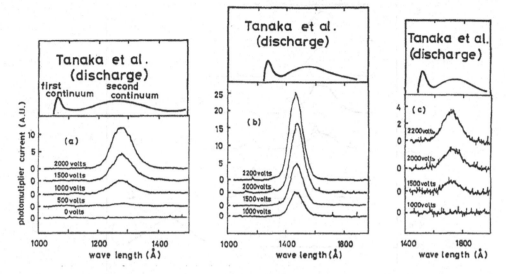

Figure 5.5 Secondary emission spectra of argon (a), krypton (b) and xenon (c) at atmospheric pressure measured with a scintillation proportional counter at different voltages (Suzuki and Kubota, 1979). By kind permission of Elsevier.

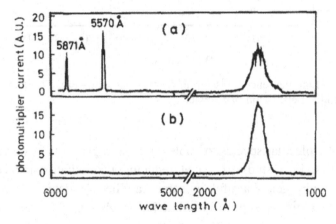

Figure 5.6 Krypton emission spectra at 400 torr in discharge (a) and at 560 torr in the proportional scintillation regime (b) (Suzuki and Kubota, 1979). By kind permission of Elsevier.

Addition to a rare gas of even small amounts of another species with smaller excitation potentials gradually shifts the emission to longer wavelengths, towards that of the additive, as shown in Figure 5.7 and Figure 5.8 for argon–xenon and argon–nitrogen (Takahashi *et al.*, 1983); with nitrogen, the emission occurs in the visible, making it a rather convenient internal wavelength shifter for the use of standard photomultipliers as sensors.

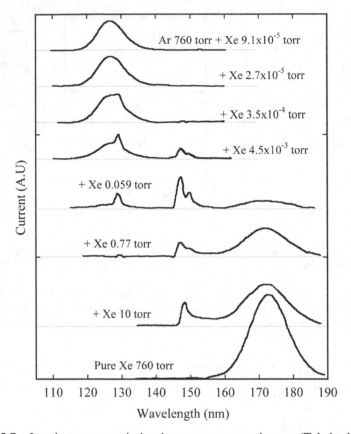

Figure 5.7   Luminescence emission in argon–xenon mixtures (Takahashi *et al.*, 1983). By kind permission of Elsevier.

Addition of molecular species to noble gases strongly reduces these secondary emission effects. In a study aimed at reducing the radiation-induced primary scintillation of nitrogen, harmful in some families of detectors, Morii *et al.* (2004) have measured the primary scintellation photon yield as a function of various additives in N_2; the results for oxygen are shown in Figure 5.9.

Photons emitted by primary or avalanche-induced excitation processes may be absorbed in the gas or by the electrodes; if their energy exceeds the corresponding ionization potentials, they can create secondary electrons adding up to the primary ionization. Depending on the counter geometry and gain, this may result in a spread of the charge, or approach a divergence situation leading to discharges.

Systematic measurements of photon emission by avalanches in heavily quenched proportional counter gas mixtures have been done to provide a quantitative understanding of the secondary photon-mediated processes leading to the appearance of streamers and discharges at high values of the electric field.

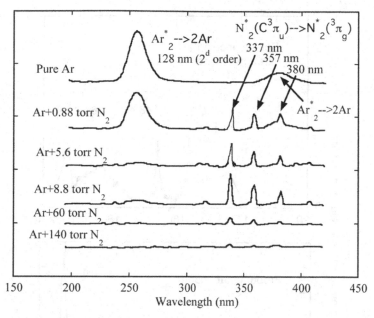

Figure 5.8 Luminescence emission in argon–nitrogen mixtures (Takahashi *et al.*, 1983). By kind permission of Elsevier.

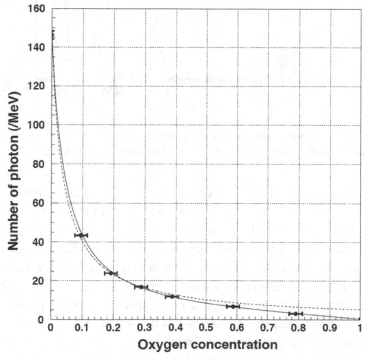

Figure 5.9 Primary scintillation photon yield in nitrogen as a function of oxygen content (Morii *et al.*, 2004). By kind permission of Elsevier.

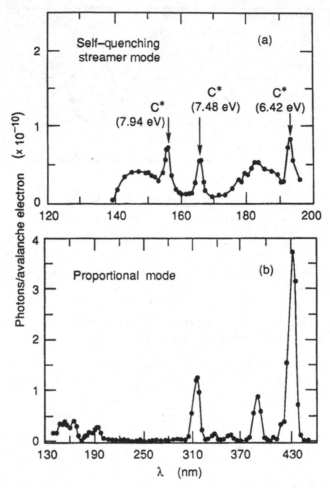

Figure 5.10   Secondary photon yields in argon–methane 90–10 in the SQS and proportional modes (Fraga *et al.*, 1992). By kind permission of Elsevier.

Measured with a single-wire counter coupled to a UV-monochromator, Figure 5.10 and Figure 5.11 show a comparison of emission spectra for argon–methane 90–10 and argon–CO_2 mixtures in the proportional and high gain conditions of the self-quenched streamer (SQS) operation (see Section 7.8); similar measurements are reported for other mixtures, together with attempts to identify the molecular species and excitation levels responsible for the multi-line emissions (Fraga *et al.*, 1992). The photon emission yield in the spectral range of the measurement is normalized to the number of electrons in the avalanche; as can be seen, it is almost two orders of magnitude lower for methane as compared to carbon dioxide, explaining the observed much better photon quenching properties of organic mixtures.

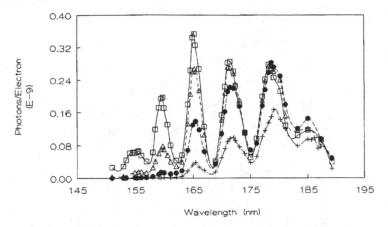

Figure 5.11   Spectral emission of argon–carbon dioxide mixtures in SQS discharges; the CO_2 concentrations vary from 21% (higher peaks) to 100% (lower peaks) (Fraga *et al.*, 1992). By kind permission of Elsevier.

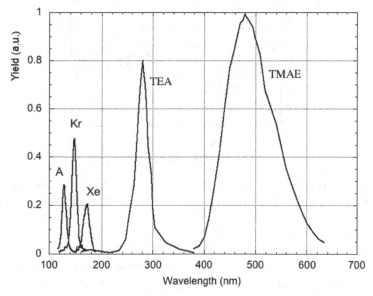

Figure 5.12   Secondary emission spectra of A, Kr, Xe, TEA and TMAE (Breskin *et al.*, 1988). By kind permission of Elsevier.

The fluorescence emission wavelength is longer the lower the excitation potential. Not surprisingly, the vapours used in photosensitive detectors, having low ionization potentials, fluoresce from the near ultra-violet to the visible; Figure 5.12 is a compilation of emission spectra for noble gases and two low-ionization potential vapours, triethyl amine (TEA) and tetrakis dimethyl amino ethylene (TMAE) (Breskin *et al.*, 1988). All spectra have been measured at pressures near

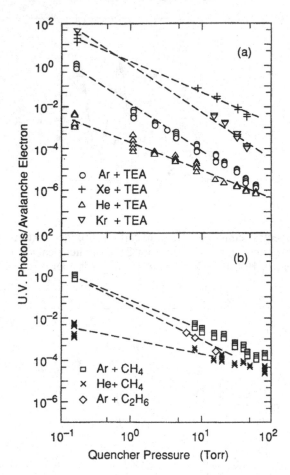

Figure 5.13   High-field secondary photon emission yield for mixtures of noble gases, TEA and some organic quenchers (Fonte *et al.*, 1991). By kind permission of Elsevier.

atmospheric and electric field conditions of charge multiplication. The secondary photon yield in these mixtures is orders of magnitude larger than for standard methane-quenched proportional counter gases, Figure 5.13 (Fonte *et al.*, 1991); due, however, to the longer emission wavelengths, these photons are harmless for the counter itself and are exploited for detection and imaging of radiation with optical means (Chapter 15).

Due to its interest as a fast, non-polymerizing filling in gaseous counters, the photon emission spectra of carbon tetra fluoride (CF_4), pure or in mixtures, have been studied by many authors in various pressure and field conditions (Pansky *et al.*, 1995a; Fraga *et al.*, 2001). As an example, Figure 5.14 (Morozov *et al.*, 2011) shows the scintillation spectrum at 5 bars and for two values of electric field; while the general shape remains the same, the emission at shorter wavelengths is

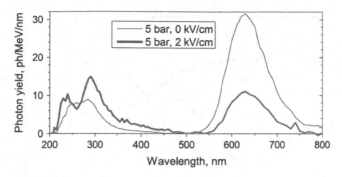

Figure 5.14 Emission spectra of CF_4 at two values of field (Morozov *et al.*, 2011). By kind permission of Elsevier.

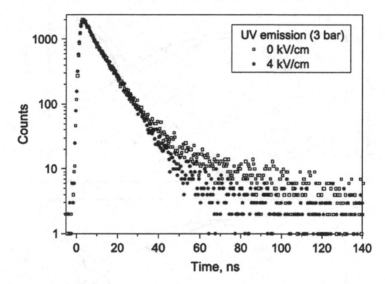

Figure 5.15 CF_4 scintillation decay times at two field values (Morozov *et al.*, 2011). By kind permission of Elsevier.

enhanced at high field. The emission is also rather fast, with typical decay times of a few tens of ns, as shown in Figure 5.15 (Morozov *et al.*, 2011).

Detailed studies of primary scintillation in rare gases show a field-dependent correlation between the primary charge and the luminescence yields induced by ionizing radiation, due to a mechanism of electron–ion recombination (Policarpo *et al.*, 1970; Suzuki, 1983; Saito *et al.*, 2008); Figure 5.16, from the second reference, shows the relative light and charge yield measured as a function of reduced electric field for argon, krypton and xenon. A simultaneous measurement of the two quantities can be exploited to improve the energy resolution of counters (Bolotnikov and Ramsey, 1999).

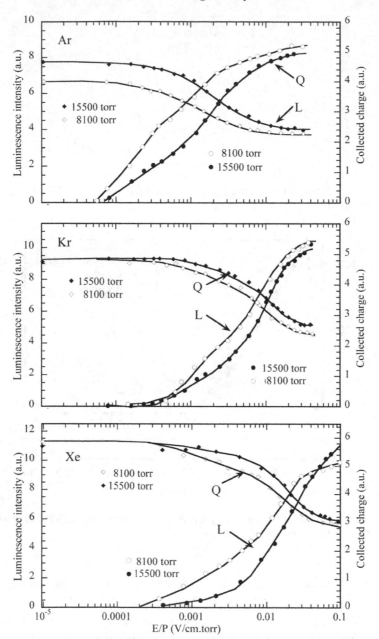

Figure 5.16   Luminescence (L) and charge yield (Q) in rare gases as a function of electric field (Suzuki, 1983). By kind permission of Elsevier.

Primary scintillation from liquid noble gases has a high photon yield and fast decay, and has been studied extensively in view of applications for gamma ray calorimetry, rare event and dark matter searches (Aprile *et al.*, 2002); as the emission is in the far and vacuum ultra-violet, appropriate wavelength shifters or

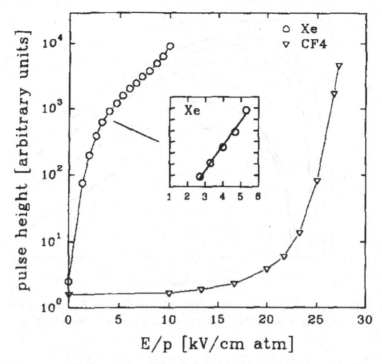

Figure 5.17   Relative scintillation yield of xenon and carbon tetrafluoride as a function of electric field. For CF$_4$, no secondary photon enhancement is observed before charge multiplication (Pansky *et al.*, 1995). By kind permission of Elsevier.

internal CsI photosensitive layers have to be used for detection. A summary of results for pure liquid rare gases and their mixtures and can be found in Doke and Masuda (1999).

Motivated by the problems connected to the primary photon emission from carbon tetrafluoride-filled radiators in Cherenkov counters, the scintillation yield of CF$_4$ in the UV region has been measured and compared with that of xenon, Figure 5.17 (Panski *et al.*, 1995).

5.3 Ionization and charge multiplication

When the energy of the electron accelerated by the electric field exceeds the ionization potential of the atom or molecule, bound electrons can be ejected, leaving behind a positive ion. Depending on the energy transfers involved and on the charge density, multiple ionized states can be produced, although in the conditions normally met in proportional counters, with electron energies below a few tens of eV, the creation of singly charged ions is the most likely process.

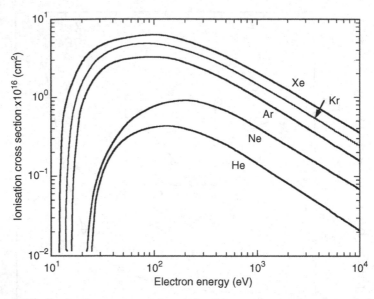

Figure 5.18   Ionization cross sections of noble gases as a function of electron energy. Simplified plot from data in Rapp and Englander-Golden (1965).

If not lost by capture or absorption in the walls, primary and secondary electrons continue their path, and can further ionize the gas molecules. The ionization cross section increases rapidly above the threshold and for most gases has a maximum at an energy above 100 eV, as shown in Figure 5.18 (Rapp and Englander-Golden, 1965).

The mean free path for ionization λ is defined as the average distance an electron has to travel before having an ionizing collision; its inverse, $\alpha = \lambda^{-1}$, is the ionization or first Townsend coefficient, and represents the number of ion pairs produced per unit length of drift; it relates to the ionization cross section through the expression:

$$\alpha = N\,\sigma_i, \tag{5.1}$$

where N is the number of molecules per unit volume.

As for other quantities in gaseous electronics, the Townsend coefficient is proportional to the gas density and therefore to the pressure P; the ratio α/P is a sole function of the reduced field E/P, as shown in Figure 5.19 for noble gases (Druyvesteyn and Penning, 1940).

Addition to a pure noble gas of even small quantities of a different species not only modifies the energy distribution of electrons, but can also open new channels for radiative and ionizing transitions. If the ionization potential of one species is lower than the excitation potential of the other, a very effective process of

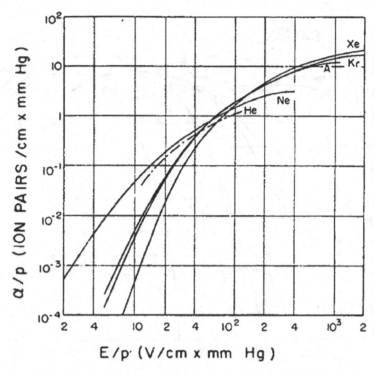

Figure 5.19   First Townsend coefficient as a function of field for the rare gases (Druyvesteyn and Penning, 1940) By kind permission of the American Physical Society.

collisional transfer can take place, resulting in the increase of ionized states (Penning effect):

$$A^* + B \rightarrow A + B^+ + e.$$

A well-known illustration of this effect results from the addition of small fractions of argon ($E_i = 15.8$ eV) to neon ($E_x = 16.6$ eV), that leads to a very effective increase of the ionization yield as compared to its value for both argon and neon alone, Figure 5.20 (Druyvesteyn and Penning, 1940). Penning mixtures are widely used in proportional counters to improve their energy resolution (Sipilä, 1976).

The process of successive ionizations by collision results in charge amplification in proportional counters. Consider an electron released in a region of uniform electric field. After a mean free path $1/\alpha$, one electron–ion pair will be produced, and the two electrons continue their drift generating, after another mean free path, two more ion pairs and so on. If n is the number of electrons in a given position, the increase in their number after a path dx is $dn = n\,\alpha dx$; integrating over a path length x:

$$n = n_0 e^{\alpha x} \text{ or } M = \frac{n}{n_0} = e^{\alpha x}, \tag{5.2}$$

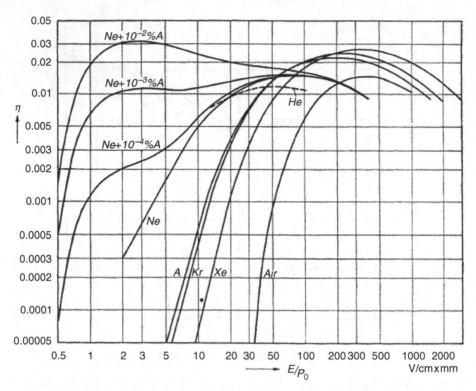

Figure 5.20   Ionization coefficient in rare gases and mixtures, illustrating the Penning effect (Druyvesteyn and Penning, 1940). By kind permission of the American Physical Society.

where M represents the charge multiplication factor. In the more general case of a non-uniform electric field, $\alpha = \alpha(x)$, the expression has to be modified as follows:

$$M = e^{\int_{x_1}^{x_2} \alpha(x)\,\mathrm{d}x}, \tag{5.3}$$

where x_1 and x_2 are the initial and final coordinates of the multiplication path. Due to the very large difference in mobility and diffusion of electrons and ions, discussed in the previous chapter, the process of charge multiplication results in a characteristic drop-like charge distribution, or avalanche, with all electrons at the front and a slow ions trail.

The multiplication factor for arbitrary field geometry can be computed by knowing the dependence of the Townsend coefficient on the electric field. Several analytic expressions exist for α, valid in different regions of E; for a summary see for example Palladino and Sadoulet (1975). A simple approximation, due to Korff (1955) is given by:

Table 5.2 *Parameters in Korff's expressions.*

Gas	A (cm^{-1}torr)	B (V cm^{-1}torr^{-1})	k (cm^{2}V^{-1})
He	3	34	0.11×10^{-17}
Ne	4	100	0.14×10^{-17}
Ar	14	180	1.81×10^{-17}
Xe	26	350	
CO$_2$	20	466	

$$\frac{\alpha}{P} = Ae^{-\frac{BP}{E}}, \tag{5.4}$$

where A and B are phenomenological constants whose approximate values are given in Table 5.2 for several gases (Korff, 1955). In the same field region, the coefficient can be assumed to be proportional to the average electron energy:

$$\alpha = kN\varepsilon. \tag{5.5}$$

Useful for qualitative considerations, the double exponential dependence of α on the field may induce large errors in the estimate even for small differences in the value of the parameters used.

With the opening of the new channels of interaction, the shape of the energy distribution is also substantially modified. With proper parameterization of the contributing cross sections, the transport theory outlined in the previous chapter can be extended to cover the high field inelastic phenomena. In Figure 5.21, experimental data for argon (Kruithof and Penning, 1937; Jelenak *et al.*, 1993) are compared with Korff's approximation, and the predictions of the program MAGBOLTZ (Biagi and Veenhof, 1995b).

Figure 5.22 provides the values computed for pure argon, methane and a 90–10 mixture of the two, at NTP. One can see that the field required to obtain a given value of α is about twice as large in methane than in argon, a clear consequence of the electron 'cooling' effect of molecular gases. The computed ionization coefficients in other gases at NTP are given in Figure 5.23.

The ionization coefficients can be measured using an ionization chamber with a parallel plate multiplication structure; Figure 5.24 is an example of measured values for argon–isobutane mixtures at NTP (Sharma and Sauli, 1992).

The MAGBOLTZ program can be used also to estimate the dependence of the multiplication factor on pressure and temperature. The computed gain variation for a 70–30 argon–carbon dioxide mixture as a function of the ratio *T/P* shows a good agreement with measurements, Figure 5.25 (Altunbas *et al.*, 2003). At constant temperature, a 1% pressure increase results in a gain reduction of about 10%, a parameter to take in due consideration when designing or operating gaseous detectors.

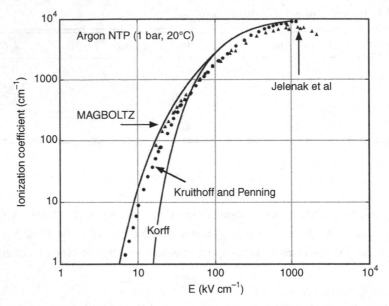

Figure 5.21   Comparison between experimental data of the Townsend coefficient for argon, full points (Kruithoff and Penning, 1937; Jelenak *et al.*, 1993), Korff's approximation (Korff, 1955) and MAGBOLTZ calculation by the author.

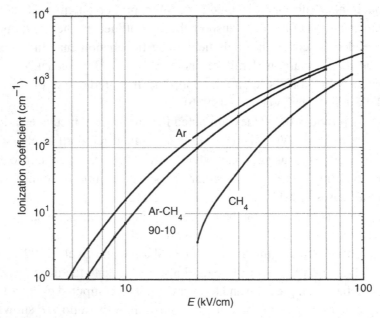

Figure 5.22   Computed Townsend coefficient for argon, methane and their mixture (NTP).

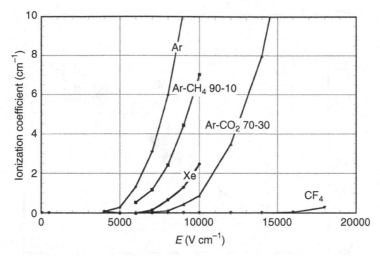

Figure 5.23   Computed Townsend coefficient for several pure gases and mixtures at NTP.

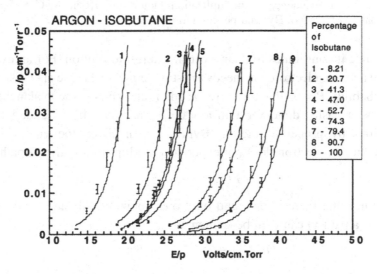

Figure 5.24   Experimental measurements of the Townsend coefficient for argon–isobutane mixtures (Sharma and Sauli, 1992). By kind permission of Elsevier.

5.4 Avalanche statistics

The considerations in the previous section refer to the average avalanche development in uniform fields, described by the Townsend coefficient α. However, statistical variations in the individual ionizing collisions paths during the avalanche development generate dispersions around the average. The problem of avalanche fluctuations has been extensively studied in the early developments of the gaseous

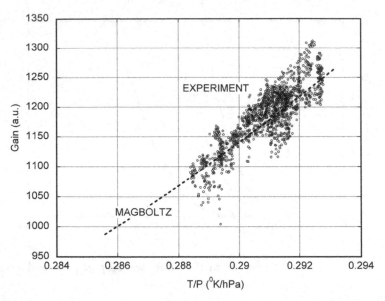

Figure 5.25   Dependence of multiplication factor on T/P in Ar-CO$_2$ 70–30 (Altunbas *et al.*, 2003). By kind permission of Elsevier.

counters, since it contributes to their intrinsic energy resolution. For an avalanche initiated by a single electron, one needs to estimate the avalanche size distribution, i.e. the probability $P(n, s)$ that there are exactly n electrons in the avalanche after a path s; this is easily done for uniform fields, and can be extended to other configurations by proper averaging (Byrne, 2002). Given the value of α, the probability that an electron does not experience multiplication after a path s is:

$$P(1, s) = e^{-\alpha s};$$

the probability that there is one, and only one, ionizing collision (two electrons present after a path x) can then be written as:

$$P(2, s) = e^{-\alpha s}(1 - e^{-\alpha s});$$

and, in general, to have n and only n electrons:

$$P(n, s) = e^{-\alpha s}(1 - e^{-\alpha s})^{n-1},$$

defining $\bar{n} = e^{\alpha s}$ as the average number of electrons in the avalanche (average avalanche size):

$$P(n, s) = \frac{1}{\bar{n}}\left(1 - \frac{1}{\bar{n}}\right)^{n-1} \approx \frac{e^{-n/\bar{n}}}{\bar{n}}, \tag{5.6}$$

the approximation, corresponding to the first term in a series expansion of the exponential, is valid for $\bar{n} \gg 1$ and is often referred to as Furry's law. The

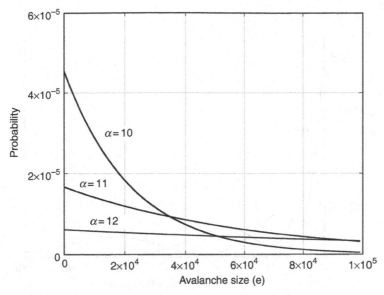

Figure 5.26 Avalanche size probability for several values of the Townsend coefficient.

probability has the particularity to decrease exponentially from a maximum at $n = 1$; surprisingly, the highest probability is for electrons to not multiply at all over the gap. The mean and the variance of the distribution are equal to the average avalanche size $\bar{n}$; the distribution does not depend explicitly on the gap thickness s (although the average size does). Figure 5.26 shows the computed avalanche size distribution in a unit length gap for three values of average size.

For an avalanche starting with N initial electrons, the corresponding size distribution can be obtained as a convolution of N independent exponentials (Breyer, 1973):

$$P(n, N) = \frac{1}{\bar{n}} \left(\frac{n}{\bar{n}}\right)^{N-1} \frac{e^{-\frac{n}{\bar{n}}}}{(N-1)!}, \tag{5.7}$$

where $\bar{n}$ is the average avalanche size for one electron. The expression reduces to (5.4) for $N = 1$, and is functionally identical to a Poisson distribution in the parameter $n/\bar{n}$. The average avalanche size, obtained by integration of the expression, is of course $N\bar{n}$.

In Figure 5.27, computed avalanche size distributions are shown for equal gain at increasing values of N. For a large number of primary multiplying electrons ($> \sim 10$) the distributions tend to have a Gaussian shape centred around the average total charge $N\,e^{\alpha s}$ and with a variance $\sqrt{N}$, independently of the gain; the same

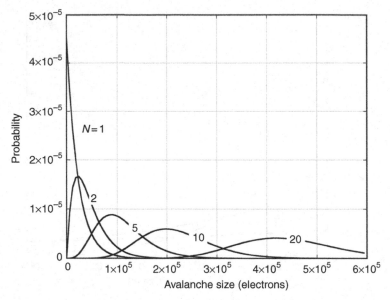

Figure 5.27   Avalanche size probability distributions for increasing values of the number of primary electrons.

property holds for the corresponding Poisson distribution for large values of the parameter in the exponential.

Expression (5.7) can be conveniently rewritten and renormalized as a function of the reduced variable $x = n/\bar{n}$:

$$P(x, N) = \frac{x^{N-1}}{(N-1)!} e^{-x}. \tag{5.8}$$

Figure 5.28 shows the renormalized function computed for several values of N; for $N = 10$, a Gaussian distribution with average $N\bar{n}$ and variance $\sqrt{N}$ is also plotted for comparison.

It has been observed experimentally that the single electron avalanche distribution can evolve from a pure exponential into a peaked shape at very high values of gain. First described by Curran (Curran and Craggs, 1949), the appearance of a peak in the distribution was theoretically analysed by many authors. Considering that for large gains, and therefore small values of the mean ionization length, the electrons use a non-negligible part of their path to acquire a sufficient energy to ionize, one can deduce the following general expression (also named the Polya distribution) for the single electron avalanche size probability (Byrne, 1969):

$$P(x, \theta) = [x(\theta + 1)]^{\theta} e^{-x(\theta+1)}, \tag{5.9}$$

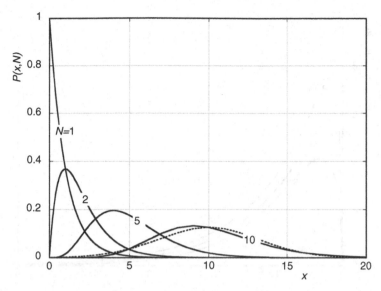

Figure 5.28   Normalized avalanche size probability. The dashed curve is a Gaussian distribution for $N = 10$.

where $x = n/\bar{n}$ is the number of electrons in the avalanche, with average value $\bar{n}$, and θ a parameter; for $\theta = 0$ the expression reduces to the simple exponential (5.6); Figure 5.29 shows the distributions for several values of θ.

Figure 5.30 is an experimental verification of the transition of the single-electron avalanche size distribution from exponential to a peaked Polya shape at increasing values of fields, measured with a methylal-filled counter (Schlumbohm, 1958).

Strictly valid only for the avalanche development in uniform fields, the Polya formulation is widely used to describe measurements obtained with proportional counters and other non-uniform field structures.

Avalanche and gas gain fluctuations can be assessed using Monte Carlo calculations based on the MAGBOLTZ program described in the previous section; the flexibility and wide data base of the program permits one to compare performances of different detector geometry and gas filling (Schindler *et al.*, 2010).

5.5 Streamer formation and breakdown

The multiplication factor cannot be increased at will. Secondary ionization processes in the gas and on the walls produced by photons emitted in the primary avalanche spread the charge over the gas volume; the space charge-induced distortions of the electric field, strongly increasing in front of and behind the

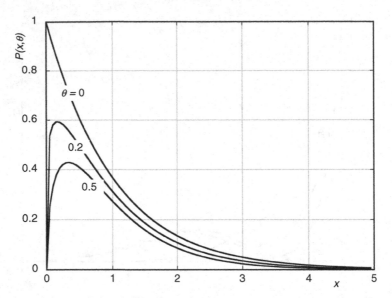

Figure 5.29   Polya single-electron avalanche size probability for several values
of the parameter θ.

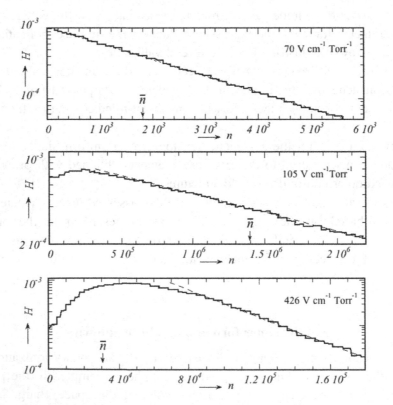

Figure 5.30   Evolution of the avalanche size from exponential to a Polya distribution at increasing values of field (Schlumbohm, 1958). By kind permission of
Springer Science+Business Media.

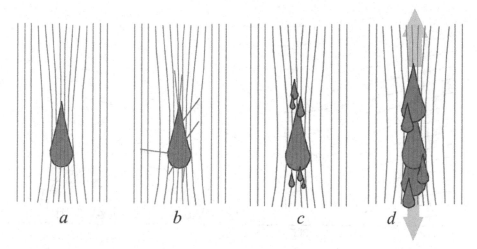

Figure 5.31   Schematic representation of the transition from avalanche to streamer.

multiplying charge, enhance the formation of secondary avalanches and may cause the transition of the proportional avalanche to a streamer and eventually to a spark breakdown. A qualitative description of the main processes leading to the avalanche–streamer transition is shown in Figure 5.31. At the onset of charge multiplication, the high density of ions and electrons in the avalanche modify the original electric field, increasing it in front of and behind the avalanche (*a*); photons emitted isotropically from inelastic collisions in the avalanche front create secondary electrons by photo-ionization of the gas molecules (*b*). Secondary avalanches then develop from electrons created in the regions of higher field (*c*) and the process continues with a forward and backward propagation of the charges, starting a streamer (*d*).

If not damped by the detector geometry or by a reduction of field, the streamer can propagate through the whole gas gap, leading to a spark breakdown. A phenomenological limit to the proportional avalanche multiplication before the transition is given by the empirical condition (Raether's limit):

$$\alpha x < 20, \tag{5.10}$$

or a total avalanche size, product of initial ionization and gain, of around 10^8. However, the statistical distribution of the energy of electrons, and therefore of the gain, generally does not allow one to operate gaseous counters at average gains above around 10^6 to avoid breakdowns.

The conditions and outcomes of the streamer transition have been extensively studied and are described in many textbooks (Meek and Cragg, 1953; Loeb, 1961;

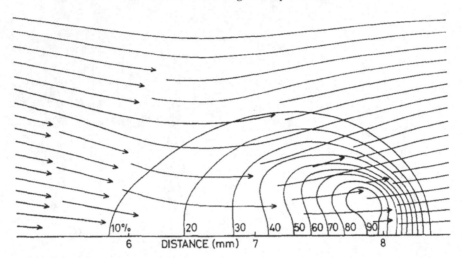

Figure 5.32   Two-dimensional plot of the electron density and electric field near
the end of the avalanche development process (Evans, 1969). By kind permission
of Elsevier.

Raether, 1964). More recently, in the course of the development of the gaseous
detector, terms describing short-distance photo-ionization have been included in
the transport equations for the atomic species involved, leading to numerical
solutions of the streamer and breakdown transition (Evans, 1969; Fonte, 1996).
Figure 5.32 and Figure 5.33, from the first reference, show a two-dimensional map
of the electron charge distribution and the field at a certain point of the avalanche
development, and respectively the longitudinal distribution of electron density and
axial field at successive times. The field enhancements in the front and back of the
avalanche are visible, as well as the reduction within the avalanche itself due to the
high density of ions and electrons.

   The presence of a precursor, a fast and short pulse induced by the streamer
transition before the main signal of the full discharge, described by Raether, has
been observed by several authors and is well reproduced by the breakdown
simulation model (Fonte *et al.*, 1991; Anderson *et al.*, 1994). The precursor
provides in fact the faster signal after the ionizing encounter, and is exploited in
detectors providing the best time resolutions (see Chapter 12).

   For high events rates, the accumulation of charges results in a rate-dependent
reduction of maximum gain, the invariant being the total charge of the avalanche,
as seen in Figure 5.35, providing the maximum event rate before breakdown for
several values of the avalanche size (Fonte *et al.*, 1999). The measurements were
realized with parallel plate chambers having different electrode resistivity; the solid

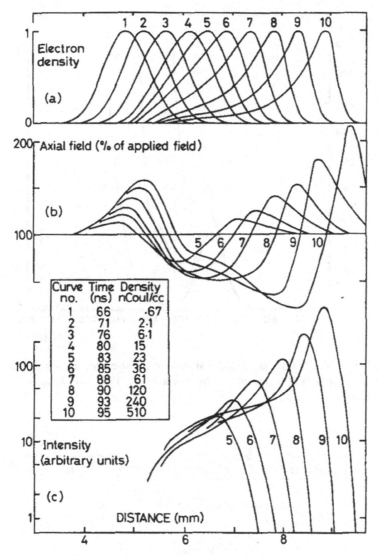

Figure 5.33 Computed axial distributions of electron density (a) and field (b) at ten successive positions in the avalanche propagation. The relative intensity of photon emission is given in (c) (Evans, 1969). By kind permission of Elsevier.

black line is the so-called metallic limit, corresponding to the use of conductive electrodes.

A naive calculation assuming typical values for the avalanche size (~ 1 mm^2) and a positive ion clearing time for a small gap, high field device (~ 1 μs) suggests that the probability of two avalanches overlapping in time at a flux of 10^5 Hz/mm^2 is only 10%, making obscure the observed collective effect. A more rigorous calculation, however, taking into account the statistical fluctuations in the

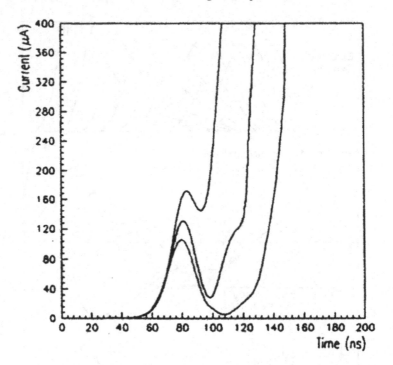

Figure 5.34   Model simulation of the signals at the streamer transition, with fast discharge precursors at increasing gains (Fonte *et al.*, 1991). By kind permission of Elsevier.

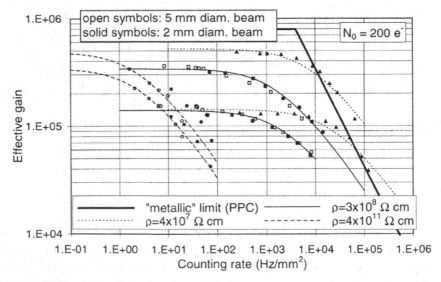

Figure 5.35   Effective gain of parallel plate counters as a function of counting rate (Fonte *et al.*, 1999). By kind permission of Elsevier.

avalanche size, reproduces rather well the observed gain–rate dependence (Peskov and Fonte, 2009).

The mechanisms, outcomes and methods of prevention of discharges in parallel plate counters have been extensively studied in the course of the development of resistive plate chambers, and are discussed in the Chapter 12, dedicated to this family of detectors.

6

Parallel plate counters

6.1 Charge induction on conductors

A positive charge placed between two conducting electrodes induces on both a negative charge distribution with shape and amplitude that depend on geometry. Figure 6.1 shows schematically the induction profile for two positions of a positive charge placed at equal distance (left) or asymmetrically between two electrodes (right). Intuitively, and as can be inferred from the image charges model, the induction profile is narrower and higher when the charge is closer to the electrode.

Electrons and ions released within a gaseous counter by ionizing events drift towards anode and cathode, respectively, under the effect of the applied electric field. For moderate fields, before the appearance of inelastic collision processes and in the absence of capture and recombination, the final outcome is the collection at the electrodes of all charges; during the drift, induced signals appear on the electrodes, with polarity and time structure that depend on the counter geometry, field strength and mobility of the charges.

The motion of ions and electrons towards the electrodes increases their surface charges, the cathode towards more negative and the anode towards more positive values; the signals detected on a load are, however, opposite in polarity. Suppose indeed that the cathode is connected to the voltage through a resistor. An increase of the negative charge on the electrode implies an inflow of electrons, or, from the definition of electric current, an outflow of positive charges; the detected voltage difference is therefore positive. This is often confusing, so the reader could memorize the golden rule of induced signal formation: 'A positive charge moving towards an electrode generates an induced positive signal; if it moves away from the electrode, the signal is negative', and similarly for negative charges, with opposite signs.

Methods to compute the currents induced by moving charges on conductors of arbitrary shape were developed to describe the signal formation in vacuum tubes

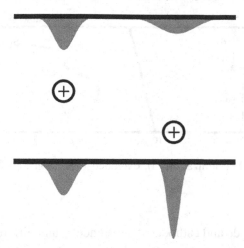

Figure 6.1   Profiles of charges induced on electrodes by a positive charge in two different positions.

(Shockley, 1938; Ramo, 1939); named the Ramo theorem, the basic equation can be written as:

$$i = E_v ev,$$

where i is the instantaneous current flowing in a given electrode due to the motion of a single electron of charge e and velocity v and E_v is the component in the direction of the electric field which would exist at the electron's position with that electron removed, the electrode raised to unit potential and all other electrodes grounded.

Standard methods of electrostatics can be used to compute analytically the induced signal distributions on electrodes with a given geometry (Durand, 1966; Morse and Feshbach, 1953; Mathieson and Smith, 1988). A general expression can be written to estimate the time development of signals for the charges induced by the system of ions located in known positions, then computing the signal development as the increment (or decrement) of induced charge on each electrode during the motion along the field lines of the various primary charges, with the appropriate velocity. For complex geometries this becomes rather tedious, and approximate numerical solutions are generally preferred; web-based programs are also available for this calculation (Veenhof, 1998; Veenhof, 2002).

6.2 Signals induced by the motion of charges in uniform fields

For parallel electrode geometry, expressions for the induced signals can be obtained from simple energy conservation principles. Consider a system composed

Parallel plate counters

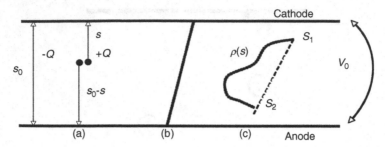

Figure 6.2   Point-like ionization (a), extended ionization trail (b) and localized trail (c).

of two electrodes, anode and cathode, at a distance s_0 and difference of potential V_0 (Figure 6.2).

Assuming that the work done by the field to move the charge Q by ds is equal to the change in electrostatic energy stored by the system, one can write for the incremental induced charge, see for example Franzen and Cochran (1956):

$$dq = Q\frac{dV}{V_0} = Q\frac{ds}{s_0}. \tag{6.1}$$

For a charge Q drifting in a uniform electric field with constant velocity w, the current delivered by the source of potential is then:

$$i = \frac{dq}{dt} = \frac{Q}{s_0}\frac{ds}{dt} = \frac{Q}{s_0}w \tag{6.2}$$

and the corresponding charge is:

$$q(t) = \frac{Q}{s_0}wt. \tag{6.3}$$

At the same time, a current equal and opposite in sign appears on the anode; the motion of a negative charge has the same effect with inverted sign. The currents cease at the moment the charges reach an electrode; integration of expression (6.1) provides the total induced charge:

$$q = Q\frac{s}{s_0}, \tag{6.4}$$

reducing to $q = Q$ if the charge is released on one electrode surface and drifts through all the gap.

Suppose now that a pair of charges of opposite sign is released at a distance s from the cathode (Figure 6.2a). At the moment of creation, and in fact forever in the absence of motion, the induced charges on each electrode, equal and opposite, cancel out. In the presence of a field, and because of their different mobility, electrons and ions are collected at a different rate; the relative contributions to the total induced signal depend on the initial position of production. In the case of an ionization event releasing at time $t = 0$ and position s equal amounts of electrons and ions of charge $-Q$ and $+Q$, the time development of the induced charge (from (6.3)) can be written as:

$$q(t) = Q\frac{w^+t}{s_0} + Q\frac{w^-t}{s_0}, \tag{6.5}$$

where w^+ and w^- are drift velocities of ions and electrons; the corresponding voltage difference across the capacitor is given by $V(t) = q(t)/C$. Both terms in expression (6.5) have values restricted to the maximum time of drift before collection, T^+ and T^- for ions and electrons respectively. The total induced charge after full collection is then:

$$q_T = Q\frac{w^+T^+}{s_0} + Q\frac{w^-T^-}{s_0} = Q\frac{s}{s_0} + Q\frac{s - s_0}{s_0} = Q.$$

Figure 6.3 gives an example of the time development of the induced charge and current on the cathode on a parallel plate ionization counter for equal amounts of electrons and ions released mid-way between two electrodes. For illustration purposes the mobility of electrons has been assumed to be only ten times that of ions (they differ in fact by three orders of magnitude). Signals on anodes are equal and opposite in sign.

The previous expressions permit us to compute the induced charge and current for the case of extended tracks, as those produced by charged particles traversing the gap (Figure 6.2b). Suppose the ionized trail to be perpendicular to the

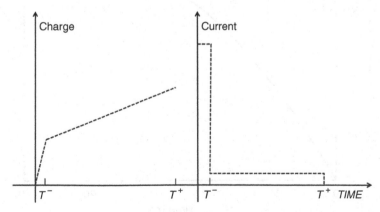

Figure 6.3   Charge and current induced on cathodes by the drift of electrons and ions.

electrodes, with uniform charge density $\rho = Q / s_0$. Considering for the time being only one species of released charges, after a time t a segment of track of length wt will have disappeared, collected by the electrode. The induced current at time t can thus be written (in analogy with expression (6.2)):

$$i = \frac{w}{s_0} \int_0^{s_0 - wt} \rho \, ds = \frac{w}{s_0} \rho \, (s_0 - wt) = \frac{wQ}{s_0}\left(1 - \frac{w}{s_0}t\right), \tag{6.6}$$

and the corresponding charge:

$$q(t) = \int_0^t i \, dt = \frac{wQ}{s_0} t \left(1 - \frac{wt}{2s_0}\right). \tag{6.7}$$

The induced charge signal rises to a maximum equal to $Q/2$, reached at the time of full collection $T = s_0/w$ (Fig 6.4); charges of opposite polarity induce a signal described by expressions (6.6) and (6.7), with the same sign and with the appropriate value of the drift velocity; the two contributions add up as in previous examples. In the case of inclined tracks, the induction process is described by the same expressions, substituting for the real charge density its projection on the normal to the electrodes.

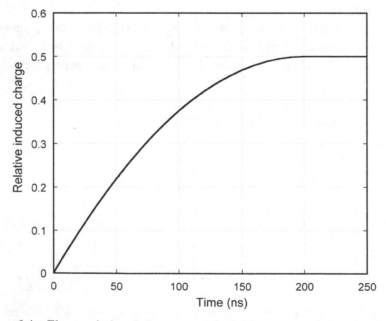

Figure 6.4   Electron-induced charge on cathodes by an extended track.

The induced charge signal can be computed in the general case of non-uniform ionization. Consider, for example, a charged particle (say, a photoelectron) produced and coming to rest between two positions at distance s_1 and s_2 from one electrode, and producing a trail with ionization density (projected on the coordinate s perpendicular to the electrodes) given by $\rho(s)$ (Figure 6.2c). The charge induced on the electrodes from the time of production to the moment the first electron reaches the anode is given by (see (6.3)):

$$q(t) = \frac{wt}{s_0} \int_{s_1}^{s_2} \rho(s)\,ds,$$

and the corresponding current:

$$i = \frac{w}{s_0} \int_{wt}^{s_2} \rho(s)\,ds,$$

where one can see that the slope of the induced signal before collection depends only on the total ionization charge (and not on its detailed distribution). From the moment collection starts, $t_1 = s_1/w$, and before collection of the last electron the induced current is:

$$i = \frac{w}{Cs_0} \int_{wt}^{s_2} \rho(s)\,ds$$

and:

$$\frac{di}{dt} = \frac{w^2}{s_0}\rho(wt),$$

where $\rho(wt)$ is the density of ionization at a distance wt-s_1 from the end of the track, measured along s; the observation of the induced charge profile is in principle providing the (projected) ionization energy loss density along the track. The same considerations apply to the signals induced by positive ions, although at a much longer time scale.

6.3 Analytical calculation of charge induction

A powerful method for evaluating the static distribution of induced charges on electrodes is based on Green's reciprocation theorem, also called the Gauss identity, stating that when a set of charges $Q_a, Q_b, \ldots (Q'_a, Q'_b, \ldots)$ is placed on insulated conductors $a, b, \ldots$, they acquire the potentials $V_a, V_b, \ldots (V'_a, V'_b, \ldots)$ so that the following relationship is satisfied:

$$\sum Q_i' V_i = \sum Q_i V_i' \qquad (6.8)$$

with the sum extended to all conductors. It is assumed that the variation of potentials in all electrodes due to deposition or induction of charges is negligible compared to the ones previously existing; this is verified in practice for the charge densities encountered in ionization and proportional chambers. A solution to the induction problem can then be found by writing expression (6.8) for a set of suitably chosen initial conditions.

As an application of Green's theorem, consider the case of a charge q placed at a point P close to a single conductor of arbitrary shape. Let V be the potential at the point P, and Q the induced charge to be evaluated on the conductor assumed at potential $V = 0$. As a second equilibrium state, consider the case of $q = 0$, induced charge Q' and potentials V' and V_0' at P and on the conductor respectively. The two sets of conditions can be written as follows:

$$(q, Q); (V, 0),$$

$$(0, Q'); (V', V_0');$$

from expression (6.8) one has then:

$$qV' + QV_0' = 0$$

and therefore:

$$Q = -\frac{V'}{V_0'} q. \qquad (6.9)$$

The induced charge on the conductor at ground potential can then be computed, knowing the potential V' produced, in the point P and with q removed, by the conductor raised at the potential V_0'. This is simple for the case of a charge at distance r from a spherical conductor of radius R:

$$V' = \frac{R}{r} V_0',$$

and therefore:

$$Q = -\frac{R}{r} q, \quad r > R.$$

For an infinite plane conductor, $Q = -q$, as can be deduced from the previous expression for $r, R \to \infty$.

Similarly, consider a charge $+q$ placed between two parallel and infinite electrodes at potentials V_1 and V_2, at distances s and $s_0 - s$ respectively (s_0 being the total gap, see Figure 6.5). We can choose the following three sets of conditions:

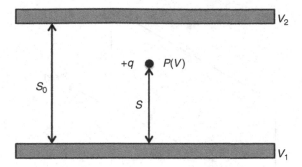

Figure 6.5   Positive charge placed between two parallel electrodes: parameter definition.

$$(q, Q_1, Q_2);(V, 0, 0),$$

$$(0, Q_1', Q_2');(V', V_1', 0),$$

$$(0, Q_1'', Q_2'');(V'', 0, V_2'');$$

applying Green's theorem to pairs of sets:

$$Q_1 = \frac{V'}{V_1'}q = -\frac{s_0 - s}{s_0}q,$$

$$Q_2 = \frac{V''}{V_2''}q = -\frac{s}{s_0}q,$$

(6.10)

where the linear dependence of the potential between two parallel electrodes, in the absence of the charge q, has been taken into account:

$$\frac{V - V_1}{V_2 - V_1} = \frac{s}{s_0}$$

and similar expressions for the other cases.

Note that in the limiting case of a charge placed infinitely close to one conductor $s \to 0$, $Q_1 \to -q$ and $Q_2 \to 0$. Also, if one brings to infinite distance the electrode 2 ($s_0 \to \infty$), as expected $Q_1 \to -q$ and $Q_2 \to 0$.

For infinite parallel electrodes, the time development of induced charges can be easily computed from energy conservation principles. In the general case of complex geometries, for example with segmented electrodes (strips or wires), and for distributions of charges, one has to resort to more sophisticated electrostatic algorithms. The method of image charges can be used to get an intuitive picture of the induction process, and in some simple cases to compute the distribution of

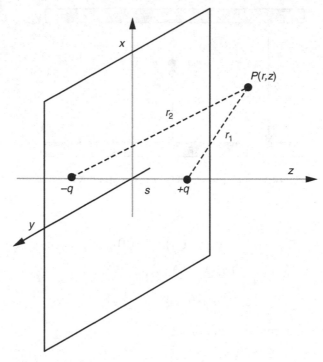

Figure 6.6   Image charge method: definition of geometry.

induced charges. Consider the case of a positive point charge q placed at a distance s from an infinite plane conductor. Elementary electrostatics suggests that one can replace the conductor with an image charge $-q$ placed symmetrically to q (Figure 6.6); indeed, the central plane in this case is equi-potential.

The potential on any point in space of coordinates *(x, y, z)* can then be written as:

$$
\begin{aligned}
V &= \frac{q}{4\pi\varepsilon_0}\left(\frac{1}{\sqrt{x^2+y^2+(z-s)^2}} - \frac{1}{\sqrt{x^2+y^2+(z+s)^2}}\right) \\
&= \frac{q}{4\pi\varepsilon_0}\left(\frac{1}{\sqrt{\rho^2+(z-s)^2}} - \frac{1}{\sqrt{\rho^2+(z+s)^2}}\right),
\end{aligned}
$$

(6.11)

with $\rho = \sqrt{x^2+y^2}$. The corresponding surface charge density on the conductor is then given by:

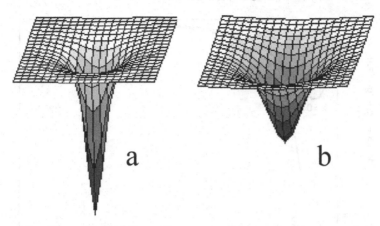

Figure 6.7   Charge induction profile on a conducting plane for two values of the distance of a charge from the plane.

$$\sigma(x,y) = \sigma(\rho,\theta) = -\varepsilon_0 \left(\frac{\partial V}{\partial z}\right)_{z=0} = -\frac{qs}{2\pi(\rho^2 + s^2)^{3/2}} = -\frac{qs}{2\pi l^3},\qquad (6.12)$$

where l is the distance of a point on the conductor from the charge q. Peaked in front of q, the induced charge distribution has a shape that depends on the distance s from the plane (Figure 6.7).

The total induced charge is obtained by integration of the previous expression and is obviously equal and opposite in sign to the original charge:

$$Q = \int \sigma(x,y)\mathrm{d}x \ \mathrm{d}y = -q.$$

The charge density in an arbitrary point (x, y) on the plane can be estimated from expression (6.12), as a function of s, as shown for example in Figure 6.8, computed for $y = 0$ and several values of x. Note that the induced charge density decreases towards negative values at the approach of q to the plane, but returns to zero when the charge reaches it ($s = 0$). For $x = 0$, the expression of charge density has a singularity, with the corresponding function tending to infinity (however, its integral equals $-q$ as previously stated). This behaviour corresponds to the general statement that, however large the induced charge on a surface element is, its final value after full collection of the moving charge is different from zero only if some charge is collected by that element.

Differentiation of (6.12) with respect to s provides the following expression, which can be interpreted as the current density at the point (x, y) generated by the uniform approach of the charge q to the conductor:

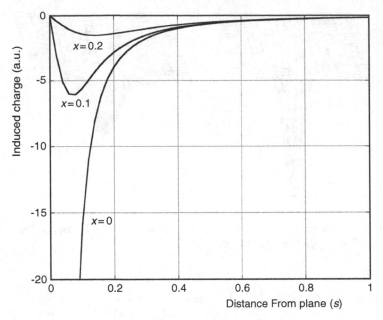

Figure 6.8   Charge induction profile as a function of distance of the charge from the (x, y) plane, for several values of x at $y = 0$.

$$i(s) = -\frac{q}{2\pi} \frac{(x^2 - 2s^2)}{(x^2 + s^2)^{5/2}}.$$

In the case of a segmented electrode, the fraction of charge induced on each segment can be computed by integration of the previous expression between the appropriate boundaries. Let us first consider the case of an electrode made of concentric rings, centred on the origin. The charge induced on the ring between r_1 and r_2 is then:

$$Q_{12} = \int_0^{2\pi} d\theta \int_{\rho_1}^{\rho_2} \sigma(\rho, \theta) d\rho = -\frac{q}{s}\left(\frac{\rho_2}{\sqrt{\rho_2^2 + s^2}} - \frac{\rho_1}{\sqrt{\rho_1^2 + s^2}}\right).$$

In Figure 6.9 the charge induced on concentric rings of unitary thickness is shown as a function of their radius, for three values of the distance s. For values of s larger than the minimum ring radius, the induced charge profile is peaked at increasing distances from the origin, reflecting the larger solid angle of those rings.

Expression (6.12) can be integrated on strips parallel to the y coordinate, between the coordinates x_1 and x_2 :

$$Q_{12} = \int_0^{\infty} dy \int_{x_1}^{x_2} \sigma(x, y) dx = -\frac{q}{\pi}\left(\arctan\frac{x_2}{s} - \arctan\frac{x_1}{s}\right),$$

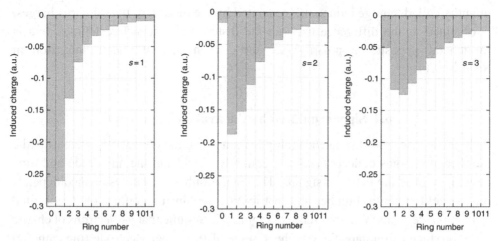

Figure 6.9 Charge induction on concentric rings, for three values of the distance *s* of the charge.

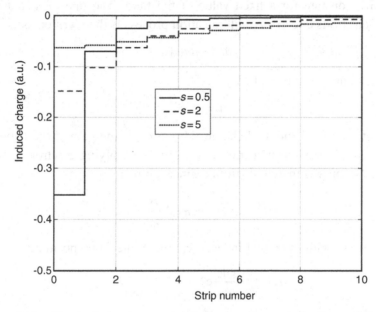

Figure 6.10 Charge induction on parallel strips, for several values of the distance of the charge.

which represents the incremental charge induction on a set of parallel strips; Figure 6.10 shows the charge induction profile on ten adjacent strips of unitary length, computed for three values of the distance *s*.

It is appropriate to note that all expressions and figures given in this section provide the static charge distributions on the electrode surface; the observable quantity called 'charge induction', generated by the motion of the primary charges, is obtained as the difference between the distributions at different times, and is in general opposite in polarity (see the golden rule of charge induction in Section 6.1).

6.4 Signals induced by the avalanche process

At high electric fields avalanche multiplication sets in, exponentially increasing the number of charges collected and consequently modifying the amplitude and time development of the detected signals. The total number of charges created depends on the field strength, number and distribution within the counter of the original ionization electrons. With reference to Figure 6.2, assume that n_0 electrons of charge e are produced simultaneously at the surface of the upper electrode (the cathode) and start moving towards the anode with a mean free path for multiplication given by α^{-1}, α being the first Townsend coefficient (number of ionizing collisions per unit path length), constant for a fixed value of the field. The rate of increase in the number of electrons due to the multiplication process is then written as:

$$\mathrm{d}n = n\,\alpha\,\mathrm{d}s, \tag{6.13}$$

and their total number after a path s:

$$n = n_0 \mathrm{e}^{\alpha s}, \tag{6.14}$$

the ratio between final and total charge, $M = n/n_0$, is the gain of the counter.

The differential charge induction due to the multiplying electrons when they reach the position s is therefore (from expression (6.1)):

$$\mathrm{d}q^- = e n_0\, \mathrm{e}^{\alpha s} \frac{\mathrm{d}s}{s_0};$$

an integration provides the total induced charge at the same position:

$$q^-(s) = \frac{e n_0}{\alpha s_0}(\mathrm{e}^{\alpha s} - 1) \cong \frac{e n_0}{\alpha s_0}\mathrm{e}^{\alpha s}, \tag{6.15}$$

or, as a function of the time from the start of the avalanche:

$$q^-(t) = \frac{e n_0}{\alpha s_0}(\mathrm{e}^{\alpha w^- t} - 1) \cong \frac{e n_0}{\alpha s_0}\mathrm{e}^{\alpha w^- t}, \tag{6.16}$$

where w^- is the drift velocity of the avalanching electrons. The corresponding current is obtained by differentiation:

$$i^-(t) = \frac{e n_0 w^-}{s_0}\mathrm{e}^{\alpha w^- t} = \frac{e n_0}{T^-}\mathrm{e}^{\alpha w^- t}\quad (t \le T^-) \tag{6.17}$$

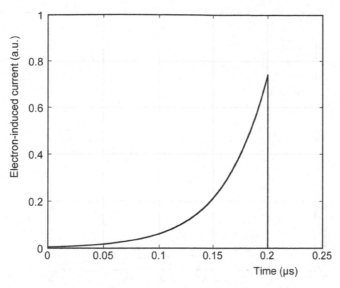

Figure 6.11 Electron-induced current by an avalanche on the cathode.

where T^- is the total collection time for the electrons in the avalanche. One can see from expression (6.15) that after full collection ($s = s_0$), the total charge induced by the electrons, the so-called fast signal in an avalanche counter, corresponds to only a fraction of the total charge generated in the avalanche:

$$q^- (T^-) \cong \frac{en_0}{\alpha s_0} e^{\alpha s_0}. \tag{6.18}$$

It is a common mistake, when using amplifiers with short differentiation constants (therefore only detecting the fast electron component of the signal), to underestimate the gain of the counter, neglecting the factor αs_0 in the previous expression.

The time evolution of the current induced by the electron component in the avalanche is shown in Figure 6.11 for a typical choice of parameters.

The contribution to the induced charge due to positive ions can also be computed by taking into account the exponential growth in the number of ions during the avalanche process followed by the drift and gradual collection of ions at the cathode. The following expressions can be derived, see for example Raether (1964):

$$i^+(t) = \frac{en_0}{T^+} \left(e^{\alpha w^- t} - e^{\alpha w^+ t} \right), \quad (0 \le t \le T^-),$$

$$i^+(t) = \frac{en_0}{T^+} \left(e^{\alpha s_0} - e^{\alpha w^* t} \right), \quad (T^- \le t \le T^- + T^+), \tag{6.19}$$

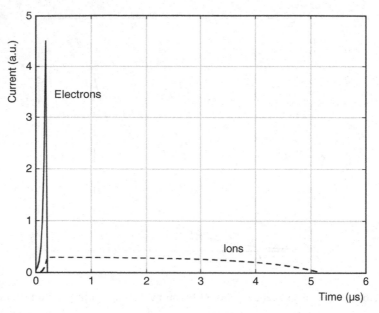

Figure 6.12   Avalanche-induced current on cathodes by electrons and ions.

with

$$\frac{1}{w^*} = \frac{1}{w^+} + \frac{1}{w^-}.$$

The electron and ion induced currents on the cathode are shown in Figure 6.12; the corresponding time evolution of the charge can be computed by integration:

$$q^+(t) = \frac{en_0}{\alpha T^+}\left(\frac{e^{\alpha w^- t}}{w^-} - \frac{e^{\alpha w^* t}}{w^*} + \frac{1}{w^+}\right), \quad (0 \le t \le T^-), \qquad (6.20)$$

$$q^+(t) = \frac{en_0}{T^+}\left((t - T^-)e^{\alpha s_0} - \frac{e^{\alpha w^* t} - e^{\alpha w^* T^-}}{\alpha w^*}\right), \quad (T^- \le t \le T^- + T^+).$$

Figure 6.13 shows the total charge induction due to electrons and ions, computed for an unphysical difference by a factor of ten between electron and ion mobility.

In the previous calculations it has been assumed that the initial electron charge released at the surface of the cathode, therefore getting full amplification. In the case of an extended distribution of charge within the gap, as produced by a charged particle traversing the chamber, each electron undergoes an avalanche multiplication by a factor $e^{\alpha s}$, s being its distance from the anode; the resulting signal depends on the density and space distribution of the charges released in the gas. In the particular case of a uniform ionization produced between anode and cathode

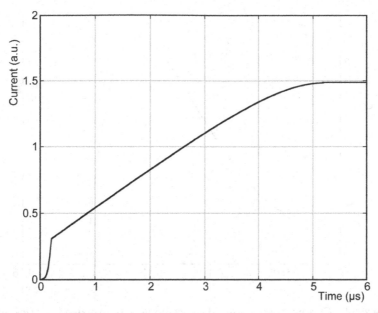

Figure 6.13   Total cathode induced charge by the avalanche multiplication.

with density r, the current induced by the electron component in the avalanche can then be written as (Raether, 1964):

$$i^-(t) = \frac{e\rho s_0}{T^-} e^{\alpha w^- t}\left(1 - \frac{t}{T^-}\right), \quad 0 \le t \le T^-; \qquad (6.21)$$

the maximum signal is obtained after a time $t_{max}^- = T^-(1 - 1/\alpha s_0)$ and has a value $i_{max}^- = (e\rho/\alpha s_0)e^{\alpha s_0 - 1}$, as shown in Figure 6.14. Since the avalanche development is much faster than the ions' collection time, the induced ion signal is very similar in shape (but not in amplitude) to the case described previously.

6.5 Grid transparency

Internal wire grids or mesh electrodes mounted between anode and cathode are used often in gaseous detectors for different purposes: screening of the pickup electrode from long signals induced by charges moving in the main drift volume of the detector (the so-called Frisch grid in ionization chambers), intermediate electrodes in the multi-step chamber (Charpak and Sauli, 1978), cathodes of the multi-wire chamber end-cap detectors in time projection chambers (Nygren and Marx, 1978), separation from the low drift and the very high avalanche field in Micromegas (Giomataris *et al.*, 1996). It is therefore instructive to estimate the

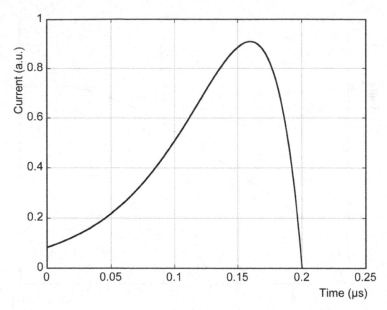

Figure 6.14   Current signal induced on cathodes by an extended track under avalanche multiplication.

transmission efficiency, or electrical transparency, of the electrodes, both for ions and electrons, and the fraction of signal induced through the mesh.

An analytical expression for the electrical transparency of a grid of parallel wires can be obtained by computing the number of field lines intercepted by a mesh inserted between two regions of field E_1 and E_2 (Bunemann *et al.*, 1949):

$$T(E_1, E_2, \rho)$$
$$= 1 - \frac{1}{\pi E_1}\left[(E_1 + E_2)\sqrt{\rho^2 - \left(\frac{E_2 - E_1}{E_2 + E_1}\right)^2} - (E_2 - E_1)\cos^{-1}\left(\frac{E_2 - E_1}{E_2 + E_1}\frac{1}{\rho}\right)\right],$$

(6.22)

with $\rho = 2\pi r/a$, where r and a are, respectively, the radius and distance apart of the wires in the mesh. The equation is only valid in the range of fields where the square root and the inverse cosine are real, namely:

$$\frac{1 - \rho}{1 + \rho} < \frac{E_2}{E_1} < \frac{1 + \rho}{1 - \rho}.$$

The quoted work, aimed at optimizing the Frisch grid ionization chamber performance, provides also the electrostatic signal shielding efficiency of a mesh. Figure 6.15 shows the transparency of a mesh as a function of the ratio of fields E_2/E_1 and several values of the parameter ρ. As can be seen, the transparency

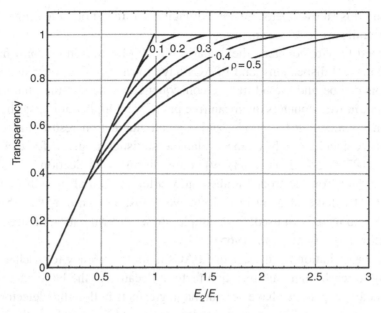

Figure 6.15 Charge transparency of a parallel wire mesh as a function of field ratio and geometry.

increases towards unity with the increasing field ratio; for large values of ρ (thicker or denser wires) full transmission is only reached at large ratios of fields. It should be noted that this calculation does not take into account the diffusion of charges in their drift through the structure, and the consequent capture losses; experimental values tend therefore to be smaller that the calculation.

No theoretical calculations seem to exist for the transparency of a crossed wire mesh, which remains an experimental issue. In a coarse approximation, it can be assumed that the electron transparency in this case is given by the square of the value for the one-dimensional grid.

6.6 Applications of parallel plate avalanche counters (PPACs)

Gaseous counters exploiting the avalanche multiplication between parallel elec-
trodes are easy to build and operate. In the simplest structure with two metal foil electrodes, the avalanche size depends on the position of the primary ionization, and therefore the response of the counter is not proportional to the initial charge. As discussed in Section 6.2, however, the collected charge is proportional to the primary ionization for extended tracks. The cathode foil in a PPAC can be replaced with a semi-transparent mesh, separating the low field drift region from the multiplication gap; the ionization clusters produced in the drift volume are then amplified uniformly, with the detected charge proportional to the energy

deposition; this allows detection and imaging of radiation releasing localized ionization trails, as X-rays or neutrons.

Compared to wire counters, the avalanche multiplication in uniform fields has smaller statistical dispersions, and PPACs have intrinsically a better energy resolution. This can be understood to be a consequence of the fact that, the total gain being equal, in wire counters the mean free path for multiplication gets shorter and shorter on approaching the anode, and a small fluctuation results in large gain variations. A detailed study of the avalanche statistics is given for example by Alkhazov (1970) and Alkhazov (1969) and is discussed in Section 7.5. Since the signal is induced on electrodes immediately after the initial growth of the avalanche, PPACs have also intrinsically a very fast response with sub-ns time resolution, and have been used, for example, in the determination of short nuclear states lifetimes (Krusche *et al.*, 1965).

The major limitation for the use of PPACs is their tendency to discharge when attempting to reach high multiplication factors; because of the large energy stored in the detector, a spark can have very damaging effects both to the detectors and to the readout electronics. The exponential dependence of the gain from the field, and therefore from the gap thickness, demands a very strict tolerance in construction, and sets limits to the size of detectors. Small imperfections on the cathode may

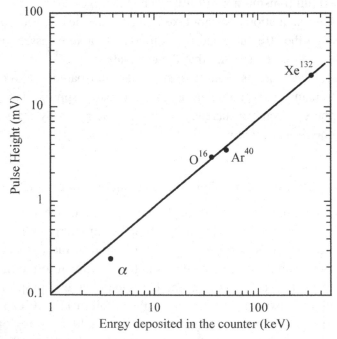

Figure 6.16   Pulse height dependence on energy loss in a PPAC (Stelzer, 1976). By kind permission of Elsevier.

cause the spontaneous emission of electrons, starting an avalanching growth that can exceed the Raether limit and leading to a discharge; although there are examples of use of PPACs for detection of fast particles (Urban *et al.*, 1981), their most common application has been for detection of highly ionizing ions.

The use of resistive electrodes is an effective way to limit the energy flowing in a discharge and allows for safer operation, yet preserving the major performances of the detector (Chtchetkovski *et al.*, 2000); this subject is covered in Chapter 12.

Measured with a medium-size PPAC with a low pressure hydrocarbon gas filling, Figure 6.16 shows the relative pulse height as a function of energy and Figure 6.17 the voltage dependence of gain for ions of different mass and energies (Stelzer, 1976). With low pressure isobutane, time resolutions between two and three hundred ps FWHM have been measured for low energy protons and α particles, Figure 6.18 (Breskin and Zwang, 1977).

The good linearity and proportionality of the detectors, combined with the excellent time resolution, permit the identification of the ions in spectrometers, see for example Sernicki (1983), Hempel *et al.*, (1975) Prete and Viesti (1985). Other applications include the detection of fission fragments and, with an internal converter, of neutrons (Nishio *et al.*, 1997; Jamil *et al.*, 2012).

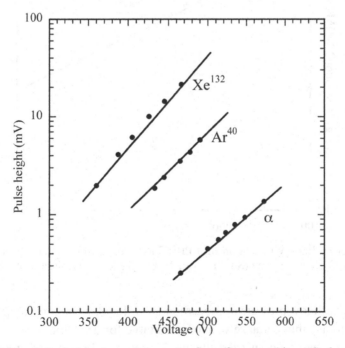

Figure 6.17   Pulse height dependence on voltage for several ions (Stelzer, 1976). By kind permission of Elsevier.

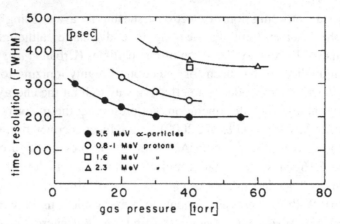

Figure 6.18   PPAC time resolution as a function of i-C_4H_{10} pressure for several ions (Breskin and Zwang, 1977). By kind permission of Elsevier.

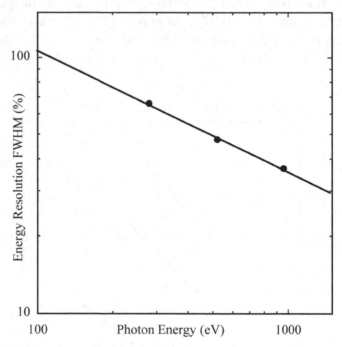

Figure 6.19   Energy resolution of a PPAC as a function of photon energy (gas filling Ar-C_2H_6 at 28 torr) (Smith *et al.*, 1990). By kind permission of Elsevier.

Operation with low pressure hydrocarbon gas fillings, while providing high gains, permits one to use thin entrance windows, suitable for detection of soft X-rays; this detector design has been successfully used, with excellent energy resolution, for fluorescence analysis, as shown for example in Figure 6.19 (Smith *et al.*, 1990).

Used as end-cap detector for large volume drift and time projection chambers, the PPAC has the advantage of reducing the positive ion feedback and increasing the rate capability, as compared to conventional MWPC readout (Peisert, 1983). The Micromegas detector, described in Section 13.3, is of similar conception and, thanks to the use of very thin multiplication gaps, solves many of the gain uniformities problems encountered in standard PPACs.

Parallel electrode multiplying structures are also used in scintillating proportional counters, reaching the statistical limit in energy resolution, see Section 7.6.

7

Proportional counters

7.1 Basic principles

A relativistic charged particle releases around 100 electron–ion pairs per cm in a gas at NTP. In a simple two-electrode parallel plate structure with a capacitance of a few tens of pF, the collection of this charge would produce a voltage signal of a few μV, below the possibility of detection with simple electronics means. As discussed in the previous chapter, application of a high electric field between the electrodes results in avalanche multiplication, boosting the signal amplitude by many orders of magnitude. Thanks to their simple construction, parallel plate or avalanche counters have been used in the past, and with several improvements are still widely used in particle physics to cover large detection areas (see Chapter 12); they suffer, however, from one basic limitation: the detected signal depends on the avalanche length, i.e. from the point of release of the primary ionization, implying a lack of proportionality between the energy deposit and the detected signal. Also, due to the large statistical fluctuations in the avalanche size, the Raether condition is occasionally met even at moderate gains, leading to breakdown.

Developed in the early years of the last century (Rutherford and Geiger, 1908), a cylindrical coaxial geometry for the counter overcomes these limitations and provides an amplified signal proportional to the initial ionization (Figure 7.1). A thin metal wire is stretched on the axis of a conducting cylinder and insulated from it, so that a difference of potential can be applied between the electrodes, with the central wire (the anode) positive in respect to the outer cylinder (the cathode).

The electric field in the tube is highest at the surface of the wire, and rapidly decreases towards the cathode (Figure 7.2); using thin wires, very high values of field can be obtained close to the anode. In most of the volume where the charges are produced by the primary interaction processes, the electric field only makes electrons and ions to drift respectively towards the anode and cathode.

182

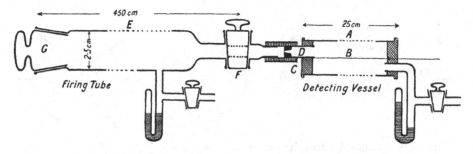

Figure 7.1   The original design of the proportional counter. Ionization electrons released by a radium sample are let into the detecting vessel, multiplied and detected (Rutherford and Geiger, 1908).

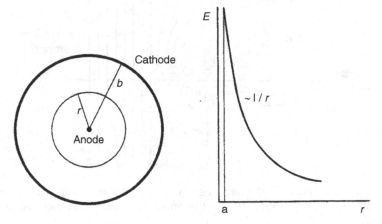

Figure 7.2   Schematics of a proportional counter and its field.

On approaching the anode, normally at a distance of a few wire radii, electrons begin experiencing a field strong enough to initiate charge multiplication; an avalanche develops, with electrons in the front and ions behind, as in the parallel plate case, but with a fast decreasing value of the mean free path for multiplication. Because of lateral diffusion and the small radius of the anode, even at moderate gains the avalanche tends to surround the wire, as shown schematically in Figure 7.3; due to the large difference in drift velocity, electrons in the front are quickly collected, while positive ions slowly drift towards the cathode. As will be seen later, most of the detected signal is due not to the collection of electrons, but to the initially fast motion of the ions receding from the anode, where most of them are produced.

The design and operation of wire proportional counters have been extensively described in articles and textbooks (Rossi and Staub, 1949; Curran and Craggs, 1949; Wilkinson, 1950; Korff, 1955).

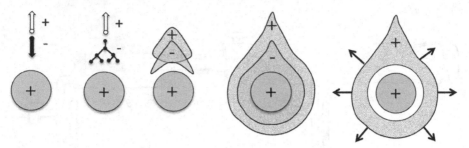

Figure 7.3   Avalanche growth around a thin wire.

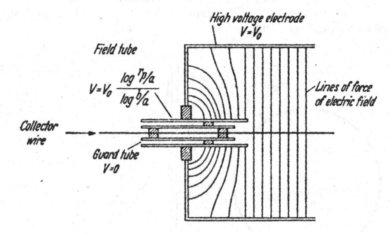

Figure 7.4   Wire holder and guard ring structure on a proportional counter edge (Cockroft and Curran, 1951). By kind permission of the American Institute of Physics.

To prevent edge discharges and delimit the active volume of the counter to the region of uniform response, one or more additional electrodes or guard rings are usually added at the edges of the counters; Figure 7.4 shows a classic configuration (Cockroft and Curran, 1951). A thin anode wire is mounted within a metallic tube, insulated from it and kept at the same potential. A second coaxial tube of larger diameter, insulated from the previous, receives a voltage corresponding to the potential present in that position within the active volume of the counter. The guard rings collect charges released in the region close to the wire's end, and ensure a uniform response of the sensitive volume.

Figure 7.5 (Montgomery and Montgomery, 1941) shows a typical dependence of detected charge on anode voltage in a proportional counter, for two sources of radiation differing in primary ionization density. At very low voltages, charges start being collected, but recombination is still a dominant process; at higher fields, full charge collection begins, and the counter is said to operate in the ionization chamber

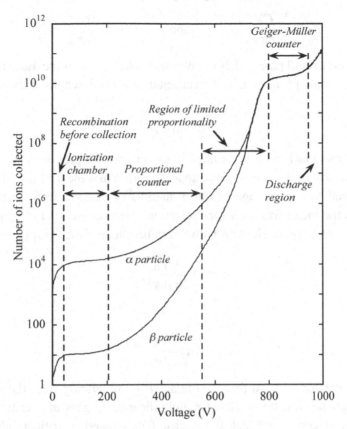

Figure 7.5   Voltage dependence of detected charge in a proportional counter (Montgomery and Montgomery, 1941). By kind permission of Elsevier.

mode. Increasing the voltage further, and above a threshold value V_T, the electric field close to the surface of the anode gets large enough to initiate a process of charge multiplication; gains in excess of 10^5–10^6 can be obtained, with the detected charge proportional, through the multiplication factor, to the original deposited charge. At even higher voltages, however, this proportionality is gradually lost, as a consequence of the electric field distortions due to the large charge density building up around the anode; the region of limited proportionality ends in a region of saturated gain, where the detected signal is independent of the original ionization density.

When the voltage is further increased, various secondary processes set in, resulting in breakdown or, under special conditions (low pressure, poorly quenched gases) in a photon-mediated propagation of the multiplication over the full length of the anode, the so-called Geiger–Müller operation.

If a and b are the radii of anode and cathode the electric field and potential at a distance r from the centre of the counter are:

$$E(r) = \frac{CV_0}{2\pi\,\varepsilon_0}\frac{1}{r} \qquad \text{and} \qquad V(r) = \frac{CV_0}{2\pi\,\varepsilon_0}\ln\frac{r}{a}. \tag{7.1}$$

$V_0 = V(b)$ is the overall potential difference, with $V(a) = 0$; ε_0 is the dielectric constant (for gases $\varepsilon_0 = 8.86$ pF/m), and C the capacitance per unit length of the counter:

$$C = \frac{2\pi\,\varepsilon_0}{\ln(b/a)}. \tag{7.2}$$

Following Rose and Korff, the multiplication factor of a proportional counter can be computed, within the limits of the approximation that the Townsend coefficient is proportional to the average electron energy, expression (5.5) (Korff, 1955). Since $1/\alpha$ is the mean free path for ionization, the average energy acquired from the electric field E by an electron between collisions is E/α; from (7.1):

$$\varepsilon = \sqrt{\frac{CV_0}{2\pi\,\varepsilon_0 kN}\frac{1}{r}},$$

therefore:

$$\alpha(r) = \sqrt{\frac{kNCV_0}{2\pi\,\varepsilon_0}\frac{1}{r}}. \tag{7.3}$$

The gain of the counter can then be obtained by integrating over the useful field region. Assuming that the avalanche multiplication begins at a critical distance r_c from the wire centre at which the electric field exceeds a critical value E_c:

$$M = e^{\int_a^{r_c} \alpha(r)\mathrm{d}r}, \tag{7.4}$$

recalling the definition of threshold voltage V_T:

$$E_C = \frac{CV_T}{2\pi\varepsilon_0}\frac{1}{a} \qquad \text{and} \qquad \frac{r_c}{a} = \frac{V_0}{V_T}; \tag{7.5}$$

substituting (7.3) in (7.4), integrating and using (7.5):

$$M = e^{2\sqrt{\frac{kNCV_0 a}{2\pi\varepsilon_0}}\left(\sqrt{\frac{V_0}{V_T}}-1\right)}. \tag{7.6}$$

For $V_0 \gg V_T$, the multiplication factor is seen to depend exponentially on the charge per unit length $Q = CV_0$:

$$M = K\,e^{HCV_0}. \tag{7.7}$$

Once the threshold voltage has been determined, the multiplication factor can be computed using the values of k given in Table 5.2. Figure 7.6 (Staub, 1953), shows

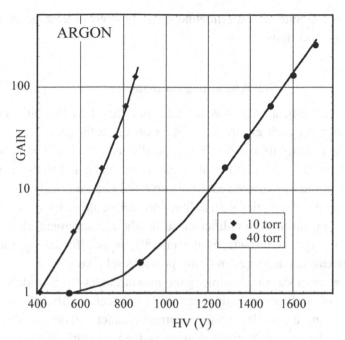

Figure 7.6 Comparison between the gain computed with Korff's approximation and measurements. Data from Staub (1953). By kind permission of J. Wiley & Son.

the computed and measured gains for argon at two values of pressure (in cm Hg): the agreement is excellent for moderate gains.

Many authors have developed alternative formulations to the original Rose and Korff expression, aiming at improving the agreement with experimental data; a summary and extended bibliography can be found in Bambynek (1973).

The expressions are adequate for a qualitative understanding of the proportional counters' operation. For gain values above 10^4, however, the approximations used for α are not justified; a summary of more detailed calculations by different authors is given in Palladino and Sadoulet (1975).

Methods of calculation of the electric field and signal induction in cylindrical counters with wire cathodes are discussed in Szarka and Povinec (1979), and for counters having rectangular cross section in Tomitani (1972); the results can be used to optimize the design of the counters, and easily extended to multi-element detectors consisting of wire-delimited identical adjacent cells as implemented in volumetric multi-wire drift chamber arrays.

A concise parametric model of the avalanche processes in gas counters is given in Bateman (2003); the discussion covers a range of detectors of different geometry. Approximate expressions of gain for cylindrical wire counters, micro-strip chambers, point anode and parallel plate counters are provided, and permit one to analyse experimental data obtained in a range of gas fillings and operating conditions.

7.2 Absolute gain measurement

The detector multiplication factor, or gain, is defined as the ratio between the collected charge on each electrode, positive on the cathode and negative on the anode, and the primary ionization charge. As discussed in the next section, due to the large difference in the mobility of electrons and ions, and to the characteristics of the amplifier, the fast component of the detected charge is only a fraction of the total. Although an electronics gain calibration can be done by injecting a known amount of charge into the amplifier-pulse height measurement chain, the result depends on the time constants of the circuits, it is affected by the generally unknown detector capacitance, and only provides relative values.

Alternative methods of absolute gain measurements are widely used. The first consists of a direct measurement of the detector current, under uniform irradiation, making a complete voltage–current characterization, as the one shown in Figure 7.5: the gain at a given voltage is then the ratio between the current recorded at that voltage and the one corresponding to the ionization chamber plateau. The method requires high source intensities and/or heavily ionizing particles in order to measure the ionization current accurately.

Alternatively, one can realize simultaneous or consecutive measurements of the current I and rate R under exposure to radiation with a known ionization yield, as for example a soft X-ray source. The absolute gain M at a given voltage is then:

$$M = \frac{I}{NeR},$$

where the average ionization yield for each event, N, can be estimated from expression (2.3). Requiring efficient detection of each ionizing event for the estimate of R, the method can only be used when sufficiently high gains can be attained.

7.3 Time development of the signal

As indicated, charge multiplication begins at a few wire radii, i.e. typically less than 50 μm from the anode surface. Assuming a value of 5 cm/μsec for the drift velocity of electrons in this high field region, it appears that the whole process of multiplication takes place in less than 1 ns: after this time, all electrons in the avalanche have been collected on the anode, and the positive ion sheath drifts towards the cathode at decreasing velocity. The detected signal, negative on the anode and positive on the cathode, is the consequence of the change in energy of the system due to the movement of charges. The simple electrostatic considerations discussed in Chapter 6 show that if a charge Q is moved by dr in a system of total capacitance LC (where L is the length of the counter), the induced signal is:

$$dv = \frac{Q}{LCV_0}\frac{dV}{dr}dr. \tag{7.8}$$

Electrons in the avalanche are produced very close to the anode (half of them in the last mean free path); therefore their contribution to the total signal is very small: positive ions, instead, drift across the counter and generate most of the signal. Assuming that all charges are produced at a distance λ from the wire, the electron and ion contributions to the signal on the anode are, respectively:

$$v^- = -\frac{Q}{LCV_0}\int_a^{a+\lambda}\frac{dV}{dr}dr = -\frac{Q}{2\pi\varepsilon_0 L}\ln\frac{a+\lambda}{a},$$

$$v^+ = \frac{Q}{LCV_0}\int_{a+\lambda}^{b}\frac{dV}{dr}dr = -\frac{Q}{2\pi\varepsilon_0 L}\ln\frac{b}{a+\lambda}.$$

The total maximum signal induced on the anode is then:

$$v = v^+ + v^- = -\frac{Q}{2\pi\varepsilon_0 L}\ln\frac{b}{a} = -\frac{Q}{LC}$$

and the ratio of the two contributions:

$$\frac{v^+}{v^-} = \frac{\ln(a+\lambda) - \ln a}{\ln b - \ln(a+\lambda)}.$$

Substituting in the previous expression typical values for a counter, $a = 10\ \mu m$, $\lambda = 1\ \mu m$ and $b = 10\ mm$, one finds that the electron contribution to the signal is about 1% of the total; it is therefore, in general, neglected for most practical purposes.

The time development of the signal can be deduced assuming then that ions leaving the surface of the wire with constant mobility are the only contribution. In this case, integration of (7.8) gives for the signal induced on the anode the expression:

$$v(t) = -\int_0^t dv = -\frac{Q}{2\pi\varepsilon_0 L}\ln\frac{r(t)}{a}. \tag{7.9}$$

From the definition of mobility, it follows that

$$\frac{dr}{dt} = \mu^+ \frac{E}{P} = \frac{\mu^+ C V_0}{2\pi\varepsilon_0} \frac{1}{r}$$

and therefore

$$\int_a^r r\,dr = \frac{\mu^+ C V_0}{2\pi\varepsilon_0} \int_0^t dt,$$

$$r(t) = \sqrt{a^2 + \frac{\mu^+ C V_0}{\pi\varepsilon_0} t}.$$

Substituting in (7.9):

$$v(t) = -\frac{Q}{4\pi\varepsilon_0 L} \ln\left(1 + \frac{\mu^+ C V_0}{\pi\varepsilon_0 a^2} t\right) = -\frac{Q}{4\pi\varepsilon_0 L}\left(1 + \frac{t}{t_0}\right), \tag{7.10}$$

and the corresponding current:

$$i(t) = LC\frac{dv(t)}{dt} = -\frac{QC}{4\pi\varepsilon_0} \frac{1}{t_0 + t},$$

$$t_0 = \frac{\pi\varepsilon_0 a^2}{\mu^+ C V_0}.$$

The current is maximum for $t = 0$:

$$i_{\mathrm{MAX}} = i(0) = -\frac{\mu^+ Q C^2 V_0}{4\pi^2 \varepsilon_0^2 a^2}. \tag{7.11}$$

The total drift time of the ions, T^+, is obtained from the condition $r(T^+) = b$:

$$T^+ = \frac{\pi\varepsilon_0 (b^2 - a^2)}{\mu^+ C V_0},$$

$$v(T^+) = -\frac{Q}{LC}.$$

Due to the high initial velocity of the ions in the field around the anode, the signal growth is very fast at short times: about one half of the signal develops in a fraction of time corresponding to the ratio of the anode to cathode radius:

$$v\left(\frac{a}{b} T^+\right) \approx \frac{Q}{2LC}.$$

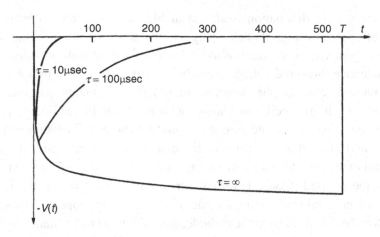

Figure 7.7   Time evolution of the voltage pulse on the anode, for several values of the time constant of the load.

With the aim of increasing the rate capability of the counter, it is normal practice to terminate the anode with a resistor R, such that the signal is shaped with a time constant $\tau = RC$; short pulses can be obtained using low impedance terminations, at the cost of a reduced amplitude.

Figure 7.7 is an example of time development of the signal computed for an argon-filled counter at STP, with $a = 10$ μm, $b = 8$ mm, $C = 8$ pF/m (from (7.2)) and $\mu^+ = 1.7$ cm^{2}s^{-1}V^{-1} (from Table 4.2); for a typical operational voltage $V_0 = 3$ kV, the total length of the pulse is $T = 550$ μs.

Substituting the numerical values of the previous example, and assuming $Q = 10^6$ e (where e is the electron charge) one gets $I_{MAX} \sim 10$ nA, a typical value for the operation of proportional counters.

7.4 Choice of the gas filling

Avalanche multiplication occurs in all gases; virtually any gas or gas mixture can be used in a proportional counter. In most cases, however, the specific experimental requirements restrict the choice to selected families of compounds; low working voltage, high gain operation, good proportionality, high rate capabilities, long lifetime and fast recovery are examples of sometimes conflicting requirements. In what follows, the main properties of different gases in the performance of proportional counters are briefly outlined; a more detailed discussion can be found in the literature (Curran and Craggs, 1949; Korff, 1955; Franzen and Cochran, 1956).

As discussed in Chapter 5, avalanche multiplication occurs in noble gases at much lower fields than in complex molecules: this is a consequence of the many non-ionizing energy dissipation modes available in polyatomic molecules. Convenience of operation suggests then using a noble gas as the main filling; addition of other components, for the reasons to be discussed below, will of course increase the multiplication threshold voltage. The choice within the family of noble gases is then dictated, at least for the detection of minimum ionizing particles, by the requirement of a high specific ionization; with reference to Table 2.1, and disregarding xenon or krypton for economic reasons, the choice falls naturally on argon.

Excited and ionized atoms form in the avalanche process; the excited noble gases return to the ground state only through a radiative process, and the minimum energy of the emitted photon (11.6 eV for argon) is well above the ionization potential of any metal constituting the cathode (7.7 eV for copper). Photoelectrons can therefore be extracted from the cathode, and initiate a new avalanche very soon after the primary; an argon-operated counter generally does not allow gains in excess of a few hundred without entering into permanent discharge.

Positive ions produced in the avalanche migrate to the cathode and are there neutralized by extracting an electron; the balance of energy is either radiated as a photon, or by secondary emission of electrons from the metal surface. Both processes result in a delayed spurious avalanche: even for moderate gains, the probability of the processes discussed is high enough to induce a permanent regime of discharge. Polyatomic molecules have a different behaviour, especially when they contain more than four atoms; the presence of non-radiative excited states (rotational and vibrational) allows the absorption of photons in a wide energy range (see Chapter 3). For methane, for example, absorption is very efficient in the range 7.9 to 14.5 eV, which covers the range of energy of photons emitted by argon. This is a property of most organic compounds in the hydrocarbon and alcohol families, and of several inorganic compounds like freons, CO_2, CF_4 and others; these molecules dissipate the excess energy by elastic collisions or by dissociation into simpler radicals. The same behaviour is observed when a polyatomic ionized molecule neutralizes at the cathode: secondary emission is very unlikely. Even small amounts of a polyatomic quencher added to a noble gas change the operation of a counter, because of the lower ionization potential that results in a very efficient ion exchange process. Good photon absorption and suppression of the secondary emission allows gains of 10^5–10^6 to be reached before discharge.

Argon–methane in the volume percentages 90–10 (the so-called P10 mixture) is widely used in wire proportional counters. The quenching efficiency of a polyatomic gas increases with the number of atoms in the molecule; isobutane (i-C_4H_{10}) is often used for high-gain, stable operation. Secondary emission has been observed, although with low probability, when using as quenchers simpler molecules like carbon dioxide, which may occasionally result in discharges.

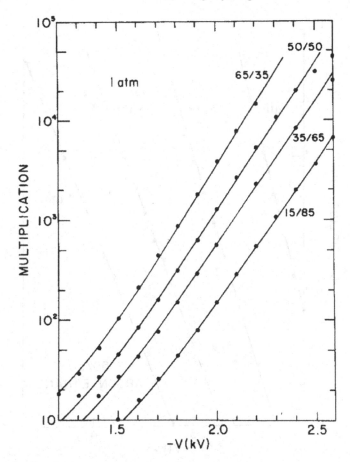

Figure 7.8 Proportional counter gain in argon–ethane mixtures (Behrends and Melissinos, 1981). By kind permission of Elsevier.

Comparative studies of gain and energy resolution of proportional counters with argon–methane and argon–ethane gas fillings in a range of conditions can be found, for example, in Behrends and Melissinos (1981). Figure 7.8 and Figure 7.9 are examples of measured gain in argon–ethane in a range of percentages and pressures; solid lines are fits using the Rose and Korff expression.

Addition of small quantities of electro-negative gases, like freons (CF_3Br in particular) or ethyl bromide (C_2H_5Br) allows one to reach a saturated mode of operation in the so-called 'magic gas', see Section 8.6 (Bouclier *et al.*, 1970). Aside from their photon-quenching capability, electro-negative gases capture free electrons, forming negative ions that cannot induce avalanches in the field values normally met in a proportional counter. If the mean free path for electron capture is shorter than the distance from anode to cathode, electrons liberated at the cathode by the described processes have very little probability of reaching the anode, and

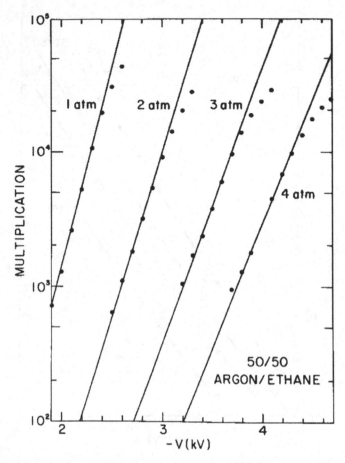

Figure 7.9 Proportional counter gain in a range of pressures in a 50–50 argon–ethane mixture (Behrends and Melissinos, 1981). By kind permission of Elsevier.

gains around 10^7 can be safely obtained before discharge or breakdown. To preserve detection efficiency, however, only limited amounts of electro-negative gases can be used (Breidenbach *et al.*, 1973).

Unfortunately, the use of polyatomic organic gases can have a dramatic consequence on the lifetime of counters, when high fluxes of radiation are detected, in a degenerative process appropriately named ageing. This subject will be covered in Chapter 16.

7.5 Energy resolution

The process of avalanche multiplication, statistical in nature, introduces fluctuations to the average charge gain of a counter. Usually of small relevance in the detection of fast charged particles, due to the large spread in the primary ionization yield, avalanche fluctuations have a dominant effect in determining the energy resolution of proportional counters in the detection of soft X-rays.

If N is the average number of electron–ion pairs released in the gas and M the counter's gain, the relative pulse height resolution can be written as:

$$\left(\frac{\sigma_P}{P}\right)^2 = \left(\frac{\sigma_N}{N}\right)^2 + \left(\frac{\sigma_M}{M}\right)^2. \qquad (7.12)$$

The terms in the expression correspond to the squared sum of the variance in the number of ionization electrons and of the multiplication factor, respectively. Averaged over the N single electron avalanches of size A_i, the gain M can be written as:

$$M = \frac{1}{N}\sum_{i=1}^{N} A_i = \overline{A}$$

and its variance

$$\sigma_M^2 = \left(\frac{1}{N}\right)^2 \sum_{i=1}^{N} \sigma_A^2 \quad \text{or} \quad \left(\frac{\sigma_M}{M}\right)^2 = \frac{1}{N}\left(\frac{\sigma_A}{\overline{A}}\right)^2. \qquad (7.13)$$

As shown in Section 5.4, for a Furry statistics the variance of the single electron avalanche fluctuation is equal to the average avalanche size, $\sigma_A = \overline{A}$, while for a Polya gain distribution it can be written as:

$$\left(\frac{\sigma_A}{\overline{A}}\right)^2 = \frac{1}{\overline{A}} + \frac{1}{1+\theta}, \qquad (7.14)$$

which reduces to the previous for $\theta = 0$ and large values of $\overline{A}$. In both cases, the avalanche fluctuations increase with the avalanche size (the gain of the counter).

As discussed in Section 3.6, for totally absorbed X-rays, the variance in the number of electron–ion pairs can be written as: $\sigma_N^2 = FN$, an expression that takes into account the statistics of energy loss for soft X-rays; the Fano factor F (≤ 1) has a value that depends on the gas (Fano, 1963). Combining the two expressions, and adding a term describing the electronics noise:

$$\frac{\sigma_P}{P} = \sqrt{\frac{1}{N}(F+b)} + f(P). \qquad (7.15)$$

Figure 7.10 shows qualitatively the dependence of the resolution on the gain of a counter, assuming an exponentially decreasing noise spectrum and $N = 100$; the best resolution is obtained at the lowest gains for which the noise contribution becomes unimportant. Figure 7.11 is an example of energy resolution for 5.9 keV

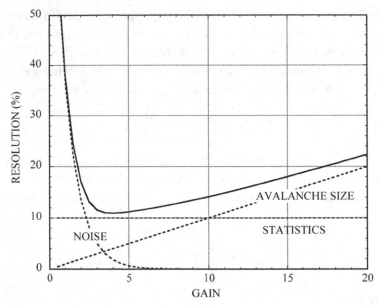

Figure 7.10   Qualitative dependence of a proportional counter resolution on gain due to various dispersive factors.

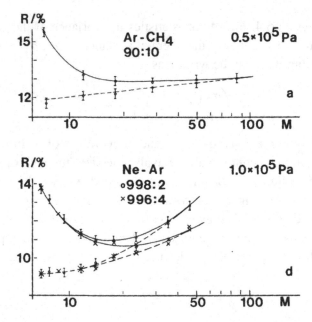

Figure 7.11   Measured energy resolutions FWHM at 5.9 keV as a function of gain for two gas mixtures (Järvinen and Sipilä 1982). By kind permission of Elsevier.

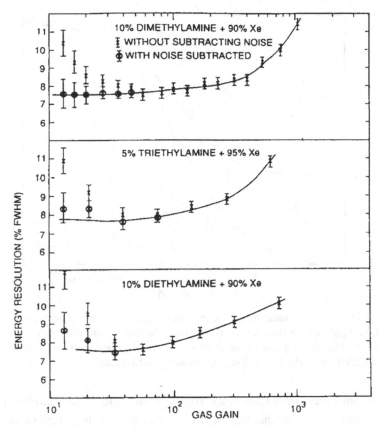

Figure 7.12 Energy resolution for 22 keV X-rays in several mixtures of xenon with low-ionization potential vapours (Ramsey and Agrawal, 1989). By kind permission of Elsevier.

X-rays, measured in the same counter with argon–methane or neon–argon fillings; the characteristic dependence of resolution on gain as well as the improvement with the second mixture, satisfying the Penning condition (see Section 5.3) are clearly seen (Järvinen and Sipilä, 1982). Detailed studies of resolution in Penning mixtures can be found in Sipilä and Kiuru (1978); Agrawal and Ramsey (1989); Agrawal *et al.* (1989); Ramsey and Agrawal (1989). Figure 7.12, from the last reference, is an example of gain dependence of energy resolution for mixtures of xenon and several photosensitive vapours, which satisfy the Penning condition thanks to their very low ionization potential. Using these quenchers has also the advantage of lowering the operating potential as compared to standard mixtures.

Many authors have studied the effect on resolution of geometrical factors in the counter construction and operation. Figure 7.13 is an example of computed dependence of resolution on the anode wire diameter and gas pressure for argon and methane, at a mean gain value of 100 (Alkhazov 1969). More detailed

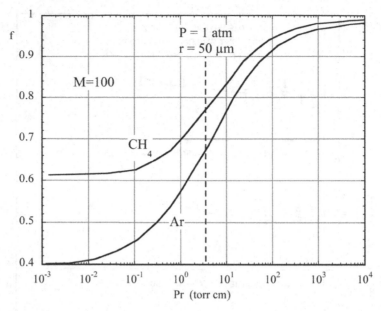

Figure 7.13   Relative variance of the gas amplification factor at an average gain of 100, computed as a function of the product between wire diameter r and gas pressure P. The dashed line corresponds to a typical choice of parameters in a counter (Alkhazov, 1969). By kind permission of Elsevier.

calculations are given in a later work by the same author (Alkhazov, 1970). It should be noted, however, that practical constraints limit the range of parameters that can be adopted for a proportional counter operation.

7.6 Scintillation proportional counters

As discussed above, the avalanche charge multiplication adds dispersions to the energy resolution, increasing with the charge gain. Detection of the charge before or very close to the onset of charge multiplication provides in principle the best resolution, close to the statistical limit; due to limitations imposed by electronics noise, this is hardly possible for small ionization yields such as those released by soft X-rays. A family of gaseous detectors, named gas scintillation proportional counters (GSPC), achieve the goal detecting, by optical means, the signals provided by secondary photon emission at fields below the onset of charge multiplication.

As discussed in Section 5.2, the inelastic collisions between electrons and molecules at moderate values of the electric field result in the creation of excited states, which return to the ground level with the emission of photons of wavelengths characteristic for each gas. The light yield increases exponentially with the

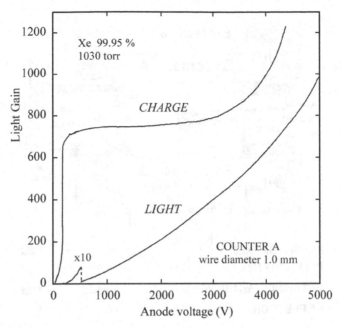

Figure 7.14   Charge and light gain measured with a scintillation proportional counter as a function of anode voltage (Policarpo *et al.*, 1972). By kind permission of Elsevier.

field, and is very copious approaching the charge multiplication threshold, as shown in Figure 7.14 (Policarpo *et al.*, 1972).

An example of a scintillation proportional counter, optimized for the detection of photons emitted by electrons approaching the anode, is shown in Figure 7.15 (Policarpo *et al.*, 1974). A spherical anode, mounted on a tip, is centred on a structure coupled to a photomultiplier through UV-grade windows on each side; the windows can be coated with a wavelength shifter to increase the quantum efficiency of the system. With a xenon filling at atmospheric pressure, the counter provides the pulse height spectra shown in Figure 7.16 for three soft X-ray lines: 227 eV, 1.49 keV and 5.9 keV, respectively (Policarpo *et al.*, 1974). The energy resolution for the 5.9 keV line is 8% FWHM, close to the statistical limit for xenon (6.5%, see Table 3.1), and can be compared with the best value of around 11% quoted in the previous section for a charge-multiplying counter using Penning mixtures.

Other scintillation counter designs have been developed, depending on experimental needs, using parallel plate or multi-wire chamber structures, operated below charge multiplication; for a review see Policarpo (1977). Recent developments include high pressure and large volume detectors; coupled to solid-state light sensors, they have the advantage of being very compact and insensitive to external

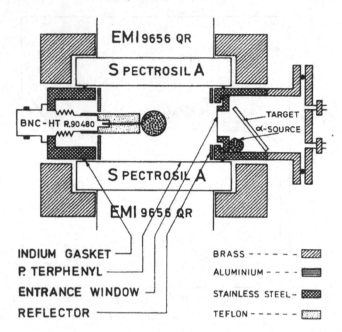

Figure 7.15   Schematics of a gas scintillation proportional counter with spherical anode and two photomultiplier light sensors (Policarpo *et al.*, 1974). By kind permission of Elsevier.

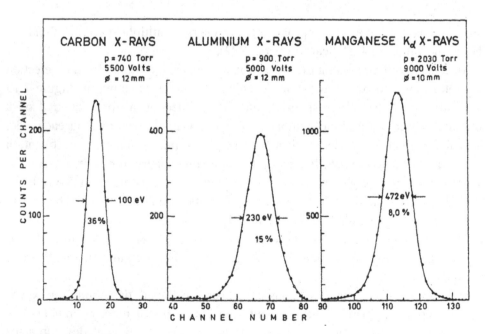

Figure 7.16   Energy resolution of a xenon-filled scintillation proportional counter for several X-ray energies (Policarpo *et al.*, 1974). By kind permission of Elsevier.

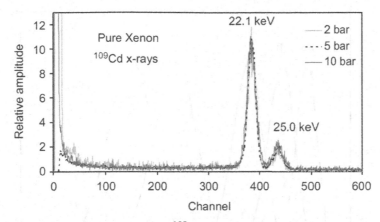

Figure 7.17   Energy resolution for [109]Cd X-rays of a Xe-filled GSPC at several pressures (Coelho *et al.*, 2007). By kind permission of Elsevier.

magnetic fields. Figure 7.17 is an example of energy resolution spectra measured in a xenon-filled GPSC with avalanche photodiode readout for [109]Cd X-rays at several pressures (Coelho *et al.*, 2007); the resolution for the 22.1 keV line is 4.5%, again close to the statistical limit (3.5%).

Methods for directly counting the number of primary ionization electrons have also been developed, making use of optical detection (Siegmund *et al.*, 1983) as well as charge detection with micro-pattern gaseous detectors, described in Chapter 13. Operation at low pressures allows spreading of the clusters by diffusion and detection of the arrival time of individual electrons; by using a dedicated algorithm to identify the peaks in the time-dispersed signal, single electron sensitivity could be reached (Pansky *et al.*, 1993); Figure 7.18 shows the energy spectra of sub-keV X-rays, measured by the electron counting method with optical detection. Results are similar for charge detection; the resolutions are close to the expected statistical values.

7.7 Space-charge gain shifts

During the avalanche development, the growth of a positive ion sheath around the wire results in a local drop of the electric field, with a consequent dynamic reduction of the amplification factor; the normal field is restored only when the ions leave the proximity of the wire and are neutralized at the cathode after several hundred microseconds. The gain is self-affected within each count, implying a loss of proportionality that reflects in a dependence of the gain-voltage characteristics from the initial ionization density, as clearly seen in the typical proportional counter response (Figure 7.5). When a counter is operated in the proportional or

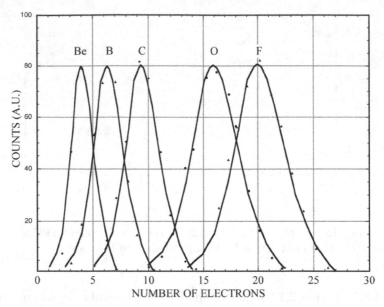

Figure 7.18   Energy resolution for ultra-soft X-rays measured with electron counting (Pansky *et al.*, 1993). By kind permission of Elsevier.

semi-proportional mode, the extension of the avalanche along the wire is rather small, between 0.1 and 0.5 mm, and therefore the field modification is confined to a small region of the counter. In Geiger–Müller operation, on the contrary, the avalanches spread all along the wire and the field in the whole counter is affected: for several hundred μsec no further detection is possible.

The proportional gain dependence on the detected charge density and production rate in wire counters has been studied theoretically and experimentally by many authors to explain observed gain non-uniformities and the appearance of distorted peaks in X-ray and α-particle spectrometry (Hendricks, 1969; Sipilä *et al.*, 1980; Mathieson 1986; Kageyama *et al.*, 1996). In the first reference, in particular, the gain shift dependence on the counter geometry (anode and cathode diameter) and operating pressure is discussed in detail; Figure 7.19 shows the computed relative gain variation as a function of anode wire radius for cylindrical counters with different cathode radii. As expected, the relative gain shift decreases with narrower tubes, due to faster ion collection; the reduction with the wire radius can be understood as due to a decrease of charge density because of the larger wire area. Practical considerations, as well as the possibility of reaching the gain required for detection, restrict the choice of geometry, however.

Following Hendrix, the approximate change of the anodic potential V_0 due to a flux R of ionizing radiation, each creating nMe charges in the avalanches is:

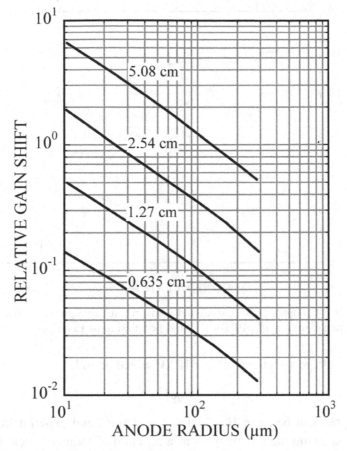

Figure 7.19 Computed relative gain shift in a cylindrical counter as a function of anode wire radius and several cathode radii (Hendricks, 1969). By kind permission of the American Institute of Physics.

$$\Delta V = V_0 - V = \frac{nMeRT^+}{4\pi^2 \varepsilon_0} = KMR,$$

$$K = \frac{neT^+}{4\pi^2 \varepsilon_0},$$

where T^+ is the total collection time of positive ions. Using the approximated expression (7.7) the gain at the rate R can then be written as:

$$M = M_0 e^{-\Delta V},$$

where M_0 is the gain at low rates; replacing in the previous expression:

$$\Delta V e^{\Delta V} = KM_0 R.$$

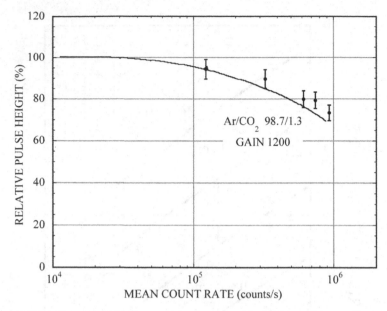

Figure 7.20 Computed (full curve) and measured relative gain as a function of counting rate (Sipilä *et al.*, 1980). By kind permission of Elsevier.

For small voltage variations, $e^{\Delta V} \approx 1$, $\Delta V \approx KM_0R$ and:

$$M = M_0 e^{-KM_0R}.$$

The good agreement between Hendricks' calculations and experiment is seen in Figure 7.20, showing the relative gain as a function of radiation measured with a single wire counter 2 cm in diameter and operated in Ar-CO_2 at a gain of 1200 (Sipilä *et al.*, 1980). In the experiment, the ^{55}Fe X-ray source used for the measurement was collimated on about 2 cm of wire; the gain is reduced to 70% of its low rate value at a flux of 5×10^4 counts/s per mm of wire.

For a given detector geometry, the relative gain reduction depends on the charge production rate, but is independent of the avalanche size, or initial gain, as seen in Figure 7.21 (Walenta, 1981); since the effect depends on the charge density, the rate is conventionally expressed per unit length of wire.

More exotic processes such as columnar recombination (Mahesh, 1976) and the formation of a sheet of polar molecules around the anode (Spielberg and Tsarnas, 1975) have been studied to explain some peculiarities of the observed gain reduction at low rates.

Despite decades of research, little progress has been done on rate capability with wire-based systems, confirming the fundamental nature of the space-charge limitation; Figure 7.22 (Aleksa *et al.*, 2000) is an example of the rate dependence of gain measured with the high-accuracy drift tubes of the ATLAS experiment at

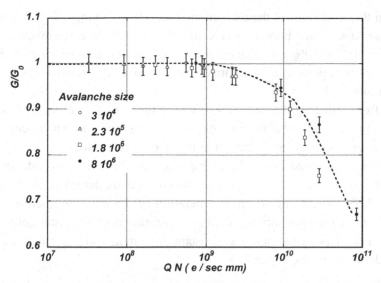

Figure 7.21 Normalized gain measured with a wire chamber as a function of particle flux, given in terms of charge production rate per mm of wire (Walenta, 1981). © The Royal Swedish Academy of Sciences. Reproduced by permission of IOP Publishing.

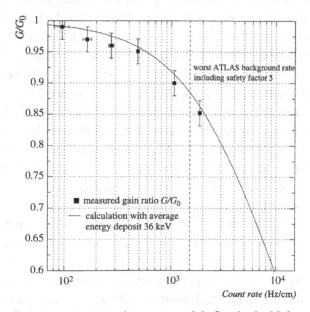

Figure 7.22 Relative gain dependence on particle flux in the high-accuracy drift tubes (Aleksa *et al.*, 2000). By kind permission of Elsevier.

CERN. In the conditions of the measurement, made with tubes 30 mm in diameter operated in Ar-CO_2 at 3 bars at a gas gain of 2×10^4, the average avalanche size is around 3×10^7; a 10% gain reduction is observed at a charge density of around 3×10^9 electrons per second and mm of wire, close to the values indicated by the plot in Figure 7.21.

Apart from reducing the gain and therefore the detection efficiency, the accumulation of ions results in distortions of the drift field and consequent deterioration of localization accuracy, as discussed in the quoted reference.

As will be discussed in the next chapters, very similar gain reductions as a function of rate are observed in all wire-based gaseous detectors, multi-wire and drift chambers; the advent of a new generation of devices, the micro-pattern gas detectors, where the geometry and field configuration permit a faster collection and neutralization of positive ions, has substantially improved the rate capability of gaseous detectors (see Chapter 13).

7.8 Geiger and self-quenching streamer operation

The detected charge in a proportional counter has the characteristic dependence on applied voltage shown in Figure 7.5, going from charge collection, through proportional amplification and to a saturated gain, before reaching the breakdown limit. In special conditions, namely poorly quenched gases and/or low pressures, a different mechanism is observed (Geiger and Müller, 1928). Instead of a localized multiplication of the original ionization charge, a photon-mediated avalanche breeding sets in, propagating the charge multiplication in both directions along the wire, and eventually involving the whole counter. The discharge stops when the operating potential, usually applied to the tube through a high value resistor, drops below a critical value for maintaining the conditions of charge multiplication. This counting mode provides very large counting pulses, typically 10^9 to 10^{10} electrons even starting from small primary ionizations, down to a single electron, and is therefore widely in use for portable radiation monitors. Due to the long time needed to clear all ions and excited molecules produced in a discharge and restore the voltage, a millisecond or more, the rate capability of a Geiger counter is rather limited. Methods of extending the dynamic range of counters with corrections of dead-time losses are discussed in Jones and Holford (1981).

For a review on the Geiger–Müller process and operation of the counters see, for example, the reference book by Wilkinson (1950) and the more recent updates of the same author (Wilkinson, 1992; Wilkinson, 1996a; Wilkinson, 1996b; Wilkinson, 1999). The detection efficiency of counters with cathodes of different thickness is discussed in Watanabe (1999).

With the use of thick anode wires and operating in hydrocarbon-rich gas mixtures, a sudden appearance of very large signals was observed in multi-wire detectors, initially attributed to a transition to a Geiger discharge, limited in extension along the wire because of the quenching effect of the gas mixture used (Brehin *et al.*, 1975). However, further work led to the deduction that the large signals were rather due to a peculiar mode of discharge, propagating perpendicular to the wires and damped before reaching the cathodes; the process was appropriately named self-quenching streamers (SQS) (Alexeev *et al.*, 1980). The mechanism of streamer propagation was described in Section 5.5; in uniform fields, it leads to a discharge between electrodes. With wire counters, the backwards propagation of the streamer towards the cathode can stop in the decreasing electric field, providing large and saturated signals; this mode of operation is exploited in the family of devices named limited streamer tubes, discussed in detail in Chapter 11.

The transition from semi-proportional to SQS occurs when the avalanches exceed a certain size, and therefore takes place at different voltages depending on the initial ionization charge; the semi-proportional and SQS modes coexist over a short range of voltages, as seen in Figure 7.23 (Koori *et al.*, 1986b).

The mechanism of propagation of SQS requires the emission in the avalanches of photons capable of photo-ionizing one of the components of the gas mixture, and has been studied by many authors in a variety of conditions and gases (Koori *et al.*, 1986b; Kamyshov *et al.*, 1987; Koori *et al.*, 1989; Ohgaki *et al.*, 1990; Koori *et al.*, 1990). A quantitative theory of the photon-mediated transition and expressions for the gain in the SQS mode are given in De Lima *et al.* (1985) and compared with experimental measurements in Ohgaki *et al.* (1990).

Due to the limited extension of the avalanche charge in the SQS mode, as compared to a full Geiger, the rate capability is increased, as shown in Figure 7.24 (Nohtomi *et al.*, 1994). Given in γ-ray radiation exposure units (mR/h), the horizontal scale is not easily converted to the counting rate, but serves as relative comparison.

It should be noted that the SQS transition was controversially observed also in mixtures where the photon feeding mechanism is energetically not allowed (Koori *et al.*, 1991).

7.9 Radiation damage and detector ageing

Permanent deterioration of performances under long-term irradiation has been observed since the early development of proportional counters. In Geiger counters, due to the extreme charge densities, this was attributed to a change in the gas composition due to the breakdown of the organic quenchers used; further studies, however, demonstrated that the culprit was a deposition of hydrocarbon polymers

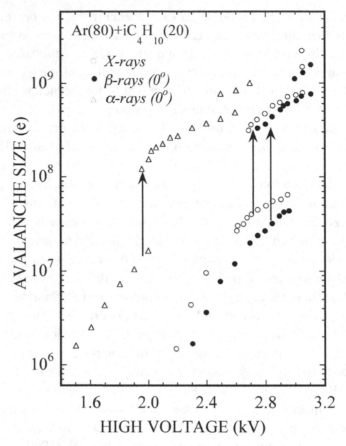

Figure 7.23   Transition from proportional to streamer regime with several ionizing sources (Koori *et al.*, 1986b). © The Japan Society of Applied Physics.

on electrodes. An example of deterioration is given in Figure 7.25, showing the
singles counting rates as a function of voltage before and after several levels of
irradiation (Farmer and Brown, 1948). Figure 7.26 is one of the first reported
observations of ageing in a proportional counter with argon–methane filling,
irradiated with a 5.9 keV X-ray source; the successive pulse height spectra, (a)
to (e), show a progressive degradation of the resolution and the appearance of a
second, lower amplitude peak (den Boggende *et al.*, 1969). By comparing the
results obtained with different counters and gases, the authors inferred a gradual
build-up of a thin, permanent and conductive deposit on the anode, which increases
its diameter. The double peak can be explained by the higher rates on the wire side
facing the source, resulting in faster ageing. A fluorescence surface analysis, and
the recovery of the counter by thermal annealing, confirmed the presence of
volatile carbon-rich deposits.

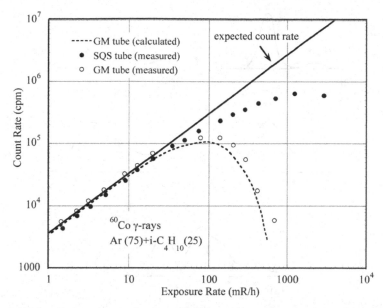

Figure 7.24 Comparison of rate capability of a counter operated in the full Geiger or the self-quenched streamer mode (Nohtomi *et al.*, 1994). By kind permission of Elsevier.

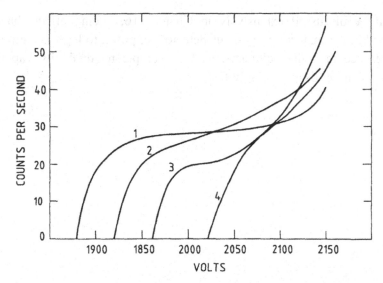

Figure 7.25 Progressive degradation of a Geiger counter with exposure to radiation. Curve 4 is the singles counting rate vs voltage measured after 10^8 counts (Farmer and Brown, 1948). By kind permission of the American Physical Society.

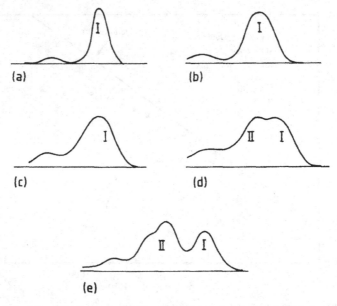

Figure 7.26   Time evolution of pulse height spectra for a counter irradiated with 5.9 keV X-rays (den Boggende *et al.*, 1969). © The Royal Swedish Academy of Sciences. Reproduced by permission of IOP Publishing.

As will be discussed extensively in Chapter 16, ageing effects have been encountered in the majority of gaseous detectors exposed to high radiation levels; in the presence of some pollutants, as for example residual oily vapours, the degradation can be disappointingly fast.

8

Multi-wire proportional chambers

8.1 Principles of operation

Single-wire proportional counters are widely used for detection and energy loss measurement of ionizing radiation; their localization capability is, however, determined by the physical size of the counter. Although some attempts were made to assemble several wire counters within the same gas volume, there was a general belief that multi-wire structures would not work properly because of the large capacitance between parallel non-screened wires, causing the signal to spread to many wires, thereby frustrating localization attempts. It was the merit of Georges Charpak and collaborators in the late 1960s to recognize that the positive signals induced on all electrodes surrounding an anode involving the avalanche largely compensate the negative signals produced by capacitive coupling; this led to the conception and successful test of the first multi-wire proportional chamber (MWPC), shown in Figure 8.1 (Charpak *et al.*, 1968).

The detector consists of a set of thin, parallel and equally spaced anode wires, symmetrically placed between two cathode planes; Figure 8.2 gives a schematic cross section of the structure. The gas gaps are typically three or four times larger than the wire spacing, although thin-gap devices have been developed and successfully operated. When symmetric negative potentials are applied to the cathodes, the anodes being grounded, an electric field develops, as seen in the equi-potentials and field lines shown in Figure 8.3.

As in a proportional counter, electrons created in the gas volume by an ionizing event drift along the field lines approaching the high field region, close to the anode wires, where avalanche multiplication can occur. Figure 8.4 shows schematically the variation of the electric field along a direction perpendicular to the wire plane, for a typical MWPC structure; notice the extended region with uniform field in the drift gap, and the decrease of the field between wires.

Figure 8.1   The first multi-wire proportional chamber, with 24 anode wires and 10×10 cm^2 of active area. Picture CERN (1967).

Analytic expressions for the electric field can be obtained using standard electrostatic algorithms, and are given in many textbooks and articles (Morse and Feshbach, 1953; Durand, 1966; Erskine, 1972; Charpak, 1970). With the geometrical definitions of Figure 8.2:

$$V(x,y) = \frac{CV_0}{4\pi\varepsilon_0} \left\{ \frac{2\pi \, l}{s} - \ln\left[4\left(\sin^2\frac{\pi \, x}{s} + \sinh^2\frac{\pi \, y}{s} \right)\right] \right\}, \tag{8.1}$$

$$E(x,y) = \frac{CV_0}{2\pi \, \varepsilon_0} \left(1 + \tan^2\frac{\pi x}{s}\tanh^2\frac{\pi y}{s}\right)^{\frac{1}{2}} \left(\tan^2\frac{\pi x}{s} + \tanh^2\frac{\pi y}{s}\right)^{-\frac{1}{2}}, \tag{8.2}$$

where V_0 is the voltage applied between anode and cathode planes, and the capacitance per unit length C is given by:

$$C = \frac{2\pi\varepsilon_0}{\frac{\pi l}{s} - \ln\frac{2\pi \, a}{s}}, \tag{8.3}$$

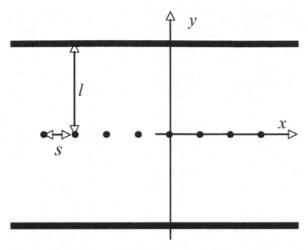

Figure 8.2   Schematics of the MWPC with definitions of the symbols used in the text.

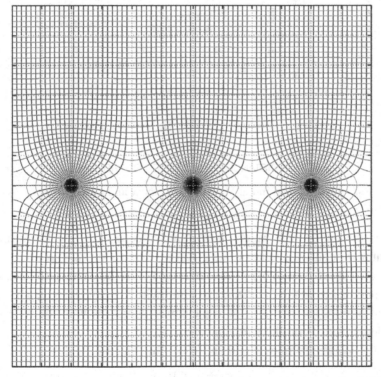

Figure 8.3   Field lines and equi-potentials near the anode wires in the MWPC.

Table 8.1 *Capacitance per unit length (in pF/m) for several MWPC geometries.*

l (mm)	2*a* (μm)	*s* (mm) 1	2	3	5
8	10	1.99	3.47	4.54	5.92
	20	2.04	3.63	4.55	6.39
	50	2.12	3.86	5.23	7.14
5	10	3.01	4.92	6.11	7.40
	20	3.13	5.24	6.62	8.15
	50	3.30	5.73	7.43	9.42

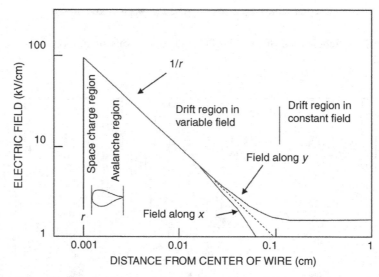

Figure 8.4   Electric field in the MWPC along the *x* and *y* directions.

where *a* is the anode wire radius. Note that, since $a \ll s$, the logarithm in the denominator is negative and the value given by expression (8.3) is always smaller than the capacity of the double plane condenser with the same surface ($2\pi\varepsilon_0 s/l$).

Computed values of the capacitance per unit wire length *C* are given in Table 8.1 for several MWPC geometries; the capacitance decreases with the wire spacing, while it depends little on the wire diameter.

Along the symmetry lines $x = 0$ and $y = 0$ the electric field is:

$$E_y = E(0, y) = \frac{CV_0}{2\varepsilon_0 s} \coth \frac{\pi y}{s},$$

$$E_x = E(x, 0) = \frac{CV_0}{2\varepsilon_0 s} \cot \frac{\pi x}{s}.$$

(8.4)

It is instructive to consider the following approximations:

$$y \ll s: \ E(x,y) \approx \frac{CV_0}{2\pi\varepsilon_0}\frac{1}{r} \quad r = \sqrt{x^2 + y^2}, \tag{8.5}$$

$$y \geq s: \ E_y \approx \frac{CV_0}{2\varepsilon_0 s}. \tag{8.6}$$

Expression (8.5) shows that the field is radial around the anode, as in a cylindrical proportional counter; the results obtained in Sections 6.2 and 6.3 can then be used to estimate the operational characteristics of a multi-wire proportional chamber, provided that the correct value for the capacitance per unit length is used. In the drift region, the field is uniform and smaller than the parallel gap value V_0/l.

8.2 Choice of geometrical parameters

The anode wire distance determines the localization accuracy in a multi-wire chamber with digital readout; spacing below 2 mm is, however, difficult to realize and operate. One can understand the reasons by inspecting the expressions for the fields (8.5) and (8.6), and the approximate multiplication factor in a proportional counter (expression (7.7)). For a fixed wire diameter, to obtain a given gain one has to keep the charge per unit length CV_0 constant, i.e. increase V_0 when s (and therefore Q) is decreased; for example, going from 2 to 1 mm spacing, V_0 has to be almost doubled (see Table 8.1). The electric field in the drift region is also doubled; the probability of a breakdown is strongly increased. Practical experience has shown that, if a chamber with 2 mm wire spacing is relatively easy to operate, 1 mm spacing chambers are hard to manufacture and operate even for small sizes. Decreasing the wire diameters helps, but this brings in mechanical and electrostatic stability problems, see Section 8.4. Down-scaling all geometrical parameters (distance and diameter of the wires and gap) is not sufficient to provide a good operation: indeed, the mean free path for ionization remains invariant, unless the gas pressure is correspondingly increased. Work in this direction has been carried out to obtain good accuracies over small areas.

The approximate expression (7.7) for the gain of a cylindrical counter, using the values for the capacitance per unit length given in Table 8.1, permits one to investigate qualitatively the influence on gain of the wire diameter for a given MWPC geometry (Sauli, 1977). Figure 8.5 shows how the gain varies, as a function of the ratio V_0/V_T, for a MWPC with $s = 2$ mm, $l = 8$ mm and several anode wire radii. Although in principle any wire diameter allows one to obtain the desired gain, the steeper slope for large diameters implies a more critical operation when taking into account mechanical and electrical tolerances (see the next

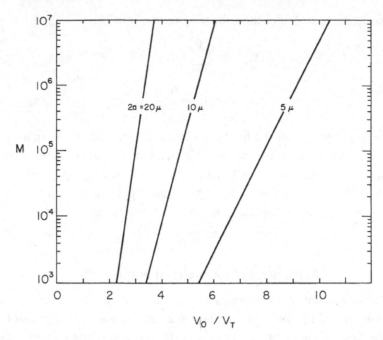

Figure 8.5   Gain dependence from voltage for a MWPC with several anode wire
diameters (Sauli, 1977). By kind permission of CERN.

section). On the other hand, thicker anode wires are easier to handle, and a
compromise has to be found; diameters around 10 μm are a practical lower limit,
while 20 μm or larger are more frequently used.

8.3 Influence on gain of mechanical tolerances

The gain of a chamber at a given operational voltage depends on the detailed shape
and value of the electric field in the multiplication region and can therefore change
along a wire or from wire to wire as a consequence of mechanical variations. The
maximum tolerable differences in gain depend on the specific application: in pulse-
height measuring devices, requirements are much more severe than in threshold-
operated chambers.

A detailed analysis of the gain variations due to different sources of mechanical
tolerances can be found in Erskine (1972). Approximate estimates can be obtained
again by making use of expression (7.7) for the multiplication factor (Sauli, 1977).
By differentiation one gets:

$$\frac{\Delta M}{M} = \ln M \frac{\Delta Q}{Q} \text{ with } Q = CV_0. \tag{8.7}$$

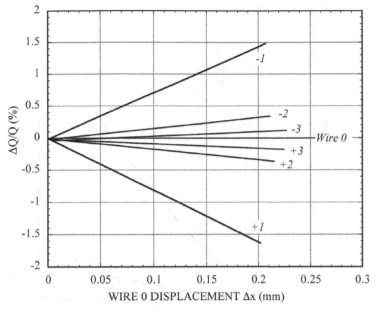

Figure 8.6   Relative charge change on adjacent wires due to wire 0 displaced in the *x* direction (Erskine, 1972). By kind permission of Elsevier.

The problem of gain variation is therefore reduced to a calculation of the change in the charge of the wires. From expression (8.3) for the capacitance per unit length *C* one can compute the effect of a change Δa in the wire radius and Δl in the gap:

$$\frac{\Delta Q}{Q} = \frac{C}{2\pi\varepsilon_0}\frac{\Delta a}{a} \quad \text{and} \quad \frac{\Delta Q}{Q} = \frac{Cl}{2\pi\varepsilon_0}\frac{\Delta l}{l}. \tag{8.8}$$

Consider, for example, a typical $l = 8$ mm, $s = 2$ mm chamber with $2a = 20$ μm operating at a gain of 10^5. The approximate gain variations will be, from the previous expressions:

$$\frac{\Delta M}{M} \approx 0.8\frac{\Delta a}{a} \quad \text{and} \quad \frac{\Delta M}{M} \approx 12\frac{\Delta l}{l};$$

a typical diameter variation around ~10%, as measured on standard 20 μm wires, results then in a ~10% change in gain, while a 1 mm difference in the gap thickness determines a ~50% gain change.

The displacement of a wire from the nominal position also results in a change of the charge per unit length for displaced wire as well as that of its neighbours. Computed for a standard MWPC geometry (2 mm wire spacing, 8 mm gaps and 20 μm wire diameter), Figure 8.6 and Figure 8.7 show the relative change in the charge per unit length of neighbouring wires, as a function of the displacement of a central wire

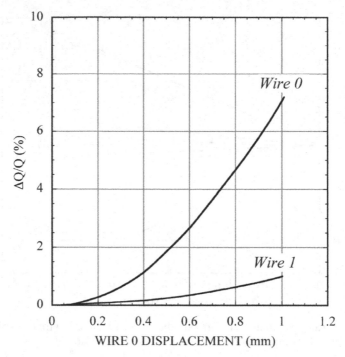

Figure 8.7   Relative charge change due to wire 0 displaced in the y direction (Erskine, 1972). By kind permission of Elsevier.

0 along the x and y directions (Erskine, 1972; Charpak, 1970).[1] A 10% (0.2 mm) displacement of a wire in the wire plane results in a 1.5% change in the charge of the two adjacent wires, and from (8.7), a 20% gain change at gains around 10^5. This is obviously a critical parameter for the construction of multi-wire chambers.

Expression (8.7) can also be used to estimate the variation of the multiplication factor with the operational voltage V_0; a 1% change in V_0 results in an increase in relative gain of about 12% at $M = 10^5$. Experience shows that, with normal mechanical tolerances and for medium-size chambers, over-all gain variation around 20–30% can be expected over the active area, and this has to be taken into account when designing the detection electronics.

8.4 Electrostatic forces and wire stability

In a multi-wire structure, the anode wires are not in a stable equilibrium condition when a difference of potential is applied between anodes and cathodes. Indeed, if

---

[1] Note that in Figure 16 of the quoted reference the scale of the relative change for a wire displacement in the x direction is wrong by a factor of ten.

one wire is displaced from the middle plane, it is attracted more to the side of the displacement and less to the opposite, and the movement would continue indefinitely if there was no restoring force (the mechanical tension on the wires). It has been observed in large chambers that, above a certain value of the applied voltage, the wires become unstable, moving off the middle plane and reaching a new equilibrium with all wires alternately displaced towards the cathodes. Due to the reduction of the distance of the wires from the high voltage electrodes, a discharge may occur, propagating to the whole chamber and leading to spectacular failures, with most of the anode wires broken, as experienced in one of the first large MWPCs (Schilly *et al.*, 1970).

The critical length of a wire for stability can be computed from the balance of forces in the MWPC geometry (Trippe, 1969). Assuming the electric field to be radial, given by the approximate expression (8.5) and not modified by a small displacement δ off the median anode plane, the force per unit length between two equal linear charges CV_0 at a distance r is:

$$F(r) = \frac{(CV_0)^2}{2\pi\varepsilon_0}\frac{1}{r};$$

approximating the tangents to the arcs, the total force per unit length on a given wire, in the direction normal to the wire plane, is:

$$\sum F_\perp = 2\frac{(CV_0)^2}{2\pi\varepsilon_0}\left[\frac{1}{s}\frac{2\delta}{s} + \frac{1}{3s}\frac{2\delta}{3s} + \cdots\right] \approx \frac{\pi(CV_0)^2}{4\varepsilon_0}\frac{\delta}{s^2}. \tag{8.9}$$

If T is the mechanical tension of the wire, the restoring force, in the direction perpendicular to the wire plane and per unit length, is:

$$R = T\frac{d^2\delta}{dx^2}.$$

$\delta = \delta(x)$ is the displacement of the wire along its length, with the conditions $\delta(0) = \delta(L) = 0$, where L is the wire length. At equilibrium, the restoring force should be equal and opposite to the force on the displaced wire, expression (8.9):

$$T\frac{d^2\delta}{dx^2} = -\frac{\pi(CV_0)^2}{4\varepsilon_0}\frac{\delta}{s^2}.$$

The equation has the solution:

$$\delta(x) = \delta_0 \sin\left(\frac{CV_0}{2s}\sqrt{\frac{\pi}{\varepsilon_0 T}}x\right),$$

and from the boundary condition $\delta(L) = 0$:

Table 8.2 *Maximum tension before deformation for tungsten wires of different diameter.*

	T_M	
$2a$ μm	N	g
5	0.04	4
10	0.16	16
20	0.65	66
30	1.45	146

$$\frac{CV_0}{2s}\sqrt{\frac{\pi}{\varepsilon_0 T}}L \quad \text{or} \quad T = \frac{1}{4\pi\varepsilon_0}\left(\frac{CV_0 L}{s}\right)^2.$$

For tensions larger than this value, no solution is possible other than $\delta(x) = 0$, implying that the wires remain stable. The required stability condition is, therefore:

$$T \geq T_C = \frac{1}{4\pi\varepsilon_0}\left(\frac{CV_0 L}{s}\right)^2$$

or, if T_M is the maximum tension allowed by the elasticity module of a given wire, the critical stability length is:

$$L_C = \frac{s}{CV_0}\sqrt{4\pi\varepsilon_0 T_M}. \tag{8.10}$$

Table 8.2 provides, for tungsten wires of several diameters, the maximum mechanical tension that can be applied before inelastic deformation.[2] For a MWPC with 2 mm spacing, 20 μm diameter tungsten wires and 8 mm gap, operated at 5 kV, expression (8.10) gives a maximum stable wire length of 90 cm. When larger detector sizes are needed, a mechanical support has to be used for the wires, at intervals shorter than the critical length (see Section 8.12).

Similar studies exist on the stability of cylindrical MWPCs (Bologna *et al.*, 1979), jet multi-wire chambers (Burckhart *et al.*, 1986) and single-wire drift tubes (Hammarström *et al.*, 1980b).

Another consequence of the electrostatic forces in a multi-wire chamber is the inward attraction of the outer electrodes towards the anode plane, and therefore an inflection of the cathode planes with a reduction of the gap width in the centre of a large chamber. Since the multiplication factor is rather sensitive to the gap width, this can be a problem for large chambers. Taking into account all charge distributions in the multi-wire chamber structure, one can compute the electrostatic force

---

[2] Mechanical wire tensions are often given in grams-weight (1 N — 102 g).

per unit surface, or pressure, on each cathode. For thin-foil electrodes, the deformation of the structure can then be estimated using standard mechanical tools, replacing weight with the electrostatic force.

An approximate estimate of the deflection can be obtained from expression (8.6) for the field in the drift volume. The field at the surface of the cathodes is:

$$E_S = \frac{E_y}{2} = \frac{CV_0}{4\varepsilon_0 s};$$

simple charge balance shows then that the average charge per unit surface on each cathode is $CV_0/2s$, as can also be deduced from the expression taking into account the basic electrostatic relationship $E = \sigma/\varepsilon_0$. The electrostatic pressure on each cathode is:

$$P = \frac{(CV_0)^2}{8\varepsilon_0 s^2}. \tag{8.11}$$

The limits of validity of this expression can be found in the requirement that the field (and therefore the charge distribution) on the surface of the cathodes is constant; from inspection of Figure 8.3 one can see that this assumption holds for $l > s$.

The maximum inwards deflection of a square foil of side H, stretched with a linear tension T and subject to a pressure P, is then given by:

$$\Delta y = \frac{P}{T} \frac{H^2}{8}. \tag{8.12}$$

Combining the two expressions, the maximum deflection for a given chamber geometry, or the maximum size for a chamber if a limit is set for the deflection, can be deduced. As an example, one can compute the maximum gap reduction in an $s = 2$ mm, 8 mm gap chamber with $H = 3$ m, operated at 4.5 kV. From Table 8.1, $CV_0 = 1.5 \times 10^{-6}$ C/m and, from (8.11), $P = 0.8$ N/m^2. For a typical aluminium foil electrode, 20 µm thick, having a Young's modulus of 2×10^8 N/m^2, the maximum stretching tension T is 4×10^3 N/m; from (8.12) the maximum deflection is then 220 µm, and the corresponding gain increase in the centre of the chamber is about 35% at gains of around 10^6 (see Section 6.3), in general not acceptable for correct operation. To reduce the gain non-uniformity, a mechanical gap-restoring support or spacer is necessary; several kinds of spacers and pillars have been developed but are a source of local inefficiency (see Section 8.12).

8.5 General operational characteristics: proportional and semi-proportional

A general discussion on gases used in proportional counters was given in Chapter 7; only a summary of experimental observations specific to multi-wire structures will be discussed here. A stable proportional or semi-proportional

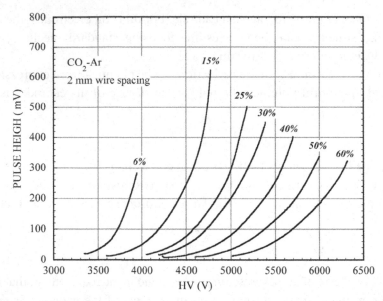

Figure 8.8   Peak pulse height measured on a MWPC on a 100 kΩ load for 5.9 keV X-rays as a function of voltage and for several argon–carbon dioxide mixtures (Bouclier *et al.*, 1970). By kind permission of Elsevier.

operation can be achieved in mixtures of a noble gas with a molecular gas having photon-quenching properties: carbon dioxide, hydrocarbons (methane, ethane, isobutane), ethylene, dimethylether, carbon tetrafluoride and many others. Gains around 10^5 can generally be obtained before breakdown. For the typical energy loss of fast particles in a thin gap of around 2 keV, this implies a total charge signal around 1 pC, or 100 mV on a 10 pF load (Table 2.1). Measured with a 2 mm wire spacing, 8 mm gap MWPC operated at atmospheric pressure, Figure 8.8 and Figure 8.9 show compilations of the average pulse height measured for ^{55}Fe 5.9 keV X-rays on a large load impedance (100 kΩ) for mixtures of argon–carbon dioxide and argon–isobutane (Bouclier *et al.*, 1970); curves stop at the appearance of discharges. Note the tendency to reach higher gains for mixtures containing 30–40% of quencher. For most applications, requiring a fast detector response, the total charge (that corresponds to the full collection of electrons and ions) is differentiated using a low load resistance, substantially reducing the signal amplitude (see the discussion in Section 7.3).

Recorded on 1 kΩ loads in similar conditions, Figure 8.10 shows the pulse-height distributions for 5.9 keV X-rays and minimum ionizing particles, having a most probable energy loss in the 2×8 mm gap filled with a 60–40 argon–isobutane mixture also around 6 keV: charged particles have the characteristic Landau distribution of pulse height; in both plots, the horizontal scale is 1 mV/division. Full detection efficiency for minimum ionizing particles can be obtained

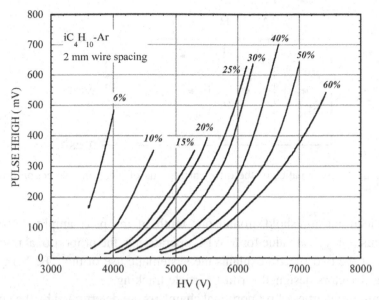

Figure 8.9   Peak pulse height measured on a MWPC on 100 kΩ load for 5.9 keV X-rays as a function of voltage and for several argon–isobutane mixtures (Bouclier *et al.*, 1970). By kind permission of Elsevier.

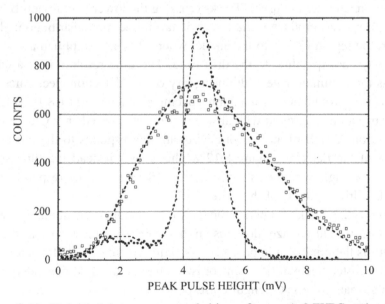

Figure 8.10   Pulse height spectra recorded in an 8 mm gap MWPC at the same voltage for 5.9 keV X-ray (the narrower peak) and minimum ionizing particles.

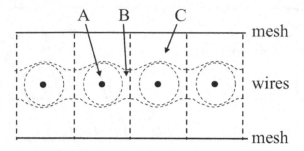

Figure 8.11   Schematics of the different regions of electrons collection around the wires.

with an electronic threshold around one tenth of the peak amplitude, or about 0.5 mV: this is a typical value for MWPC operation in the proportional region. The similarity of the two spectra explains the systematic use of the ^{55}Fe X-ray source for testing detectors designed for fast particle tracking.[3]

The timing properties of proportional chambers are determined by the collection time of the electrons produced by the ionizing tracks. The structure of the electric field around the wires allows the separation of three regions (Figure 8.11): electrons released in region A are quickly collected (typical drift velocities in these high fields are above 5 cm/μsec). Tracks crossing the lower field region B between the wires, however, produce a characteristic tail in the time distribution; electrons created in the region C drift to the anode, where they are amplified and collected with a delay corresponding to the drift time. The time resolution of a chamber, defined as the minimum gate width necessary on the detection electronics for full efficiency, is of around 30 ns for a 2 mm spacing chamber. Figure 8.12 (left) shows a time distributions measured with all wires connected together, so that each track crosses region A or B of at least one wire, and corresponds to the intrinsic time resolution of the detector. Figure 8.12 (right) shows instead the time spectrum recorded on a single wire for an inclined beam: the long tail corresponds to tracks crossing the drift region C of this wire.

When detecting tracks not perpendicular to the chamber, the number of wires hit on each track (or cluster size) depends on the timing gate on the electronics; if the gate length is the minimum imposed by the requirement of full efficiency (around 30 ns), the cluster size will be of one or two wires in an angle-dependent ratio. If, on the other hand, the gate length corresponds to the maximum drift time from region C (about 200 ns for an 8 mm gap), the cluster size depends on the tracks' angle. Figure 8.13 shows the measured average cluster size for a large gate width

---

[3] With the introduction of micro-pattern detectors, having narrower sensitive gaps, tests with the X-ray source tend to be optimistic.

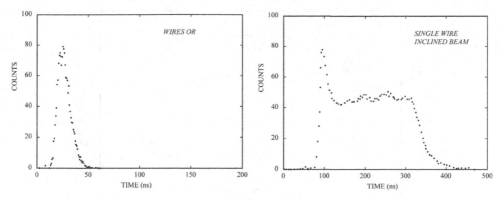

Figure 8.12   Time spectra recorded in a MWPC for minimum ionizing tracks. Left, with all wires connected together; right, for a single wire and an inclined beam.

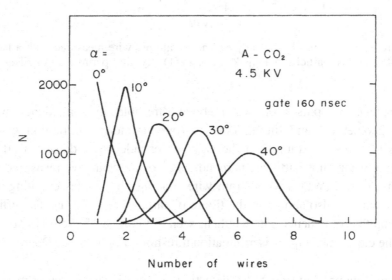

Figure 8.13   Cluster size for ionizing particle tracks at different incidence angles for a 2 mm pitch chamber (Bouclier *et al.*, 1970). By kind permission of Elsevier.

as a function of the angle of incidence ($\alpha = 0°$ corresponds to tracks perpendicular to the wires' plane) (Bouclier *et al.*, 1970).

The detection efficiency of a MWPC depends on many parameters: amount and space distribution of the ionization released by radiation, gain of the detector, electronics threshold, time resolution; a low threshold allows one to get full efficiency at lower voltages, but is prone to be more sensitive to spurious signals due to input noise.

Efficiency and timing properties for charged particles are measured by exposing the detector to a charged particle beam and recording, with appropriate electronics,

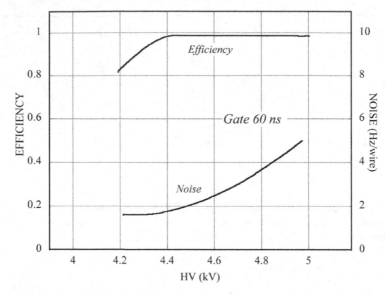

Figure 8.14   A typical efficiency and noise rate per wire measured with a large MWPC for fast particles (Charpak *et al.*, 1971). By kind permission of Elsevier.

the occurrence of pulses on wires above threshold in coincidence with an external geometrical and timing monitor (one or more scintillation counters). Figure 8.14 gives an example of the voltage dependence of detection efficiency for minimum ionizing particles, perpendicular to the chamber, measured in one of the first large MWPCs with 2 mm wire spacing; the singles counting rate, or noise per wire, is also shown in the figure (Charpak *et al.*, 1971). The efficiency depends on the coincidence gate width, as seen in Figure 8.15 (Schilly *et al.*, 1970) and on the electronics input discrimination threshold, Figure 8.16 (Duerdoth *et al.*, 1975).

For applications requiring X-ray detection, as for diffraction experiments from synchrotron radiation sources, xenon mixtures are preferred to provide higher detection efficiency. If a good energy resolution is also required, the operating voltage is a compromise between high gain and best resolution, as shown in Figure 8.17 and Figure 8.18 (Duerdoth *et al.*, 1975).

A comparative analysis of gas properties and of the criteria for a choice meeting the experimental requirements can be found in Va'vra (1992).

8.6 Saturated amplification region: Charpak's 'magic gas'

Presence in the gas mixture of electro-negative pollutants, like oxygen or halogens, results in the capture of electrons with the consequent loss of proportionality, and

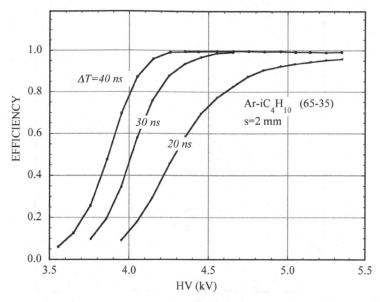

Figure 8.15   Efficiency as a function of voltage for several gate lengths (Schilly *et al.*, 1970). By kind permission of Elsevier.

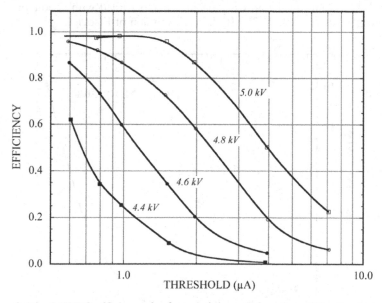

Figure 8.16   MWPC efficiency for fast particles as a function of the discrimination threshold and high voltage (Duerdoth *et al.*, 1975). By kind permission of Elsevier.

should in general be avoided. However, since the effective capture probability depends on the electrons' distance from the wires, the property can be exploited to restrict the sensitive region, thus reducing the dependence of the detected signals on the incidence angle of tracks. With 20% ethyl bromide (C_2H_5Br)

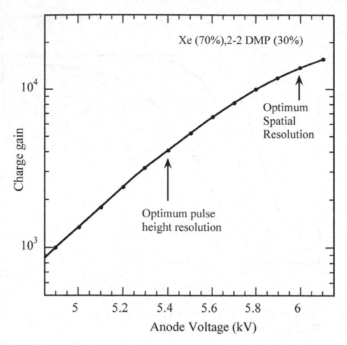

Figure 8.17   Gain of a 1 mm wire spacing chamber filled with xenon–dimethyl propane (Duerdoth *et al.*, 1975). By kind permission of Elsevier.

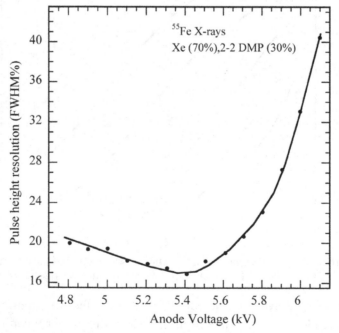

Figure 8.18   Energy resolution on 5.9 keV X-rays in a xenon-DMP filled chamber as a function of applied voltage (Duerdoth *et al.*, 1975). By kind permission of Elsevier.

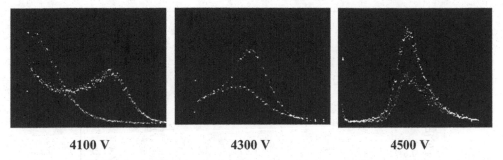

| 4100 V | 4300 V | 4500 V |

Figure 8.19   Pulse height spectra for 5.9 keV and fast particles recorded in magic gas at increasing voltages. The horizontal scale is adjusted to match the increasing pulse heights (Bouclier *et al.*, 1970). By kind permission of Elsevier.

added to the mixture, the sensitive region around the anode wires is reduced to about 1 mm, at the same time permitting one to reach high gains (Grunberg *et al.*, 1970).

As first noticed by Charpak and collaborators, addition of freon-13B1(CF_3Br) permits one to reach a fully saturated operation, i.e. a pulse-height distribution independent of the number of primary ionization charges (Bouclier *et al.*, 1970). The appearance of gain saturation in the so-called magic gas, argon–isobutane–freon in the volume proportions 70/29.6/0.4 is illustrated in Figure 8.19, showing pulse-height spectra for minimum ionizing electrons and 5.9 keV X-rays recorded at increasing voltages; owing to the large increase in the gain, the horizontal scales in the plots differ. Due to the presence of the electro-negative addition, spectra are degraded due to electron capture, but the evolution from a quasi-proportional to a fully saturated regime is apparent. Under these conditions, a single photoelectron provides fully saturated signals (Breidenbach *et al.*, 1973).

The amount of electro-negative gas that can be used in a chamber is limited by the requirement of full efficiency: the mean free path for electron capture, λ_C, should not be smaller than the wire spacing. Figure 8.20 shows the efficiency for a 2 mm wire spacing chamber measured at increasing concentrations of freon; under reasonable assumptions on the detection and capture mechanisms, the experimental points are well approximated by the relationship $\lambda_C = 0.63/p$, where p is the freon percentage and λ_C is given in mm (Breidenbach *et al.*, 1973).

An important consequence of the presence of an electro-negative gas in the mixture is a different dependence of the cluster size on the gate length; electrons produced in the drift region have a very small probability of reaching the anodes, and the cluster size is limited even for long gates.

The large amplitude and reduced dynamic range of saturated pulses, easing the electronics requirements, was essential in the early development of the technology

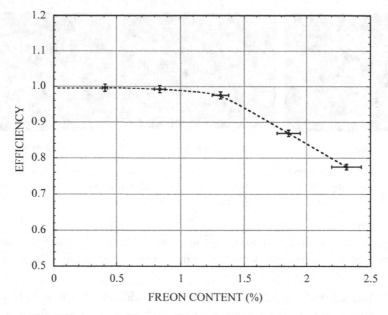

Figure 8.20   Efficiency for fast particles of a 2 mm pitch chamber as a function of freon content (Breidenbach *et al.*, 1973). By kind permission of Elsevier.

and has been adopted by many experiments (Bouclier *et al.*, 1974; Baksay *et al.*, 1976; Frieze *et al.*, 1976; Doll *et al.*, 1988).

The large, saturated gain features of the magic gas have been observed in other mixtures containing small percentages of freon, such as argon–methane and argon–ethane (Koori *et al.*, 1984; Koori *et al.*, 1986a); they should, however, not be confused with the very large signals obtained in the conditions where a transition to a self-quenching streamer is possible (see the next section).

The availability of low threshold highly integrated front-end electronics, and the moratorium on production of fluorinated gases due to ecological reasons have phased out the use of the 'magic gas'.

8.7 Limited streamer and full Geiger operation

A peculiar mode of operation has been observed in proportional chambers having widely spaced thick wires and filled with small concentrations of organic quenchers: the appearance of very large signals in what was first attributed to a transition from proportional avalanches to a damped Geiger propagation, limited in spatial extension to around 10 mm along the wire (Brehin *et al.*, 1975). Although very attractive because of the large signal pulse height that could be obtained (30–40 mV on a 50 Ω load) this mode of operation has a severe rate limitation due to the long time taken by the positive ion sheath to clear the concerned anode wire section. For an 8 mm gap chamber, the assumption that after each count a 10 mm long section of the wire is dead for about 300 μs provides a good agreement with the experimentally measured efficiency.

It was demonstrated later that the large signals were probably due to another mode of discharge, propagating perpendicular to the wires and stopping before reaching the cathodes; the process was appropriately named self-quenching streamers (SQS) (Alexeev *et al.*, 1980), and was confirmed by optical observation of the photons emitted by the avalanches (Atac and Tollestrup, 1982). This operational mode has been used for the limited streamer tube arrays, to be described in Section 12.1.

Full Geiger propagation along all the wire has also been observed in multi-wire chambers operated with a small percentage of organic quencher, like methylal or ethyl bromide added to pure argon, albeit with the long dead-times characteristic of this mode of operation. A measurement of the propagation time of the Geiger streamer along the wires has been used to provide two-dimensional images of the conversion points, making them suitable for low-rate neutral radiation imaging (Charpak and Sauli, 1971).

A large gain, saturated mode of operation has also been observed using heavily quenched gas mixtures in the so-called thin-gap chambers (TGC), developed for applications where the total thickness of the detector has to be minimized (Majewski *et al.*, 1983; Bella *et al.*, 1986). The detectors have been successfully used in the pre-sampling of the OPAL electro-magnetic calorimeter (Beard *et al.*, 1990) and the ATLAS muon spectrometer (Aloisio *et al.*, 2004), and are under development for a high-rate, large size upgrade of the muon tracker for the Super Large Hadron Collider (SLHC) (Amram *et al.*, 2011). In extended beam tests, the detectors have demonstrated reliable operation and resilience to radiation damage up to a large accumulated charge.

8.8 Discharges and breakdown: the Raether limit

The transition of the avalanche to a streamer, and the subsequent sudden appearance of discharges in uniform field counters were discussed in Section 5.5. The transition occurs when the total amount of charge in an avalanche exceed about 10^7 electron–ion pairs, the so-called Raether limit. In wire chambers, a proper choice of geometry and gas mixture permits the operation in the limited or self-quenching regimes, but on increasing the voltage further the conditions are met for full discharge propagation through the structure, often with irreversible damage. Methods of dumping the energy of a discharge by making use of resistive electrodes are discussed in Chapter 12.

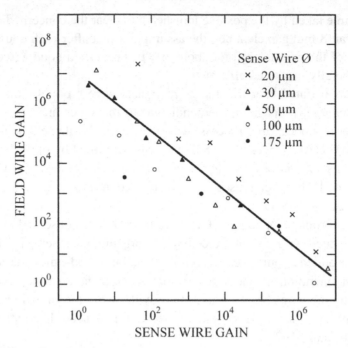

Figure 8.21   Correlation between measured gains of the anode and estimated gain of the cathode wires at discharge (Giubellino *et al.*, 1986). By kind permission of Elsevier.

In structures having non-planar cathodes (wires or strips) the critical Raether condition can be met at smaller apparent values of collected charge due to secondary emission from the high field regions in the cathode plane. A measurement of this effect is shown in Figure 8.21 (Giubellino *et al.*, 1986): in a detector having anode wires centred in a square wire cathode cell, the maximum gain at discharge for the anodes has been correlated to the gain of the cathodes, estimated from the gain the wire would have if used as anode, in a range of wire diameters. Electrons ejected from the cathode by secondary emission experience charge multiplication before getting to the anodes; the product of the two gains roughly corresponds to the Raether limit. Similar gain-limiting processes have been observed in micro-pattern gaseous detectors (see Chapter 13).

The presence of thin insulating layers on cathodes, either manufacturing residuals or created by the organic polymerization processes induced by the avalanches, described in Chapter 16, can also trigger a cascade of events leading to a discharge. Ions produced in the multiplication process can accumulate on the insulating layer, creating a high local dipole electric field; if this field exceeds a critical value, electrons can be extracted from the underlying metal electrode and

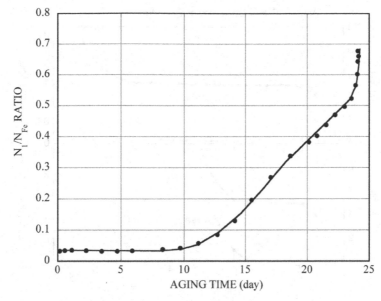

Figure 8.22   Normalized spontaneous single electron counting rate as a function of irradiation time (Boyarski, 2004). By kind permission of Elsevier.

injected into the gas gap, the so-called Malter effect (Malter, 1936). Drifting to the anode, the secondary electrons create more ions, and the process can diverge, leading to the condition of discharge.

A field-emission model taking into account the ions' accumulation and depletion rates on thin-film deposits has been developed and compared with the measured current spikes generated by field-emitted electrons for the specific conditions met in wire chambers (Boyarski, 2004). Figure 8.22 shows how the single electron counting rate, normalized to the source rate, increases with the irradiation time during an ageing test, until reaching the divergence condition; the full line is a fit of the model prediction to the experimental data.

Other breakdown mechanisms, involving long-term modifications of the charge distributions in counters, named 'cumulative' and attributed to a memory of previous occurrences in the detectors, have also been discussed (Ivanouchenkov *et al.*, 1999; Iacobaeus *et al.*, 2001).

Although not directly related to the breakdown mechanisms, an enhancement of spontaneous electron emission due to a decrease of the electronic work function on aluminium cathodes has been found to make gaseous detectors sensitive to visible light. As an example, Figure 8.23 compares the normal counting and the noise rate of a detector exposed to a radioactive source in darkness (upper plot) and in ambient light (lower plot) (Chirihov-Zorin and Pukhov, 1996).

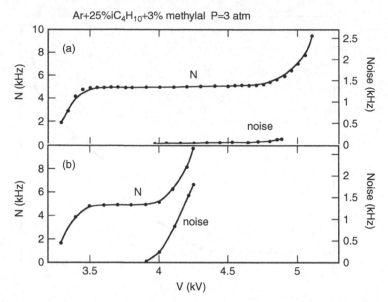

Figure 8.23   Counting and noise rates of a detector in the darkness (a) and exposed to light (b) (Chirihov-Zorin and Pukhov, 1996). By kind permission of Elsevier.

An increase of leakage current due to infra-red radiation, attributed to a person approaching the detector, was also curiously observed in the early developments of MWPCs, but remains anecdotal (Marsh *et al.*, 1979).

The survivability of a wire chamber to breakdowns depends on mechanical construction, energy and frequency of the discharges. Thin anode wires are obviously more prone to failure than thick wires or continuous electrodes. Although spark damages are a common occurrence during detectors' development and tests, only a few systematic studies have been reported on this matter. Figure 8.24 (Rotherburg and Walsh, 1993) is an example of measurements of the energy of a spark needed to break a 10 μm tungsten wire stretched at increasing mechanical tensions; curves are the predictions of a simple model discussed in the reference.

Methods for reducing the discharge energy by sectoring the high-voltage electrodes, with individual high value protection resistors, or by making the electrodes themselves resistive, have been introduced in the development of the limited streamer tubes, resistive plate chambers and micro-pattern gas counters, and will be discussed in the following chapters.

8.9 Cathode induced signals

As noted in the introduction, the avalanche formation, fast electrons collection on the anodes and the ensuing backward motion of ions, result in a distribution of

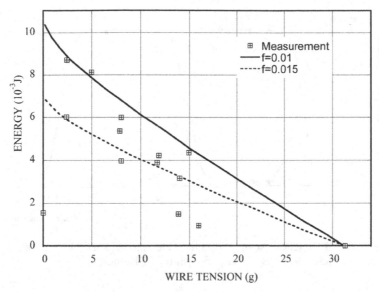

Figure 8.24   Energy of a spark needed to break a 10 μm wire as a function of wire tension (Rotherburg and Walsh, 1993). By kind permission of Elsevier.

induced positive charge on electrodes surrounding the wire, the adjacent wires and the cathodes. The possibility of exploiting a measurement of induced signals to obtain information on the position of the avalanche was discussed in the early works on MWPCs (Charpak *et al.*, 1970).

A simple signal over threshold detection on wide strips on the high voltage planes can be used to provide coarse information on the coordinates of the avalanche along the wires; Figure 8.25 is an efficiency scan through several adjacent cathode strips, 8 cm wide, for a collimated electron source (Fischer and Plch, 1972). The combined detection efficiency is close to 100%, but due to mutual cross talk some residual efficiency is measured for off-beam positions, which can be minimized by a thorough choice of operating conditions and discrimination threshold. A digital readout of wide strips on the cathode planes was used in the first large experimental setup using two-dimensional multi-wire chambers, and had the advantage of reducing the number of modules and mass of the tracker when compared to a solution deploying chambers with inclined wires (Bouclier *et al.*, 1974).

The process of cathode charge induction has been extensively analysed both experimentally and theoretically, particularly in view of its capability of two-dimensional and accurate localization of neutral events. In coincidence with the anode pulses or started by an external trigger, the positive induced charge profile is recorded on sets of readout electrodes on the high-voltage planes: groups of cathode wires or printed circuit strips, at angles with the anodes (Figure 8.26).

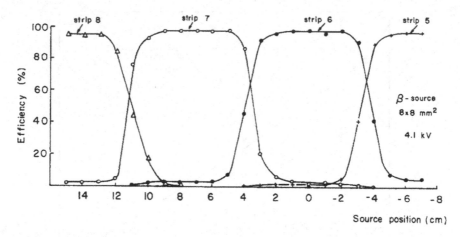

Figure 8.25 Efficiency scan through wide strips on the cathode plane (Fischer and Plch, 1972). By kind permission of Elsevier.

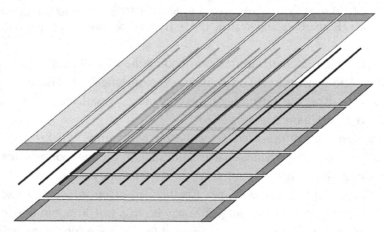

Figure 8.26 Schematics of a MWPC with readout of the cathode induced signals on two sets of perpendicular strips.

The coordinates are then deduced, in the simplest formulation, from a calculation of the centre-of-gravity (COG) of the recorded charge:

$$\bar{x} = \frac{\sum x_i X_i}{\sum X_i} \text{ and } \bar{y} = \frac{\sum y_i Y_i}{\sum Y_i}, \qquad (8.13)$$

where x_i (y_i) and X_i (Y_i) are respectively the central coordinate and the charge recorded on strip i for the projections along two coordinates x and y; the weighted average is made on the set of adjacent strips having a detected signal above some pre-determined threshold. The space–COG correlation is almost linear for the coordinate y, along the

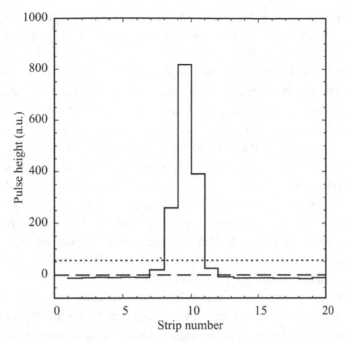

Figure 8.27 Induced charge distribution on cathodes of a 5 mm gap MWPC; the readout strip pitch is 5 mm. The dotted line indicates a bias value used in the analysis (Charpak *et al.*, 1979b). By kind permission of Elsevier.

anodes; under special conditions, to be discussed later, an exploitable non-linear correlation exists also for the coordinate x perpendicular to the wires.

Figure 8.27 is a measurement of a single event charge distribution recorded on cathode strips, 5 mm wide; the distribution has a quasi-Gaussian shape with a standard deviation close to the MWPC gap thickness of 5 mm (Charpak *et al.*, 1979b); this has become a standard rule of thumb for the cathode induction processes, discussed in detail in Section 10.4. For a modern formulation of the signal formation and induction in gaseous counters see Riegler (2004).

For the coordinate along the anode wires, a simple modification of the expression permits one to reduce the dispersive effect of noise and inter-strip cross talk, subtracting a constant bias level B to all measured charges:

$$\bar{s} = \frac{\sum (Q_i - B)\, s_i}{\sum (Q_i - B)}, \tag{8.14}$$

in which negative terms in the sum are neglected; a fit to the data provides the best estimate of the bias term B (Charpak *et al.*, 1979b). In a further refinement of the expression, choice of a bias level proportional to the measured amplitude,

$B = b\Sigma Q_i$ permits one to achieve, with a proper value of the parameter b, the best localization independently of the position of the source with respect to the strips (Figure 8.28) (Piuz *et al.*, 1982).

For charged particles, the angle of incidence to the detector plane plays a major role in determining the localization accuracy, due to asymmetric energy losses caused by delta electrons, as discussed in Section 2.4. A detailed analysis shows the correlation between localization and recorded energy loss for fast particles: while for normal incidence the position accuracy is almost independent of the pulse height, for inclined tracks the dispersion is considerably worse for large energy losses, confirming the crucial role of the delta electrons, Figure 8.29 (Charpak *et al.*, 1979b). The dispersive effect of long-range secondary electrons is manifest in the residual distribution measured with high-accuracy detectors for a perpendicular beam, shown in Figure 8.30. While the standard deviation of a Gaussian fit to the single detector resolution is 15 μm, large deviations in the tails due to delta electrons are clearly visible (Bondar *et al.*, 1983).[4]

For soft X-rays, and using a xenon-based gas mixture to reduce the physical path of the photoelectron in the gas, a localization accuracy of 35 μm rms for the coordinate along the wires could be achieved for a collimated soft X-ray source, Figure 8.31 (Charpak *et al.*, 1978).

The correlation between real position and COG measurement, continuous in the direction parallel to the anode wires, jumps through steps in the perpendicular direction, due to the discrete wire spacing. However, it was found in the early works that, at least at moderate proportional gains, the avalanche does not surround the anode, and the direction of the ions' backflow retains a memory of the original direction of approach of the ionization. The correlation between reconstructed GOG and real position is not linear, but can be unfolded with appropriate algorithms to obtain the space coordinate in the direction perpendicular to the wires, albeit not with the same accuracy as for the one along the anodes; the result can be improved further by a measurement of the induced charge on the two adjacent anode wires (Breskin *et al.*, 1977). Figure 8.32 is the image obtained by exposing a 2 mm spacing MWPC with 3 mm wide cathode strip readout to soft X-rays through a metal mask with letter cuttings $4 \times 1.5 \ mm^2$ in size (Charpak *et al.*, 1978). Using the proper correlations, the centre of gravity method provides bi-dimensional localization with accuracy better than 100 μm.

The angular spread of the avalanche around the wires and its dependence on geometry, gas mixture and gain have been studied both experimentally and theoretically by many authors (Mathieson and Harris, 1979; Fischer *et al.*, 1978;

---

[4] For charged particles, the space resolution is usually deduced from the calculation of the residuals, the difference between predicted and measured coordinates for three identical detectors; if σ_R is the standard deviation of the residuals, the single coordinate resolution is given by $\sigma_1 = \sigma_R / \sqrt{2}$.

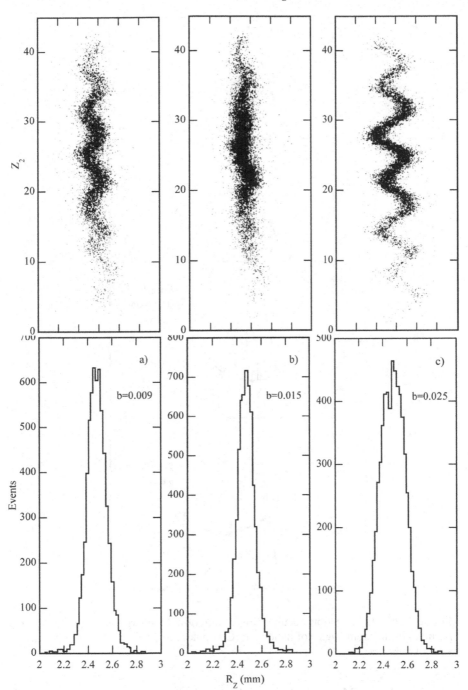

Figure 8.28   Localization accuracy along the anodes for three choices of the bias parameters (Piuz *et al.*, 1982). By kind permission of Elsevier.

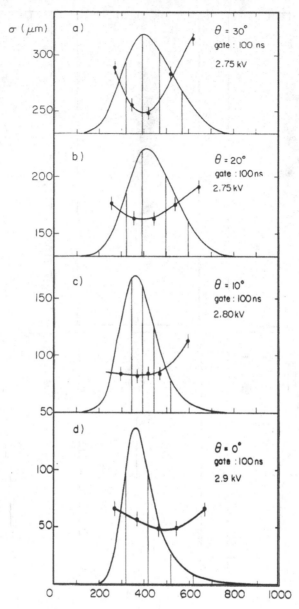

Figure 8.29   Localization accuracy as a function of energy loss for charged particles at different angles of incidence (Charpak *et al.*, 1979b). By kind permission of Elsevier.

Okuno *et al.*, 1979). Measured with a collimated [55]Fe X-ray source, Figure 8.33 is an example of the dependence of the avalanche width (FWHM) on the size, for several anode wire diameter (Fischer *et al.*, 1978). In the same references, the effects of varying the ionization density and the gas are also discussed. For small

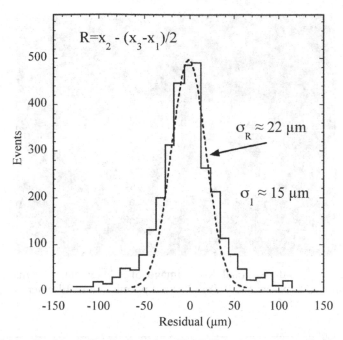

Figure 8.30   Residual distribution for 65 MeV protons perpendicular to three identical detectors. The single detector resolution is 15 μm rms (Bondar *et al.*, 1983). By kind permission of Elsevier.

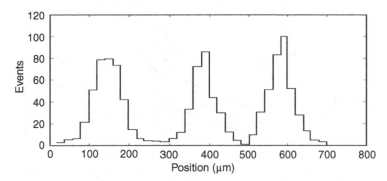

Figure 8.31   Centre of gravity position determination for three soft X-ray source position 200 μm apart in the direction of the anodes (Charpak *et al.*, 1978). By kind permission of Elsevier.

angular spreads, the difference in the signals induced on the two adjacent wires adjacent to the counting one can be exploited to resolve the right–left ambiguity, intrinsic in wire structures (Breskin *et al.*, 1977; Fischer *et al.*, 1978).

Aside from its relevance for optimizing the localization accuracy, the preferential production of positive ions along the direction of approach of the ionized trail

Figure 8.32 Two-dimensional X-ray image of a letter cut. The mask size is 4×1.5 mm^2 (Charpak *et al.*, 1978). By kind permission of Elsevier.

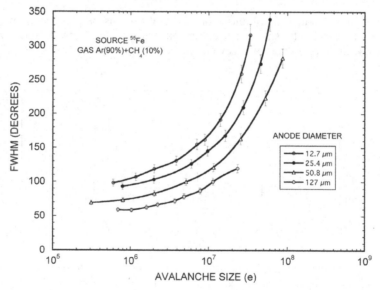

Figure 8.33 Avalanche angular spread as a function of size for several anode wire diameters (Fischer *et al.*, 1978). By kind permission of Elsevier.

affects other detector properties, as the fraction of ion feedback in time projection-like chambers (see Chapter 10).

Many authors have studied the shape of the avalanche-induced signals on anodes and cathodes in various geometries (Mathieson and Harris, 1978; Gatti *et al.*, 1979; Erskine, 1982; Chiba *et al.*, 1983; Gordon and Mathieson, 1984; Thompson *et al.*,

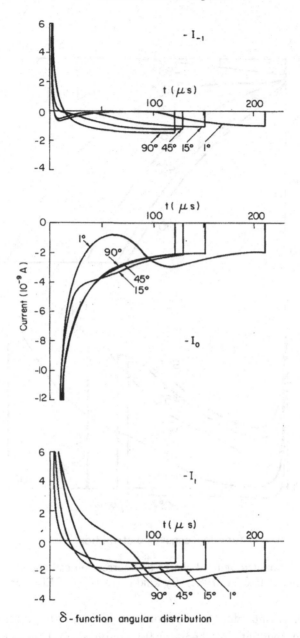

Figure 8.34   Computed signal current on three adjacent anode wires (Erskine, 1982). By kind permission of Elsevier.

1985; Staric, 1989; Landi, 2003; Riegler, 2004). Figure 8.34 and Figure 8.35 (Erskine, 1982) are examples of computed current and charge as a function of time on three adjacent anodes and on cathode strips, for a δ-function avalanche distribution centred at different angles (90° is perpendicular to the cathode planes).

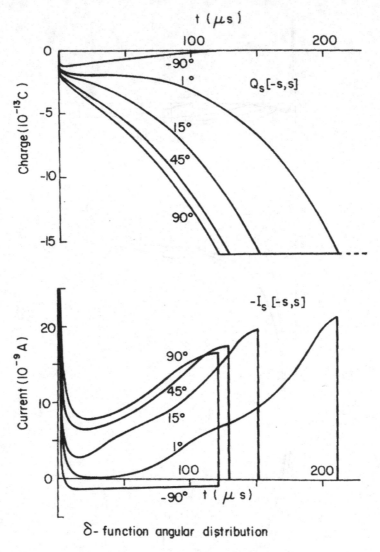

Figure 8.35   Computed induced charge and current on cathode strips facing the central anode wire (Erskine, 1982). By kind permission of Elsevier.

   The resolution limits due to electronic noise and the optimum geometry for strip cathode readout are theoretically analysed in Gatti *et al.* (1979); it is found that the optimum ratio between the cathode sensing electrode width and the anode to cathode distance is about 1, a simple rule followed in most detectors.

   With the introduction of the time projection chamber (TPC) (Nygren and Marx, 1978), the study of induced signals has been extended to devices with pad rows on the cathode plane, discussed in Chapter 10.

The induced charge readout method, while fully exploiting the MWPC performances, requires the implementation of a large number of electronic channels. Alternative methods of recording the two-dimensional coordinates of events has been introduced in the early developments of the detector, making use of electromagnetic delay lines, coupled to the signal electrodes, and inspired by similar systems used for the readout of wire spark chambers (Breskin *et al.*, 1974c). Variously manufactured with wire coils wound around an insulating core of rectangular cross section, the lines pick up the induced signals and transmit them with typical delays from a few to a few tens of ns/cm; an amplifier–discriminator circuit at each end of the line records the arrival time of the signals, thus providing the space coordinates after suitable calibrations. The limiting position accuracy achievable with delay lines depends on the ratio of signal rise time and delay; many authors have described optimizations of the design for best performances (Birk *et al.*, 1976; De Graaf *et al.*, 1979). While limited in rate capability and generally not able to disentangle multiple events, monolithic delay lines are a very cost-effective solution for moderate rate spectrometers (Breskin *et al.*, 1978; Atencio *et al.*, 1981), X-ray crystallography and biomedical imaging (Chechik *et al.*, 1996; Breskin 2000).

A lumped delay line, consisting of a multi-tap array of inductive delay cells capacitively connected to the detector cathode strips, permits one to achieve better localization and rate capability (Boie *et al.*, 1982). With a thorough optimization work on the components and of the front-end sensing electronics, the authors demonstrated in a small MWPC detector position accuracies better than 100 μm FWHM and integral non-linearity below 0.2%. Detectors based on this design have been used for high-rate X-ray synchrotron radiation experiments (Smith *et al.*, 1992).

A longitudinal localization method exploiting the current division on the two ends of the anode wires, originally introduced for the readout of MWPCs (Foeth *et al.*, 1973), has been developed mainly for the second coordinate readout in drift chambers, and is described in Chapter 9.

8.10 The multi-step chamber (MSC)

All detectors described so far have a single multiplication element, amplifying and detecting the ionization released in the gas. A double structure named the hybrid chamber, described in the early 1970s, consisted of a multiplication region delimited by wire meshes, transferring electrons into a second element, a spark chamber; the resulting delay permitted one to generate the high-voltage pulse used for triggering the main detector (Fischer and Shibata, 1972; Bohmer, 1973). The process leading to the transfer of a fraction of the electrons from a high into a low field region was, however, rather mysterious.

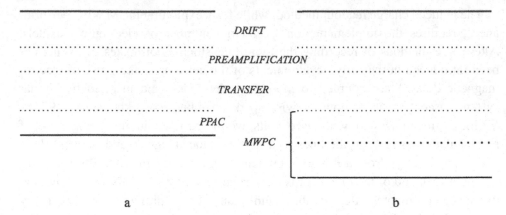

a   b

Figure 8.36   Schematics of the multi-step chamber with a parallel plate multiplier (a) and an MWPC as second amplification element (b).

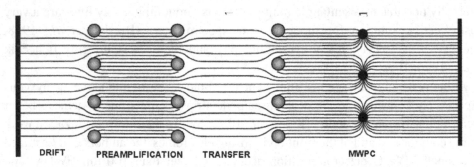

Figure 8.37   Electric field structure in the multi-step chamber with MWPC.

The pre-amplification and transfer mechanisms have been studied in the development of the multi-step chamber (MSC) (Charpak and Sauli, 1978). The structure has a high-field gap between two wire meshes, amplifying the ionization released and transferring a fraction of the electron avalanche through a transfer gap into a second element of amplification, parallel plate avalanche chamber (PPAC) or standard MWPC. Figure 8.36 shows schematically two possible implementations of the structure. Figure 8.37 is a schematic representation of the electric field lines in the second structure.

The picture in Figure 8.38 is an oscilloscope recording of the signals detected by irradiating the detector with a 5.9 keV X-ray source, converting in the drift region (Charpak and Sauli, 1978): the upper track corresponds to the pulses observed on the lower mesh of the pre-amplification gap, and has the shape expected from a negative charge multiplied and then leaving the electrode; the lower track corresponds to the delayed charge reaching the second amplification element. As seen in the picture, the energy resolution after transfer remains excellent. A measurement

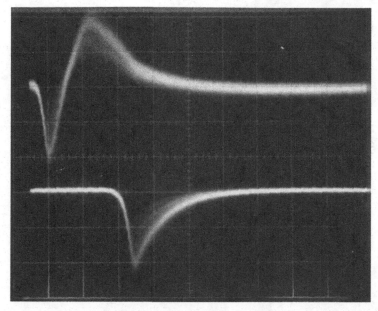

Figure 8.38   Prompt and delayed charge detected on the pre-amplification gap (top) and MWPC (bottom) (Charpak and Sauli, 1978). By kind permission of Elsevier.

of the ratio of transferred charge to the total avalanche size in the first amplification gap provides surprisingly large values of transfer efficiency, Figure 8.39; an example of total gains measured in a double step chamber with a PPAC main amplifier as a function of the voltage difference applied to the two multipliers is given in Figure 8.40 (Charpak and Sauli, 1978).

Developed initially with the aim of achieving large gains for the detection of UV photons emitted in a radiator by the Cherenkov effect, the early measurements were done using gas mixtures containing a photosensitive vapour, favouring an explanation of the observed large charge transfers based on a mechanism of UV photon emission and re-conversion in the gas. Further work demonstrated that the pre-amplification and transfer mechanism is effective in most gas mixtures and could be explained simply by the large lateral spread due to diffusion of the electrons multiplying in high fields.

Owing to the short absorption length of the additives, photons emitted in the final avalanche are reabsorbed before reaching the first multiplier, hence preventing feedback-induced divergence. Combining two amplification stages, gains in excess of 10^6 could be achieved with the MSC using photosensitive gas mixtures, permitting the detection and localization of single UV-generated photoelectrons. Figure 8.41 is an example of single electron pulse height spectra recorded at increasing gains, showing the evolution from exponential to a peaked Polya

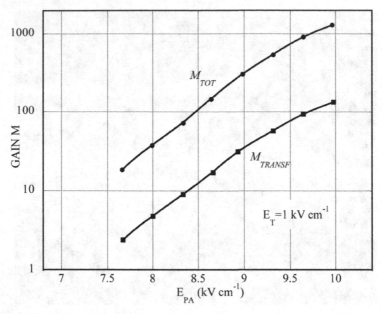

Figure 8.39   Total and transferred gain measured on the pre-amplification gap
(Charpak and Sauli, 1978). By kind permission of Elsevier.

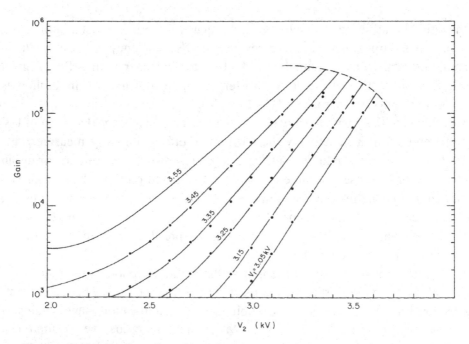

Figure 8.40   Total gain of a double parallel plate multi-step chamber
as a function of voltages (Charpak and Sauli, 1978). By kind permission of
Elsevier.

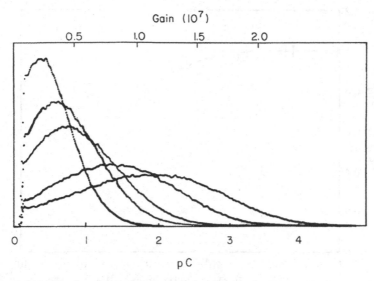

Figure 8.41 Single electron pulse height distributions recorded with a photo-sensitive multi-step chamber at increasing gains (Bouclier *et al.*, 1983). By kind permission of Elsevier.

distribution, as discussed in Section 5.4 (Bouclier *et al.*, 1983). This opened the way to application in ring imaging Cherenkov counters (RICH), described in Chapter 14. The gas electron multiplier (GEM) device, invented by the author (Sauli, 1997) and discussed in Chapter 13, is a reincarnation of the pre-amplification and transfer concept.

In a detector without the drift gap, thanks to the exponential multiplication characteristics of the pre-amplification element, the largest gain is obtained for ionization released close to the upper cathode, improving localization for low energy electron tracks, scattering in the gas; the principle has been expoited for imaging labelled anatomic samples in radiochromatography (Petersen *et al.*, 1980), and in the detection of thermal neutrons using converters (Melchart *et al.*, 1981).

8.11 Space charge and rate effects

The effect on gain of the positive space charge built up in proportional counters at high rates was discussed in Section 7.7. Since the gain reduction is a localized effect, extending perhaps one or two gap lengths around the hot spot in a chamber, substantial distortions in the detection can be induced if the rate limit is locally exceeded. Measurements of the average pulse-height as a function of rate in proportional counters were given in that section; the effect depends on the detector geometry and ion mobility, which can, however, not be varied over a very wide

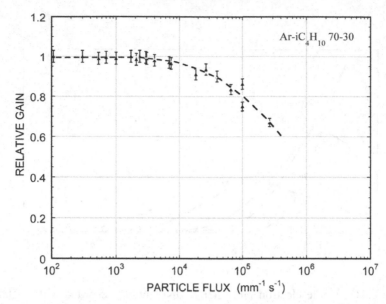

Figure 8.42   Relative gain as a function of rate measured with a drift chamber, normalized to the flux per unit wire length (Breskin *et al.*, 1974b). By kind permission of Elsevier.

range. For a given threshold of detection, inefficiency sets in when the lower part of the pulse-height spectrum distribution decreases below the threshold. High chamber gains and low thresholds are therefore needed for high rate full efficiency operation; in practice, however, at any given chamber gain the threshold value cannot safely be reduced below one tenth to one twentieth of the average pulse height, due to the noise generated by micro-discharges and other sources, and therefore almost identical efficiencies of detection versus rate have been measured in a large variety of gases and conditions.

Measured with a 50-mm anode wire spacing drift chamber, Figure 8.42 shows the gain as a function of minimum ionizing particle rate, expressed per unit length of wire (Breskin *et al.*, 1974b); Figure 8.43 provides similar measurements, realized with MWPCs having 1 mm and 0.64 mm wire pitches. Once normalized to the rate per unit length of wire, the results are consistent and show that the gain drops rapidly above a radiation flux of around 10^4 particles/mm of wire.

Depending on the electronics' detection threshold, the corresponding efficiency also drops at high rates. Figure 8.44 (Crittenden *et al.*, 1981) and Figure 8.45 (Breskin *et al.*, 1974b) compare the results of efficiency measurements for fast particles, realized with 1 and 2 mm pitch MWPC and with a 50-mm wire spacing drift chamber at several anode potentials; normalized to the rate per unit length of wire, the drop in efficiency is about the same, confirming the fundamental nature of the process, although the spatial extension of the positive charges may create some

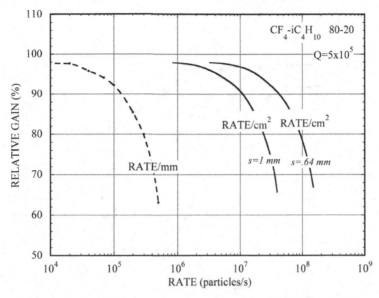

Figure 8.43   Relative gain as a function of rate for MWPCs with 1 mm and 0.64 mm wire spacing. Normalized to the flux per unit wire length, the two sets of data coincide (dashed curve) (Breskin *et al.*, 1974b). By kind permission of Elsevier.

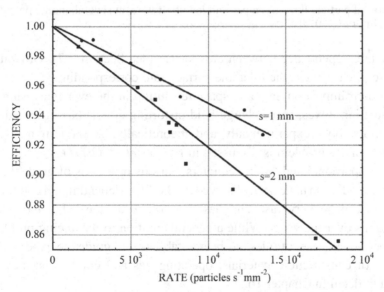

Figure 8.44   Efficiency as a function of particle rate measured in MWPCs with 1 and 2 mm wire spacing (Crittenden *et al.*, 1981). By kind permission of Elsevier.

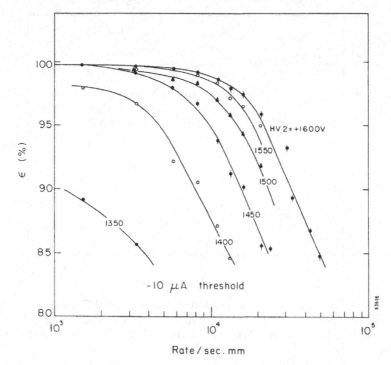

Figure 8.45 Detection efficiency of a 50 mm wire spacing drift chamber as a function of particle flux, given per unit length of wire, at several anodic potentials (Breskin *et al.*, 1974b). By kind permission of Elsevier.

degree of interdependence between close wires (see Section 7.7). Obviously, the larger wire pitch reduces the tolerable surface rate correspondingly.

Due to their impact on gain and space resolution in the ever-increasing experimental radiation fluxes, space charge field-distorting effects have been studied by many authors, both experimentally and theoretically. A modern mathematical formulation of the problem is discussed in Riegler *et al.* (2007).

As for proportional and Geiger counters, various processes of gradual deterioration under irradiation have been observed in MWPCs; depending on conditions, the degradation can be very fast, and has therefore been the subject of extensive investigations over the years. While a general solution to the ageing problems has not yet been found, a set of rules has been established to guide the experimenters in the choice of construction materials, operating gas and conditions; they will be discussed in detail in Chapter 16.

8.12 Mechanical construction of MWPCs

The basic problem for wire chamber construction is to support, on suitable frames, a succession of foils and wire planes constituting the electrodes, within strict

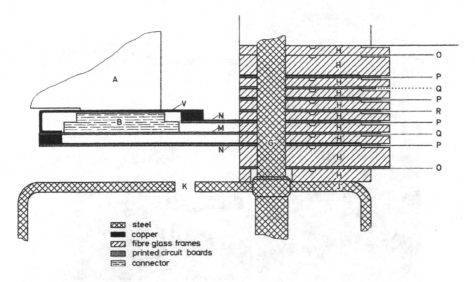

Figure 8.46   Schematic cross section of a MWPC frame assembly (Schilly *et al.*,
1970). By kind permission of Elsevier.

mechanical and electrical tolerances, and to make the whole structure gas tight and
capable of sustaining the applied voltage. Several methods of construction have
been developed; two representative examples are described here. The first, and
more frequently used, consists of fabricating a set of self-supporting insulating
frames, normally of extruded or machined fibreglass, one for each electrode. Wire
planes are in general soldered on printed circuits pasted on the frames; a complete
detector is then mounted either by gluing together the required number of frames
(in the case of small chambers), or assembled with bolts and screws traversing the
frames, the gas tightness being guaranteed by rubber joints (O-rings) embedded in
the frames; thin gas-containment windows on the two outer surfaces complete the
assembly. Figure 8.46 is a cross section of the frames used in the construction of
the first large MWPC, and Figure 8.47 is a picture of the assembled detector used
in a large aperture magnetic spectrometer experiment at CERN (Schilly *et al.*,
1970). A similar structure was used for a large MWPC prototype chamber
developed by Charpak and collaborators, Figure 8.48 and Figure 8.49 (Charpak
et al., 1971). A photograph of the fully assembled detector was shown in the
introduction (Figure 1.6), and is also the first known picture of Georges Charpak
with a MWPC. A close view of the frame with the anode wires soldered on a
printed circuit board, providing the connection to the readout electronics, is shown
in Figure 8.50.

   A drawback of the described technique lies in the rather unfavourable ratio
between active and total detector surface. In cases where the detection volume is

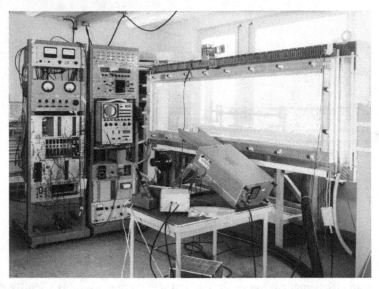

Figure 8.47   The first large MWPC (Schilly *et al.*, 1970). By kind permission of Elsevier.

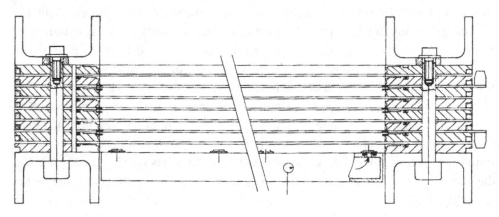

Figure 8.48   Cross section of a three-wire plane MWPC framing (Charpak *et al.*, 1971). By kind permission of Elsevier.

limited, as inside a magnet, the loss is considerable. For this reason, a different construction principle has been developed, based on the use of metal-coated self-supporting honeycomb or expanded polyurethane plates, which constitute both the cathode planes and the mechanical support of the chamber. A schematic of such self-supporting chambers is shown in Figure 8.51 for a twin-gap MWPC; narrow insulating frames (one cm or less), on which the anode wires are stretched and soldered, are glued along edges of the plates (Bouclier *et al.*, 1974). The fishbone

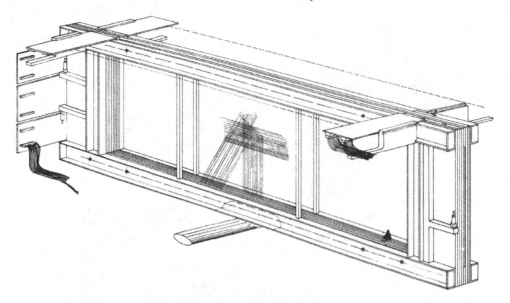

Figure 8.49   Assembled large MWPC prototype (Charpak *et al.*, 1971). By kind permission of Elsevier.

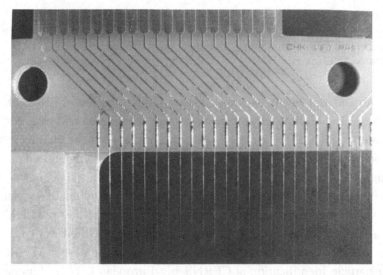

Figure 8.50   Close view of the anode wires soldered to a printed circuit board. Picture by the author (1973).

structure of the strips in the cathode plane, used to provide coarse information on the tracks' second coordinate, is also visible. The honeycomb-based structural construction has the advantage of being light, and can be tailored to particular requirements of the experiment: in the example, the structure is shaped to allow

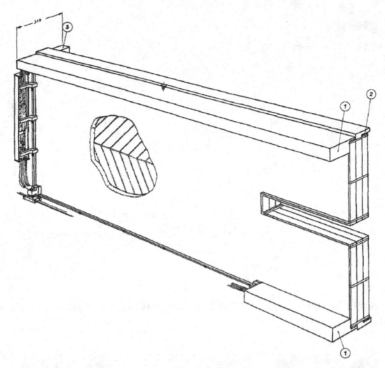

Figure 8.51 Light MWPC construction, making use of expanded polyurethane plates and thin frames for holding the anode wires (Bouclier *et al.*, 1974). By kind permission of Elsevier.

assembly around the beam vacuum tubes (in this case the CERN intersecting storage rings). The obvious advantages of a large active area and ease of construction are counterbalanced by a rather substantial increase in the chamber thickness in the active area (~ 0.6 g/cm^2 against a few mg/cm^2 for a conventional thick-frame construction).

A picture in the introduction (Figure 1.7) shows the 40-chambers system, built with self-supporting chambers as described, installed in the split field magnet detector at CERN's intersecting storage rings and operated for many years. A similar self-supporting construction was used for the external MWPC system providing particle identification at CERN's big European bubble chamber (Brand *et al.*, 1976).

Systems of large MWPCs named cathode strip chambers (CSC) are used for triggering and localization in the CMS end-cap muon detector at CERN's LHC. Built with sizes up to 3.4 m × 1.5 m, the chambers have rigid supporting cathode planes manufactured with copper-clad, 1.6 mm thick fibreglass plates; radial readout strips are milled directly on the panels (Acosta *et al.*, 2000; Ferguson *et al.*, 2005). While a single layer time resolution is not sufficient to identify the

Figure 8.52   Cathode strip chambers array for the CMS muon end-cap. Picture CERN (2007).

bunch crossing,[5] this is achieved by a hardware selection of hits on several aligned planes. Off-line analysis of the charge profiles recorded on cathode strips can provide the track position with sub-mm accuracy (Barashko *et al.*, 2008). A long-term irradiation of the detectors has demonstrated their survivability in the LHC harsh radiation environment (Acosta *et al.*, 2003). The picture in Figure 8.52 shows one layer of the cathode strip chambers system built for the CMS muon-end cap detector.

As discussed in Section 8.4, for large detector sizes, wire instabilities and over-all electrode deflections take place, degrading the operation, unless suitable mechanical support lines are used to balance electrostatic forces. Supports must be insulating if in contact with the anodes and, because they have a dielectric rigidity different from the gases, the local field modification affects the detection efficiency over a large area, a centimetre or so, around the support. The measured local efficiency drop in the region of a thin insulating wire support, perpendicular to the anode wires, is shown in Figure 8.53 (Schilly *et al.*, 1970).

Other support methods have been developed, in which an insulated wire or a conductive strip, close to the anode wires but not in contact with them, is raised to a potential high enough to, at least partially, restore the field and therefore the efficiency of detection. Figure 8.54 is an example of a vinyl-insulated support line and of the corresponding local efficiency measurement (Charpak *et al.*, 1971); the

---

[5] The time between collisions at LHC is 25 ns.

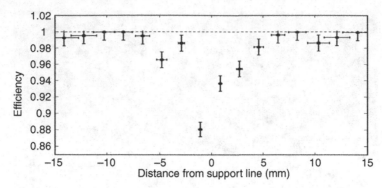

Figure 8.53   Local inefficiency induced by a thin insulating support line in contact with the anode wires (Schilly *et al.*, 1970). By kind permission of Elsevier.

picture in Figure 8.55 shows a corrugated thin kapton strip, or garland, compensating both for the anode wire instability and for the gap squeezing due to electrostatic forces (Majewski and Sauli, 1975). The structure has been used by many groups in the construction of large size multi-wire and drift chambers (Bozzo *et al.*, 1980; Hammarström *et al.*, 1980a; De Palma *et al.*, 1983).

A comparison of construction methods and materials and can be found in Veress and Montvai (1978); the note also provides numerical evaluations of the frames' deformations caused by the stretched wires for several structures.

Methods for mounting wires on the chambers' frames are mainly of three types: manual stringing, weaving machines and turn-wheel sprockets; such systems have been developed by the engineering support groups in various laboratories. The simplest, used for prototype construction or small quantities, consists of stringing by hand one wire after the other over a horizontal frame, keeping the tension with appropriate weights placed at one or both ends, and soldering the wires to printed circuit boards. The desired wire distance is obtained with the use of external reference marks or light saw-tooth-shaped rules placed over the wire mesh. An example of manual construction can be found in Staric *et al.* (1983). Limited to small productions, the method is often used for repairing defective wires.

More sophisticated, and with a conception close to their industrial siblings, dedicated winding machines have been built having a motor-driven flying chariot with a wire feeder travelling between the two sides of a heavy frame. Wires are stretched between two comb-shaped racks with teeth at the desired pitch; a stepping motor and a wire tensioning system ensure the correct position. Once the full wire plane is woven, it can be directly soldered to the final frame, or fixed to a temporary transfer frame for later assembly. Description of such wire stretching systems can be found in Alleyn *et al.* (1968) and Bock *et al.* (1994).

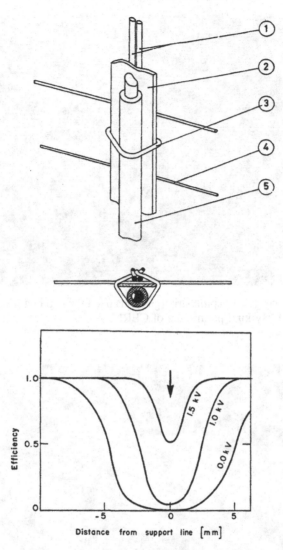

Figure 8.54   Support line with field-restoring insulated wire. With a suitable potential applied, the inefficiency loss is limited to a few mm (Charpak *et al.*, 1971). By kind permission of Elsevier.

The pictures in Figure 8.56 and Figure 8.57 are examples of waving machines of this kind used at CERN for the construction of wire chamber planes.

An alternative wiring method consists of a spinning frame, planar or cylindrical, rotating on an axis perpendicular to the wire feeder; a stepping motor, synchronized with the frame, ensures the desired wire pitch. To avoid breaking the wires during weaving, the mechanical tension is usually only a fraction of the one required; the final tension is obtained with the use of stretchable transport frames.

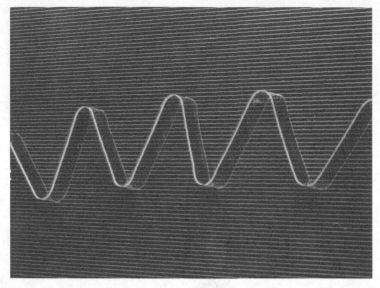

Figure 8.55   Corrugated kapton strip (garland) used for gap restoring (Majewski and Sauli, 1975). By kind permission of CERN.

Figure 8.56   Close view of a winding machine at CERN. Picture CERN (1975).

Figure 8.57   A wire winding machine used at CERN for the construction of the NUSEX chambers. Picture CERN (1981).

Figure 8.58   A rotating frame winding machine used for the wiring STAR TPC MWPC modules. Courtesy Brookhaven National Laboratory (USA).

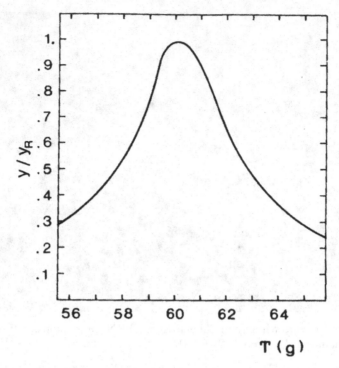

Figure 8.59   Relative resonance frequency of a wire, 1 m long, as a function of mechanical tension (Borghesi, 1978). By kind permission of Elsevier.

A system of this kind, making use of an octagonal drum, is described in Cavalli-Sforza *et al.* (1975). The picture in Figure 8.58 shows a wire weaving rotating frame used at BNL for the construction of the STAR TPC end-cap multi-wire modules. A simple table-top automatic winding and wire-soldering machine are described in Martinez *et al.* (2007).

To avoid electrostatic instability problems, after fabrication of a wire plane, it is advisable to check value and uniformity of the mechanical tension of each wire. For small detectors, a simple method consists of measuring the wire's sagging with a small weight gently put down in the centre of each wire. More sophisticated mechanical resonance systems rely on finding the natural frequency of oscillation of the wires, excited by an oscillator coupled through a metal foil or plate placed parallel to the wire plane; the resonance can be enhanced by an external magnetic field (Borghesi, 1978; Regan, 1984; Bhadra *et al.*, 1988; Mueller, 1989). Figure 8.59, from the first reference, gives an example of relative resonance frequency as a function of tension for a wire 1 m long.

Further reading

Charpak, G. (1970) Evolution of the automatic spark chambers, *Ann. Rev. Nucl. Sci.* **20**, 195.

Sauli, F. (1977) *Principles of Operation of Multi-wire Proportional and Drift Chambers* (CERN 77-09).

Charpak, G. and Sauli, F. (1984) High-resolution electronic particle detectors. *Ann. Rev. Nucl. Part. Sci.* **34**, 285.

9

Drift chambers

9.1 Single wire drift chambers

The possibility of measuring the electrons' drift time and getting additional information about the spatial coordinates of an ionizing event was suggested in the early works on multi-wire proportional chambers (Charpak *et al.*, 1968). In its basic form, a single-cell drift chamber has a narrow gap of constant electric field, with a wire counter at one end, Figure 9.1. Field-shaping electrodes, wires or strips, at decreasing potentials between the anode and a top drift electrode, provide the required electrical configuration. Ionization electrons released in the gas by a charged particle at a time t_0 migrate in the electric field and reach the anode wire where they are amplified and detected at a time t_1. The distance of the track from the anode wire is then obtained from $t_1 - t_0$ through the knowledge of the time–space correlation, measured or computed from the electric field map and the field dependence of drift velocity, $w = w(E)$. In the region of uniform field, the relative space coordinate is simply given by $s = w (t_1 - t_0)$.

Unlike MWPCs, drift chambers require a time reference signal, and can be used for the localization of charged particles exploiting an external trigger provided by one or more scintillation counters; for neutral radiation, the time reference can be obtained from the detection of the primary scintillation caused by the interaction in the gas. Albeit having reduced rate capability, due to the long electron collection time as compared to MWPCs, drift chambers permit one to reduce the number of readout channels while providing sub-mm localization accuracies, and have therefore been developed in a variety of models and widely used in experimental setups.

If a large surface of detection is required, a simple structure like the one described leads to uncomfortably large working voltages and long drift times. Nevertheless, chambers of this design reaching 50 cm drift lengths have been successfully operated, with drift voltages around 50 kV and a maximum drift time of about 10 µs (Saudinos *et al.*, 1973; Rahman *et al.*, 1981). Figure 9.2, from data

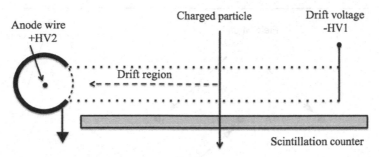

Figure 9.1   Schematic of a single-cell drift chamber.

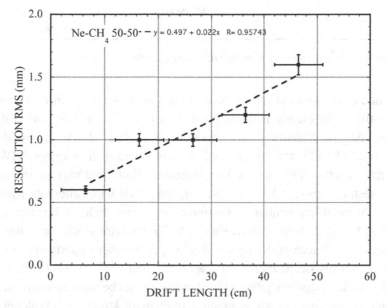

Figure 9.2   Position accuracy for a 50 cm long single cell drift chamber. Data from Saudinos *et al.*, (1973).

in the first reference, shows the position accuracy of the detector as a function of drift distance; the horizontal error bars represent the region over which the measurement was averaged.

9.2 Multi-cell planar drift chambers

Multi-cell drift structures can cover larger detection areas with shorter memory times; however, since the region of the anode wire becomes part of the active volume, the drift field is not uniform across all the active cell, and the space–time correlation is necessarily distorted. A standard multi-wire proportional chamber can be used as a drift chamber; nevertheless, especially for large wire spacing,

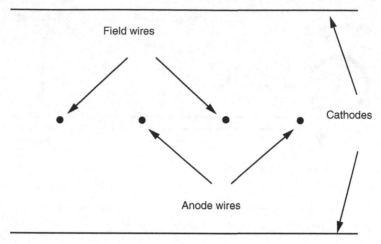

Figure 9.3   Schematics of a multi-wire drift chamber.

the presence of a low field region between the anodes results in non-linearity of the space–time relationship and can have slow electron trails at long drift times.

Introduced by H. Walenta and collaborators (Walenta *et al.*, 1971), a modification of the original MWPC structure, named the multi-wire drift chamber (MWDC) or simply drift chamber (DC) allows the reduction of low field regions in the central plane. As seen in Figure 9.3 and Figure 9.4, the anode wires alternate with thicker field wires, at suitable potentials to reinforce the electric field in the critical region. A system of drift chambers of this design, with 20 mm wire spacing and sizes up to a square metre, was successfully operated in a spectrometer experiment in the early seventies, achieving localization accuracies better than 400 μm rms (Walenta 1973). A recording of the induced signals on the field wires can be used to resolve the right–left ambiguity, intrinsic in all symmetric multi-wire structures (Walenta, 1978). Similar detectors have been built in a wide range of wire spacing and up to sizes exceeding tens of square metres (Cheng *et al.*, 1974; Bernreuther *et al.*, 1995).

The position resolution that can be achieved with multi-wire drift chambers depends on the knowledge of the space–time correlation and the intrinsic diffusion properties of migrating electrons. Due to the chamber geometry with a non-uniform electric field, the correlation is generally not linear, although an appropriate choice of the gas and operating condition can reduce the variations; the correlation depends also on the angle of incidence of the tracks. An example is shown in Figure 9.5, measured for tracks normal to a large size MWDC with 10 cm anode wire spacing (Cheng *et al.*, 1974).

Operation of the drift chambers at high pressures improves the localization accuracy, because of both the increase of ionization density and decrease in diffusion, as seen in Figure 9.6 (Farr *et al.*, 1978).

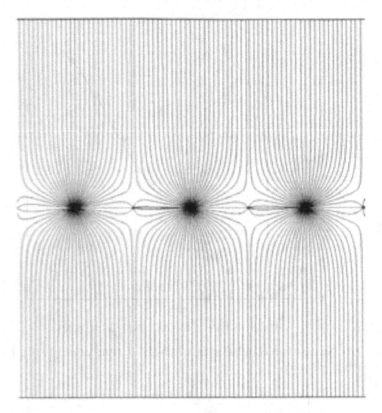

Figure 9.4   Electrical field lines in the multi-wire drift chamber.

In the presence of an external magnetic field, the electron drift lines do not coincide with the field lines, introducing distortions in the space–time correlation, which has to be measured or computed using the tools described in Chapter 4. Figure 9.7 is an example of computed drift lines and contours of constant drift times at increasing magnetic fields, parallel to the wires, and a specific choice of the gas filling and detector parameters (De Boer *et al.*, 1978).

A modification of the original MWDC, optimized to the measurement of inclined tracks and named the vertical drift chamber, is shown schematically in Figure 9.8 (Bertozzi *et al.*, 1977); the additional field wire between anodes reduces signal cross-talk between adjacent cells. A multi-cell version of this design was used for the construction of a large array of drift chambers for the L3 experiment muon detector at CERN (Becker, 1984; Viertel, 1995).

A limitation of the described design comes from the fact that, in order to obtain a relatively uniform field in the drift cells, the anode to cathode distance has to be comparable to or larger than the anode wire pitch; for the large wire spacing desirable to reduce the number of readout channels, this implies the construction of rather thick chambers, and therefore a reduced packaging density. Moreover,

Drift chambers

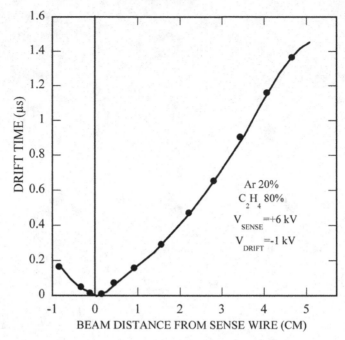

Figure 9.5 Space–time correlation in a MWDC (Cheng *et al.*, 1974). By kind permission of Elsevier.

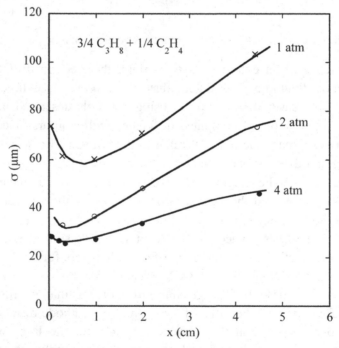

Figure 9.6 Position accuracy of a drift chamber at increasing pressures (Farr *et al.*, 1978). By kind permission of Elsevier.

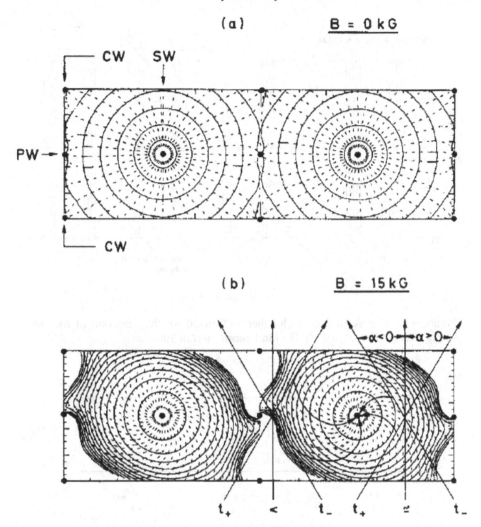

Figure 9.7   Computed electron drift lines and contours of equal drift times at increasing values of magnetic field parallel to the wires (De Boer *et al.*, 1978). By kind permission of Elsevier.

although the closest ionization electrons produce the signal used for the coordinate measurement, due to the field structure distant electrons are collected over a long time, affecting the multi-track capability.

The structure shown in Figure 9.9 (Charpak *et al.*, 1973) helps in solving both problems. Two sets of parallel cathode wire planes are connected to increasingly high negative potentials, symmetrically from the centre of the cell; the anode wire is kept at a positive potential, and additional field wires, at the potential of the adjacent cathode wires, sharpen the transition from one cell to the next. Two grounded screening electrodes protect the field structure from external influence. As shown in

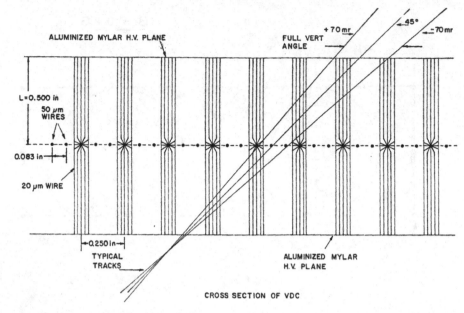

Figure 9.8 The vertical drift chamber, optimized for the detection of inclined tracks (Bertozzi *et al.*, 1977). By kind permission of Elsevier.

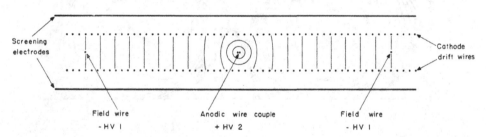

Figure 9.9 Schematics of the high accuracy drift chamber (Charpak *et al.*, 1973). By kind permission of Elsevier.

the figure, an almost equi-potential field distribution can be obtained in the median plane of the drift cell, even for gaps much narrower than the cell size; typical values of 2 × 3 mm and 50 mm have been used for the gaps and the anode wire spacing, respectively. An offset of the potentials applied on the two field-shaping meshes, Figure 9.10, permits one to operate in strong magnetic fields parallel to the wires, approximately compensating the Lorentz angle of the drifting electrons.

Owing to their quasi-uniform field, drift chambers of this design have a linear space–time correlation, almost independent of the incidence angle (Figure 9.11) (Breskin *et al.*, 1974b). They can achieve very good localization accuracy and multi-track resolution, and have therefore been appropriately named high-accuracy

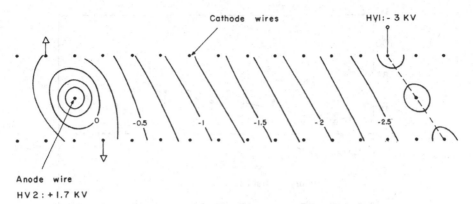

Figure 9.10 Modified equi-potentials in the high-accuracy drift chamber for operation in a magnetic field parallel to the wires (Charpak *et al.*, 1973). By kind permission of Elsevier.

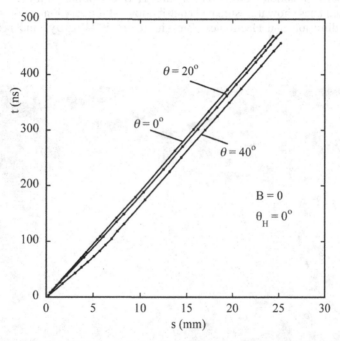

Figure 9.11 Space–time correlation in the high-accuracy drift chamber for several angles of incidence of the beam (Breskin *et al.*, 1974b). By kind permission of Elsevier.

drift chambers (HADC); an example of measured space accuracy for fast particles as a function of drift length is given in Figure 9.12 (Breskin *et al.*, 1974b). In the figure, the dashed lines show the estimated contributions of the primary ionization statistics, decreasing with the distance, of the electrons' diffusion and of the constant term due to the time measurement error.

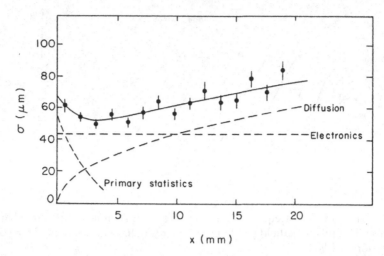

Figure 9.12 Position resolution of the HADC (points with error bars). The dashed lines show the estimated contributions of primary ionization statistics, electron diffusion and electronics (Breskin *et al.*, 1974b). By kind permission of Elsevier.

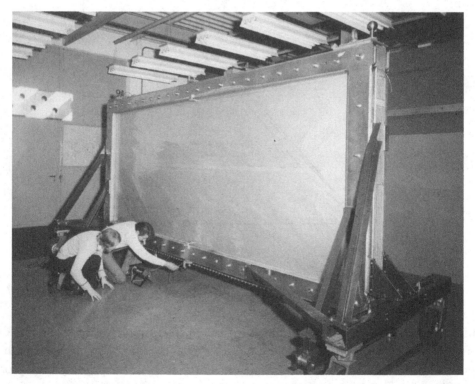

Figure 9.13 A large high-accuracy drift chamber built for the OMEGA spectrometer at CERN. Picture CERN (1976).

Multi-cell drift chambers cannot resolve the right–left ambiguity intrinsic in their symmetric construction; the reconstruction of tracks is generally obtained using several staggered planes of measurement. A method for resolving the ambiguity locally is to separate the sensitive regions on the two sides of an anode by replacing the single wire with wire doublets, at a very short distance apart (Breskin *et al.*, 1974a). The ensuing field modification results in a small (~1%) localized loss of efficiency for tracks perpendicular to the chamber, obliterated for inclined tracks. Due to the electrostatic repulsion between the pair, the wires have to be bonded together at short distances to avoid distortions; for the 100 μm distance used in the quoted work, droplets of epoxy 5 cm apart were used to stabilize the structure.

High-accuracy drift chambers of the described design have been used in experiments requiring single-track position resolutions of 100 μm or better, as in charged particle spectrometers (Filatova *et al.*, 1977; Chiavassa *et al.*, 1978), channeling experiments (Esbensen *et al.*, 1977) and other applications. The picture in Figure 9.13 shows the large HADC used for tracking in the Omega spectrometer at CERN; Figure 9.14 is a close view of the central wire plane under microscope

Figure 9.14   Close-up of the wire plane in the HADC. Picture CERN (1976).

Figure 9.15   A wire plane of the large Omega drift chamber under construction. Picture CERN (1976).

inspection. Insulating garland spacers and wire supports, described in Section 8.12, were used to ensure the stability of the large structure. The picture in Figure 9.15 shows a wire plane of the chamber before assembly, with the garlands clearly visible.

Many variants of drift chamber designs have been developed to meet the requirements of large detection areas and simplified construction, aiming at large productions. Figure 9.16 is one of the early studies: metal I-beam shaped electrodes, insulated from the (grounded) cathodes and kept at negative potential, provide structural strength to the structure and permit one to reinforce the field in the middle plane of drift (Becker *et al.*, 1975). Due to charging-up and breakdown problems around the insulator's edges, this design seems to have been abandoned.

An alternative structure, also permitting a favourable gap to pitch aspect ratio is shown in Figure 9.17 (Marel *et al.*, 1977); the drift cell includes an anode and a cathode wire in the middle plane, with additional field wires kept at appropriate potentials to define the electric field structure. Grounded electrodes, fixed to honeycomb plates on both sides of the structure, ensure mechanical strength and gas tightness. Very large (4 m diameter) hexagonal-shaped drift chambers of this design were used in the eighties to instrument the WA1 neutrino experiment at CERN, Figure 9.18, and routinely achieved sub-mm localization accuracies (Holder *et al.*, 1978).

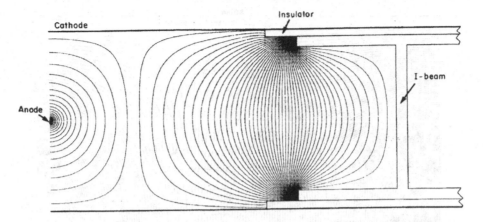

Figure 9.16   Scheme of a simple drift chamber cell with reinforcement I-beam electrodes, insulated from the cathodes, providing structural rigidity (Becker *et al.*, 1975). By kind permission of Elsevier.

Optimized for mass production, an improved design of the drift chamber described in Figure 9.16, somehow improperly named drift tube,[1] has the cell structure shown in Figure 9.19, with wires mounted between aluminium plates and I-beam profiles defining the electric field and providing mechanical stiffness (Aguilar-Benitez *et al.*, 2002; Cerminara 2010). A very large system of drift tubes has been built and operated for muon triggering, identification and tracking in the CMS barrel muon spectrometer at CERN's LHC collider; the detector is assembled in sectors around the beam, and includes 250 modules ranging between 2 and 4 m in width and 2.5 m in length (Abbiendi *et al.*, 2009).

9.3 Volume multi-wire drift chambers

The devices described above, consisting of a single plane of detection, are suitable for use in fixed target experiments such as long-arm spectrometers. Multi-cell structures have been developed to better suit the needs of experiments requiring compact detectors, particularly inside the solenoids used for magnetic analysis in storage ring colliders. Cylindrical in shape, they fill the sensitive volume with repeating cells having anodes surrounded by several field wires, and were built with a variety of schemes.

One of the first large multi-cell chambers, with simple drift cells consisting of an anode and six cathode wires, was operated for the Mark II at SLAC (Davies-White

---

[1] Arrays of proportional drift tubes with cylindrical cross section are described in Chapter 11.

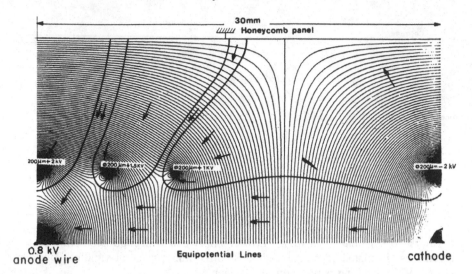

Figure 9.17   Schematics of the drift cell developed for the WA1 large drift chambers (Marel *et al.*, 1977). By kind permission of Elsevier.

Figure 9.18   The WA1 neutrino experiment; large hexagonal drift chambers alternate with scintillators and converters. Picture CERN (1976).

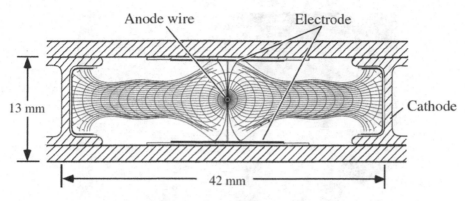

Figure 9.19   Mechanical structure and electric field in a drift tube cell (Aguilar-Benitez *et al.*, 2002). By kind permission of Elsevier.

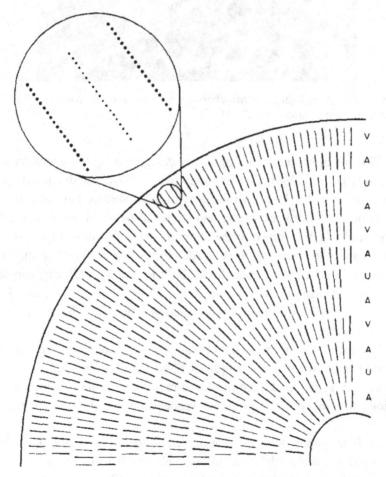

Figure 9.20   Schematics of the Mark II cylindrical drift chamber; the inset shows the cell structure (Hanson 1986). By kind permission of Elsevier.

Figure 9.21   A cylindrical drift chamber in construction for the MARK II
detector. Picture courtesy SLAC (1977).

et al., 1972), and upgraded later with a new design having the geometry shown in
Figure 9.20 (Hanson, 1986). Concentric cylindrical layers with cathode wires and
alternating anode and field wires fill the cylindrical volume between an inner and
an outer radius; the wires are strung and stretched between two disk-shaped end-
plates and soldered or crimped to the supports. The picture in Figure 9.21 shows
one such chamber in construction at SLAC, with the wires hand-strung in successive layers. Figure 9.22 is a similar device built for the VENUS experiment for the
TRISTAN storage ring at KEK (Arai *et al.*, 1983); in this case, the detector is
assembled vertically, exploiting gravity to stretch the wires between the top and
bottom end-plates.

   In the presence of a magnetic field parallel to the wires, the electron drift lines
are distorted as shown in the example of Figure 9.23, computed for the Mark II
detector at 0.45 T; the track reconstruction requires then thorough calibrations or
calculations to supply the appropriate space–time correlations (Abrams *et al.*,
1989).

   With a different approach, aimed at improving performances and reliability, the
UA1 central detector at CERN was an assembly of 40 drift chamber modules with
the geometry shown in Figure 9.24 and Figure 9.25 (Baranco Luque *et al.*, 1980;
Beingessner *et al.*, 1987). With a simultaneous recording of the track's drift time

Figure 9.22   A cylindrical multi-cell drift chamber in construction at KEK (Arai *et al.*, 1983). By kind permission of Elsevier.

and of the longitudinal coordinate from the ratio of pulse heights on the two ends of the wires, and making use of a colour display visualization, innovative at the time, the detector provided amazing three-dimensional views of the events, and was appropriately named the imaging chamber. The UA1 experiment led to the discovery of the W and Z heavy bosons, and to the assignment of the 1984 Nobel Prize to Carlo Rubbia and Simon van der Meer; Figure 9.26 is the display of a W event, recorded with the drift chambers complex.

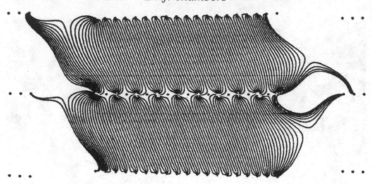

Figure 9.23   Electron drift lines in the Mark II cylindrical drift chamber in 0.45 T magnetic field parallel to the wires (Abrams *et al.*, 1989). By kind permission of Elsevier.

Similar structures have been built for many other experiments; for a summary of detectors and a comparative discussion of performances see Heintze (1978); Walenta (1983); Va'vra (1986a); Saxon (1988); Blum and Rolandi (1993).

9.4 Jet chambers

With a cylindrical geometry optimized for the detection of multiple tracks, the jet chamber (so named from the physical processes under investigation) has the structure shown in Figure 9.27 (Drumm *et al.*, 1980). Segmented radial sectors, in three consecutive shells, have a central wire plane with alternating anode and field wires, and field-shaping cathode walls to provide a near uniform electric field in the drift regions. The right–left ambiguity, intrinsic in the drift time measurement, is resolved by staggering by 150 µm the anode wires alternately right and left from the median plane; a measurement of the charge sharing between the two ends of each wire provides the longitudinal coordinate.

The computed electron drift trajectories and equal time lines for a cell, in the presence of a longitudinal magnetic field, are shown in Figure 9.28. The detector has been operated for many years in the JADE experiment on the electron–positron storage ring PETRA at DESY, consistently providing single-track radial position accuracies between 150 and 200 µm and around 10 mm along the wires (Heintze, 1982).

A larger jet chamber of similar conception, but with a single shell of radial sectors, was operated in the OPAL experiment at CERN's LEP electron–positron collider (Fischer *et al.*, 1989), achieving intrinsic radial coordinate resolutions

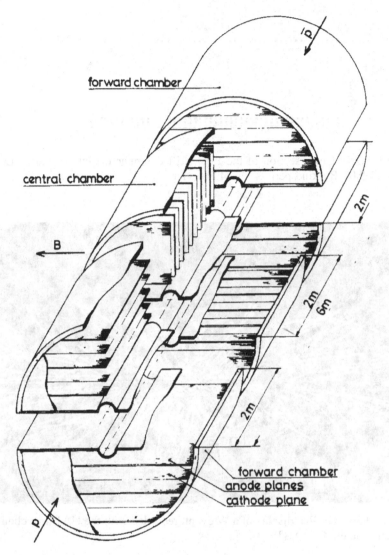

Figure 9.24   Assembly of drift chamber modules in the UA1 detector (Baranco Luque *et al.*, 1980). By kind permission of Elsevier.

between 100 and 200 μm, depending on drift distance, and an overall momentum resolution of ∼7% for particles around 50 GeV/c (Biebel *et al.*, 1992). The picture in Figure 9.29 shows the detector under construction, and Figure 9.30 shows a recorded hadronic Z_0 decay event.

Smaller cylindrical devices of similar design have been developed as vertex detectors; operated at pressures higher than atmospheric, they can reach localization accuracies of 50 μm or better (Mμller, 1986; Alexander *et al.*, 1989).

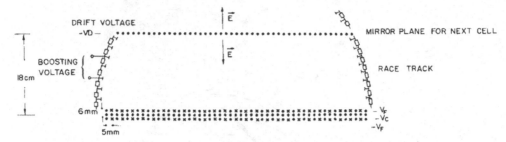

Figure 9.25   Cross section of the UA1 drift chamber module (Baranco Luque *et al.*, 1980). By kind permission of Elsevier.

Figure 9.26   On-line display of a W event, recorded with the UA1 drift chamber system. Picture CERN (1984).

9.5 Time expansion chamber

The ultimate limit in localization properties of a drift chamber depends on two major factors: the diffusion properties of electrons, with the ensuing statistical dispersion of the single cluster collection time, and the resolution of the electronics used to record the drift time. The time expansion chamber (TEC) (Walenta, 1979), with the electrode arrangement shown in Figure 9.31, is based on the choice of a field value and a gas with low diffusion and low drift velocity to reduce the error given by the electronics' resolution. As in general diffusion decreases with field while drift velocity increases, the somewhat conflicting requirements are best met in some 'cool' gas mixtures; one example is shown in Figure 9.32 for a carbon

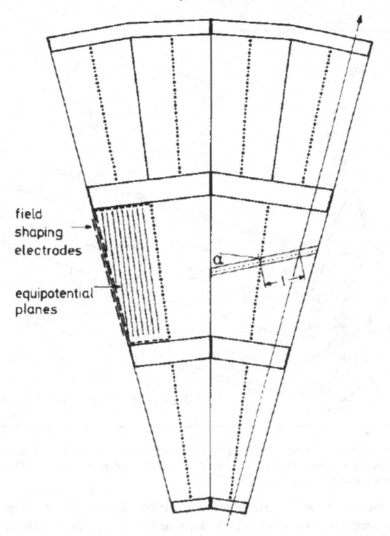

field shaping electrodes

equipotential planes

Figure 9.27   Cross section of two sectors in the JADE jet chamber (Drumm *et al.*, 1980). By kind permission of Elsevier.

dioxide–isobutane mixture, used for most the TEC developments (Commichau *et al.*, 1985). At fields around 600 V/cm, the single electron diffusion is 50 μm for 1 cm of drift, and the drift velocity 5 μm/ns, about a factor of ten lower than for a standard, saturated drift chamber gas; this corresponds to an intrinsic electronic resolution of 5 μm for a 1 ns time recording accuracy.

With a cylindrical multi-sector construction, TEC-based vertex detectors have been operated in the L3 experiment at CERN's large electron–positron collider (LEP) (Adeva *et al.*, 1990; Akbari *et al.*, 1992; Anderhub *et al.*, 2003) and the

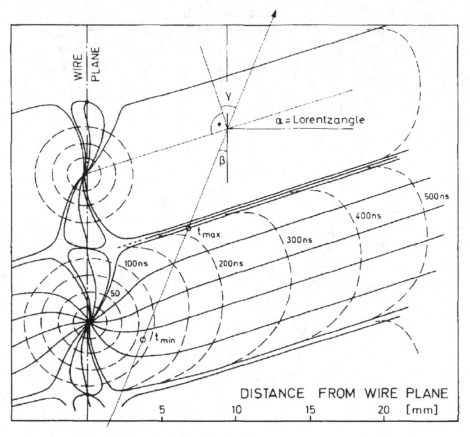

Figure 9.28   Computed electron drift trajectories and equal time lines in the jet chamber in a longitudinal magnetic field of 0.45 T (Drumm *et al.*, 1980). By kind permission of Elsevier.

Mark J detector at DESY (Anderhub *et al.*, 1986). Figure 9.33 and Figure 9.34 show, respectively, a recorded event and the single-track position resolution in the radial coordinate measured with the L3 vertex detector (Akbari *et al.*, 1992).

While satisfactorily operating at moderate particle rates, concerns about the ageing properties of hydrocarbon-based gases have suggested in later applications the use of the non-polymerizing dimethyl-ether (DME) as gas filling (Hu *et al.*, 2006). Due to the use of a non-saturated drift velocity, a good knowledge of the dependence of drift time on temperature and pressure and thorough monitoring of the ambient conditions are needed for long-term operation.

9.6 Determination of the longitudinal coordinate from current division

Drift chambers of the various designs described above provide only one coordinate, albeit with great accuracy. For unambiguous track reconstruction, particularly

Figure 9.29   The OPAL jet chamber in construction. Picture CERN (1988).

in the case of high multiplicities, a second coordinate along the wires is desirable and in some cases, as for cylindrical devices, indispensable; the methods described in the previous chapter based on signal pickup on electrodes facing the anodes are generally not possible.

The longitudinal coordinate can be obtained from a measurement of the charge at one or both ends of a wire, exploiting the attenuation due to its resistivity, a method used already for triggered spark chambers (Charpak *et al.*, 1963) and single-wire proportional counters (Kuhlmann *et al.*, 1966; Miller *et al.*, 1971).

For a wire of linear resistivity ρ and length l, the currents detected at the two ends on equal loads R for a charge generated at position x are:

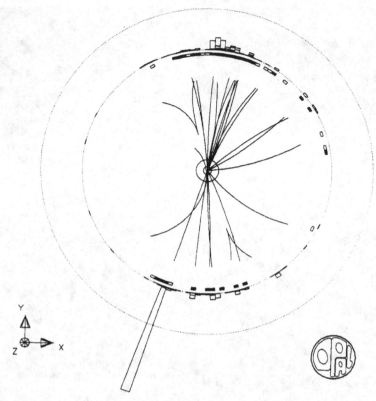

Figure 9.30 A Z_0 decay event recorded with the OPAL jet chamber; the polygons on the outer shell represent the response of the calorimeter (Baines *et al.*, 1993). By kind permission of Elsevier.

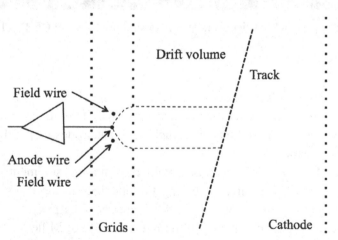

Figure 9.31 Schematics of the time expansion chamber (Commichau *et al.*, 1985). By kind permission of Elsevier.

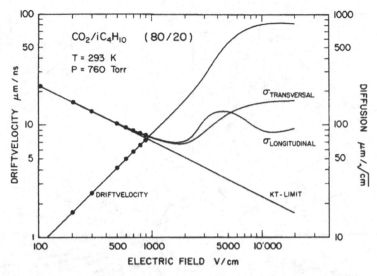

Figure 9.32   Electron drift velocity and diffusion as a function of electric field in a CO_2-iC_4H_{10} 80–20 mixture at STP (Commichau *et al.*, 1985). By kind permission of Elsevier.

$$I_1 = I_0 \frac{1 + (l - x)\rho/R}{2 + l\rho/R} \text{ and } I_2 = I_0 \frac{1 + x\rho/R}{2 + l\rho/R}, \; I_0 = I_1 + I_2.$$

A more general formulation that takes into account unequal amplifier impedances and gains is given in Buskens *et al.* (1983).

The position determination achievable with the current division method depends on the resistivity of the anode wire used; with a standard gold-plated tungsten wire 20 μm in diameter ($\rho = 1.8$ Ω.cm) an accuracy of around 4 mm rms, or ∼1% of the wire length, could be obtained (Charpak *et al.*, 1973); better localization can be obtained using higher resistivity wires (Bouclier *et al.*, 1898; Foeth *et al.*, 1973; Fischer *et al.*, 1976; Ford 1979; Biino *et al.*, 1988).

Longitudinal localization by current division has been used successfully in many large volume drift chamber systems described in the previous sections: the UA1 vertex detector (Beingessner *et al.*, 1987) and the JADE and OPAL jet chambers (Heintze, 1982; Biebel *et al.*, 1992).

9.7 Electrodeless drift chambers

A curious variation of the basic drift chamber design has been developed, in which the field needed to drift the electrons is not shaped by additional electrodes, but by the charges deposited on insulating surfaces constituting the detector walls (Allison *et al.*, 1982). Named wireless drift tubes or electrodeless drift chambers, the

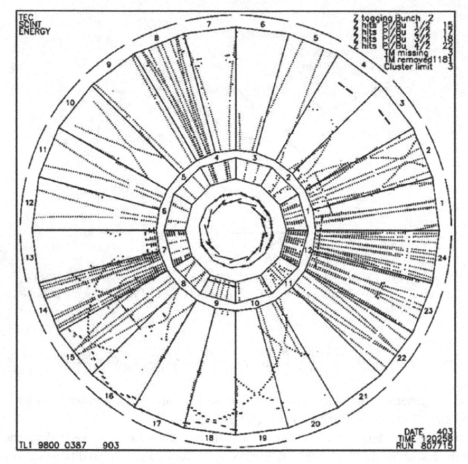

Figure 9.33   An electron–positron collision event recorded with the L3 TEC at LEP (Akbari *et al.*, 1992). By kind permission of Elsevier.

devices exploit a fundamental property of electrostatics: charges created by ionization or avalanche processes in the gas deposit on insulating surfaces with a distribution that tends to oppose further depositions, the equilibrium being reached when all field lines are made parallel to the insulator surfaces. Figure 9.35 (Zech, 1983) shows schematically the field build-up process; this in principle allows the construction of drift tubes of arbitrary shape and length.

Resulting from a dynamic equilibrium condition, the field shaping process depends on the charge production rate and distribution, and needs a 'formation' time before the counter becomes operational. These processes have been studied in detail for different detector geometry, demonstrating that reasonably high and stable efficiencies can be reached at moderate particle fluxes (Dörr *et al.*, 1985;

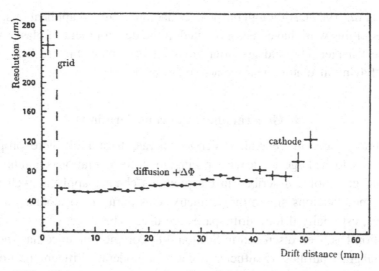

Figure 9.34   Space accuracy as a function of distance from the anode measured with the L3 TEC detector (Akbari *et al.*, 1992). By kind permission of Elsevier.

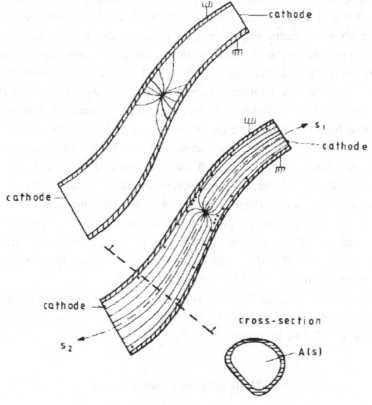

Figure 9.35   Schematics of the field-shaping build-up in the electrodeless drift tube (Zech, 1983). By kind permission of Elsevier.

Budagov *et al.*, 1987). Although the devices had only limited applications, some of the results of this work have been exploited in the development of plastic streamer tubes (see Chapter 11), and are often invoked to explain bizarre behaviours of detectors having insulators near the active electrodes.

9.8 General operating considerations

Drift chambers are, in general, easier to operate than multi-wire proportional chambers, due to the large anode wire spacing; most considerations on multiplication factors and gas choice described in the previous chapter apply as well with the necessary modifications due to the geometry. Gas purity is, of course, of primary importance, especially if long drift spaces are used. The effect of electro-negative gas pollution has been discussed in Section 4.8; common practice has shown that commercial grade purities are sufficiently good for moderate drift lengths, up to a few cm, but the gas tightness of the chamber and of the tubing have to be carefully checked and continuously monitored in long-standing experiments.

It should be emphasized that the good values of accuracy given in the examples of Figure 9.6 and Figure 9.12 are the result of local measurements, realized generally with short beam exposures. To preserve the accuracies in an actual experiment, one has to know precisely the local space–time correlations, which can be affected by the electric field strength and direction, atmospheric pressure, gas composition and temperature, external factors modifying the drift properties (electric or magnetic stray fields), and mechanical imperfections. Although it is in principle possible to take all these factors into account by calibration and monitoring, for a realistic system it is more reasonable to define limits to the tolerable variations, as a function of the desired final accuracy. The choice of a gas with drift velocity saturation decreases the dependence on the reduced electric field E/P and on temperature, as discussed in Section 4.7. The dependence on gas composition is also reduced at saturation, and has been measured in the gas mixture used in the high-accuracy drift chambers (Breskin *et al.*, 1974b): to maintain a ± 50 μm stability over 25 mm of drift, a maximum temperature variation of $\pm 7\ ^0$C and gas composition changes of $\pm 1.6\%$ should be assured.

Mechanical tolerances and electrostatic deformations contribute directly to the factors limiting accuracy; for large chambers the thermal expansion of the materials should also be taken into account.

9.9 Drift chamber construction

As for the multi-wire proportional chambers, only a brief mention is given here on the construction of drift chambers, and the reader is referred to the quoted literature for more information. Basically, the same techniques developed for the construction of proportional chambers have been used for drift chambers.

Calculation of the electrostatic forces in a chamber with a structure like the one in Figure 9.8 is relatively straightforward and shows that anodic instabilities do not appear for commonly used lengths of the wires, owing to the larger wire distance and diameter (Cheng *et al.*, 1974). On the other hand, the two cathode planes are attracted inwards and the overall deformation can be estimated following the methods outlined in Section 8.12. For the more complex high accuracy drift chamber (Figure 9.9) the critical elements are the field wires: since the adjacent cathode wires have a charge equal in sign, they are subjected to an outward force. Gap-restoring strip inserted at intervals between the cathodes and the screening electrodes, visible in Figure 9.15, have proved to be sufficient to compensate the electrostatic forces; external to the active volume of the chamber, the strips do not affect the performance of the drift chamber.

Further reading

Palladino, V. and Sadoulet, B. (1975) Applications of classical theory of electrons in gases to drift proportional chambers. *Nucl. Instr. and Meth.* **128**, 323.

Blum, W. and Rolandi, G. (1993) *Particle Detection with Drift Chambers* (Berlin, Springer-Verlag).

10

Time projection chambers

10.1 Introduction: the precursors

Various designs of drift chamber, with large sensitive gas volumes coupled to planes of multiplying wires, have been described in the previous chapter. As the event separation depends on the maximum drift time, or memory, of the detector, in many applications the drift volume thickness is kept relatively small. For low event rates, or for use at storage ring accelerators where the collision frequency is low, long drift lengths can be used; providing a full three-dimensional representation of events, large volume drift chambers have also been named imaging chambers. While this implies a strict control on gas purity and requires high voltage values, it considerably reduces the complexity of the readout device and electronics.

An early example of very large volume, single-ended readout drift chamber is the detector built for the identification of secondary particles by ionization sampling (ISIS and ISIS2), operated at CERN in the mid-1970s (Allison *et al.*, 1974) and consisting of a large gas box with field-shaping electrodes on the walls, between two cathodes and a central wire plane for amplification and detection of ionization trails, shown in Figure 10.1 (Allison *et al.*, 1984). Drift time and signal charge are recorded on each anode wire, providing the vertical coordinate and the charge for the corresponding track segments, the second coordinate is given by the anode wire number. Figure 10.2 is an example of a multi-track single event recorded in a beam exposure with ISIS2; the vertical scale corresponds to the two metre drift volume length. The device was used as part of the European Hybrid Spectrometer, providing both the particles' trajectory and the differential energy loss used for particle identification in the low momentum region.

Due to its conception, the ISIS device can be considered as a precursor of the time projection chamber, to be described in the following sections.

292

Figure 10.1 The large volume ISIS2 chamber (Allison *et al.*, 1984). By kind permission of Elsevier.

10.2 Principles of operation

Knowledgeable of the successful development of large volume drift chambers and of the coordinate localization method from cathode induced signals in MWPCs, David Nygren and collaborators proposed in 1976 a new detector concept named the time projection chamber (TPC) for the PEP-4 experimental area at SLAC (Clark *et al.*, 1976; Nygren and Marx, 1978; Aihara *et al.*, 1983).

In its basic design, the TPC consists of a large volume of gas with field shaping electrodes, terminating at one end with a multi-wire end-cap chamber plane having alternating anode and field wires; the two parts are separated by a cathode mesh permitting one to set the drift and multiplication field independently. On the outer side of the MWPC, the cathode plane consists of a printed circuit board with pad rows parallel to the anodes, shown schematically in Figure 10.3. Electrons released by ionization trails in the main volume drift towards the end-cap, cross the first

Figure 10.2   An event recorded with the ISIS drift chamber (Allison *et al.*, 1984). By kind permission of Elsevier.

mesh electrode, and are collected and amplified on the anode wires, thus producing a negative signal on the anodes and a positive charge induction profile on the corresponding pad rows, as shown schematically in Figure 10.4. A recording of the charge distribution on the pads in successive time slices provides the drift time and longitudinal coordinate of track segments, as well as the differential ionization energy loss; signals on wires can be also recorded to improve the drift time measurement.

In most TPC applications, the major information is the one projected on the MWPC plane, or azimuthal, provided by the pad rows, and therefore a large effort has been devoted to measurements and simulations of the induced signal distribution, the so-called pad response function, determining the localization properties of the detector. Operation in a strong magnetic field parallel to the drift field, with an appropriate choice of the gas filling, reduces the transverse diffusion and hence its contribution to the dispersion of the collected charge for long drift distances, with an ensuing improvement in localization accuracy as discussed in Chapter 4. Figure 10.5 (Clark *et al.*, 1976) shows the estimated reduction of diffusion for several common gases; while well known to gaseous electronics experts, this property, basic for the TPC operation, was not familiar at the time to the gas detector community.

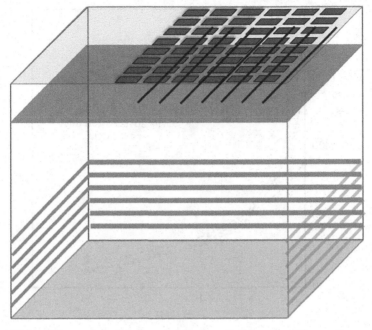

Figure 10.3 Schematics of a time projection chamber: ionizing trails produced in a large volume of gas drift to an end-cap MWPC with cathode pad rows.

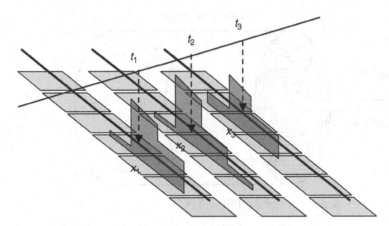

Figure 10.4 The charge induction process on the pad rows.

Adapted for operation in a solenoidal magnet, the PEP-4 TPC, as well as many similar devices implemented later, had the cylindrical shape shown schematically in Figure 10.6 (redrawn from Clark *et al.*, 1976), with two independent, back-to-back detectors having a common central high-voltage drift electrode, and two composite MWPC end-caps for amplification and signal recording (Figure 10.7). To optimize the particle identification power and momentum resolution, the PEP-4

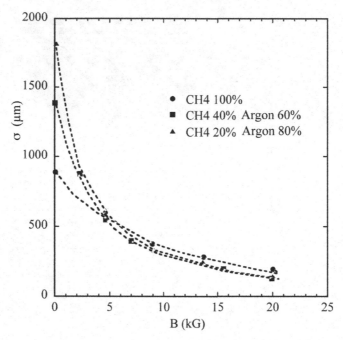

Figure 10.5   Dependence of the transverse electron diffusion on magnetic field for several gas mixtures. Data from Clark *et al.* (1976).

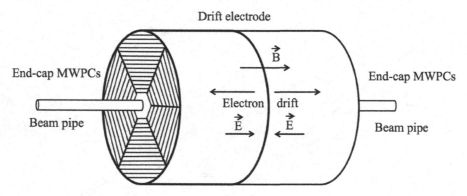

Figure 10.6   Schematics of the PEP-4 TPC.

TPC was operated at high pressures (argon–methane at 8.5 bars); for practical and security reasons, most of the later TPC systems have been operated at pressures close to atmospheric.

An example of a multi-prong event recorded with the PEP-4 TPC is shown in Figure 10.8; represented as three independent projections, the information is in fact fully three-dimensional. The particle identification power of the device, deduced from a weighted average of the 185 independent differential energy loss measurements for each track, is seen in Figure 10.9 (Nygren, personal communication).

Figure 10.7   One of the MWPC modules for the LBL PEP-4 TPC.D. Nygren, personal communication (1985).

10.3 TPC-based experiments

Time projection chambers have been built and successfully operated in many colliding beam and fixed target experiments, demonstrating excellent imaging and particle identification capabilities, particularly for large track multiplicities. Due to the long memory time, several tens of microseconds or more, TPCs find their natural use for electron–positron storage rings having comparable collision frequencies, as the now discontinued LEP at CERN and TRISTAN at KEK, low rate fixed target experiments as TRIUMF in Vancouver, very high multiplicity heavy ions collisions as NA49 at CERN and STAR at RHIC, and the gigantic ALICE TPC at CERN. Smaller devices are used in many other experiments both in high-energy and nuclear physics; cryogenic TPCs, making use of liquid noble gases, are also under development and are described in Chapter 15. For summaries of the early developments, see, for example, the proceedings of the Time Projection Chamber Conference (TPC (1983)) and the book of Blum and Rolandi (1993).

Figure 10.10 is a perspective view of the TRUMF TPC, one of the first devices of this kind to be operated after the original PEP4 development, and used for the study of muon decays (Hargrove *et al.*, 1984). The quoted reference includes also

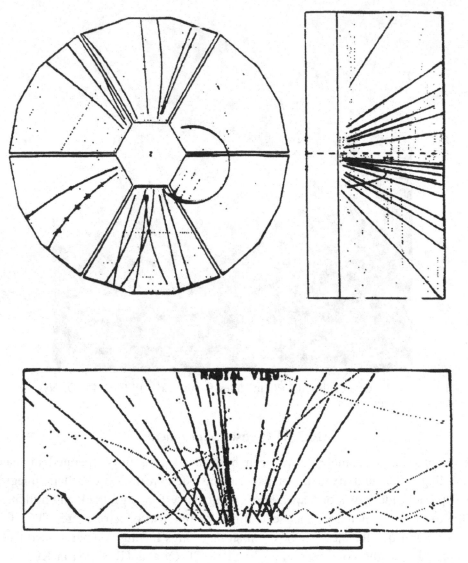

Figure 10.8   Projection on three planes of an event recorded with the PEP-4 TPC
D. Nygren, personal communication (1985).

one of the first detailed studies of the effects of magnetic field and track angle on resolution (see the next section).

Figure 10.11 shows the schematics of the ALEPH TPC, operated at LEP from its commissioning in 1989 to the end of operation of the collider in December 2000, and the picture in Figure 10.12 is a front view of the full detector with the TPC end-cap system recognizable from the composite shape of the MWPCs used for the readout (Decamp *et al.*, 1990); Figure 10.13 is an example of on-line event

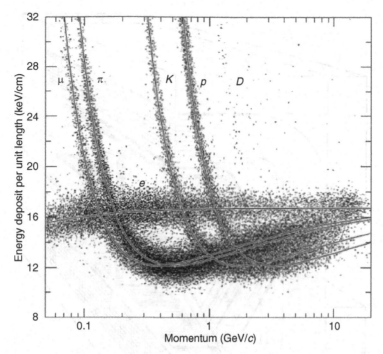

Figure 10.9 Particle identification power of the PEP-4 TPC, operated at 8.5 bars. The momentum is given by the tracks' curvature in the 0.4 T magnetic field. D. Nygren, personal communication (1985).

display for the device. Similar in conception and performances, the DELPHI TPC equipped a second collision point at LEP (Brand *et al.*, 1989).

The design of the cylindrical double back-to-back TPC, with the composite field cage and the MWPC sensors arrays at the two outer ends, has become, with small variations, a standard for detectors inserted in a solenoidal magnets around storage rings. Figure 10.14 shows the field cage and end-cap support structure of the STAR TPC at Brookhaven's RHIC heavy ions collider (STAR), and Figure 10.15 the projection on the end-cap plane of a heavily populated multi-track event in a gold–gold collision at 30+30 GeV (Anderson *et al.*, 2003). Two smaller devices with radial drift geometry, named forward TPCs, complement the main system to extend detection at lower angles to the beam axis (Ackermann *et al.*, 2003).

With an outer diameter close to 5 metres and active volume around 100 m³, the ALICE TPC installed at CERN's LHC is the largest device of this kind ever built, Figure 10.16 and Figure 10.17 (Alme *et al.*, 2010); the picture in Figure 10.18 shows Peter Glässel, the TPC technical coordinator, sitting within the central opening of the end-cap array. Used for the detection of heavy ion collisions, the device is capable of reconstructing very complex events; Figure 10.19 is an example of outcomes of a lead–lead collision at 2.76 TeV. The detector also performs

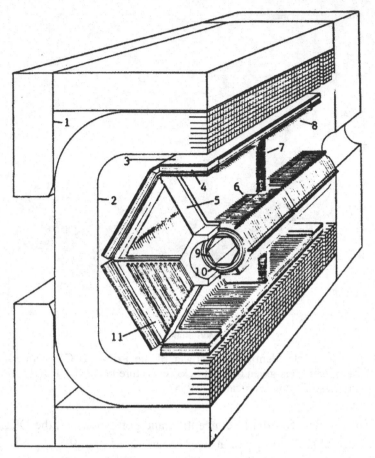

Figure 10.10   Perspective cut view of the TRIUMF TPC (Hargrove *et al.*, 1984).
By kind permission of Elsevier.

particle identification, particularly in the low momentum range, from the analysis of
the recorded differential energy loss of the tracks, Figure 10.20 (You, 2012).

A simpler, box-like TPC design is used for fixed target experiments. The NA49
large acceptance detector includes along the interacting beam line a pair of
medium-size vertex TPCs and two larger main TPCs to detect the interaction
yields; Figure 10.21 shows a view inside the main TPC field cage; Figure 10.22
is a reconstructed event resulting from a high-energy lead–lead interaction.

While in the current generation of TPCs the end-cap detector is implemented
with multi-wire chamber modules, the recent developments of micro-pattern
gas detectors have suggested the use of these devices to improve performances,
and in particular the localization accuracy and multi-track resolution (Kappler
et al., 2004; Ableev *et al.*, 2004; Radicioni, 2007). A medium-size TPC with

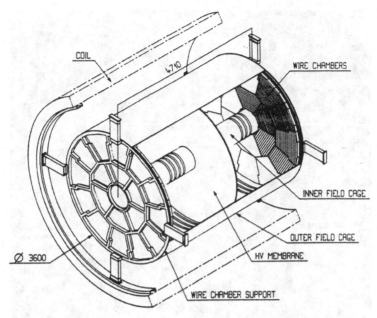

Figure 10.11   Artist's view of the ALEPH TPC assembly (Decamp *et al.*, 1990).
By kind permission of Elsevier.

Micromegas readout is already operational in the experiment T2K (Abgrall *et al.*, 2011) and both Micromegas and GEM-based readouts are investigated for the large TPC under development for the International Linear Collider (ILC) (Behnke, 2011) and the ALICE GEM-TPC upgrade (Böhmer *et al.*, 2013). Performances of these devices are discussed in Chapter 13.

10.4 Signal induction: the pad response function

As shown schematically in Figure 10.4, the multiplication process occurring on the anodes when the ionization trail reaches the wires results in the appearance on the facing pad rows of an induced charge profile around the position of the avalanches. The profile evolves with time, reaching a maximum in front of the avalanches at the time of full collection of the corresponding segment of primary ionization. The screening effect of the field wires and the separation between pad rows minimizes cross-talk problems.

As discussed in Sections 6.3 and 8.9, for a point-like avalanche the charge induction profile on the cathode plane has a two-dimensional Gaussian-like shape, building up and then decreasing with time, with a FWHM approximately equal to twice the anode-to-cathode distance, or gap G. Recording and suitable analysis of

Figure 10.12   Open-ended front view of the ALEPH detector. The spokesman of
the experiment and Nobel Laureate Jack Steinberger stands third from left. Picture
CERN (1988).

the charge induction profile on pads or strips provides the coordinate of the
avalanche along the anode wire.

Projected in the direction x parallel to the anode wire, the distribution is
approximated by the expression (see Figure 10.23 for the definition of parameters):

$$P(x) \approx e^{\frac{x^2}{2\sigma^2}}, \quad \sigma = \frac{2G}{2.34}. \tag{10.1}$$

Integration over the area of the strip provides the so-called intrinsic pad response
function (PRF), describing the dependence of the induced signal on the pad

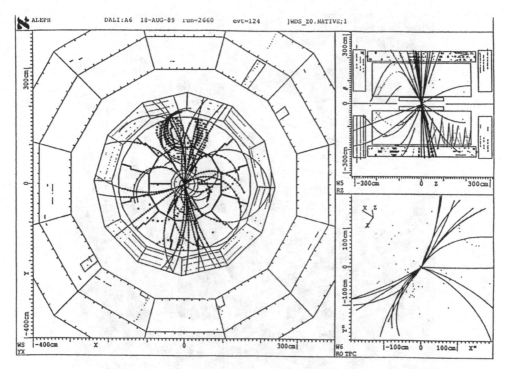

Figure 10.13 On-line display of an event recorded with the ALEPH TPC. Picture CERN (1991).

geometry and position with respect to the avalanche; for the simple case of a track parallel to the strip it can be written as (Blum and Rolandi, 1993):

$$P_0(X) = \int_{X-W/2}^{X+W/2} P(x)\mathrm{d}x, \tag{10.2}$$

where W is the pad width and X the distance between the track and the centre of the pad. An example of the agreement between computed and measured distributions for a 4 mm gap chamber with 7 mm pads is shown in Figure 10.24 (Hargrove *et al.*, 1984).

While a small pad size W provides more measurements, thus improving localization, sharing of the charge reduces the ratio signal over noise; for best performances, a good choice is a pad width close to the gap size, $W \approx G$, for which most of the induced charge is shared between three adjacent pads. In this case, a simple algorithm can be used to compute the coordinate and the Gaussian width of the avalanches along the wire from the measured charges on three adjacent pads, Q_1, Q_2 and Q_3 (Anderson *et al.*, 2003):

Figure 10.14   Field cage support structure of the STAR TPC at RHIC. Picture courtesy Brookhaven National Laboratory (USA).

$$X = \frac{\sigma^2}{2W}\ln\left(\frac{Q_3}{Q_1}\right), \qquad \sigma^2 = \frac{W^2}{\ln(Q_2^2/Q_1 Q_3)}, \tag{10.3}$$

where one should note that the charge on the central pad does not affect the coordinate measurement. The pad length L does not appear in the expressions, but determines the total value of the induced charge and affects the coordinate localization for inclined tracks and in the presence of a magnetic field, as discussed later.

In the real case of detection of the extended ionization trails released by charged particles, the electrons reaching the anode are not point-like, but spread along the wire depending on the geometry of the track and detector. For a proper description of the cathode-induced signals and the position resolution limits, the following effects have to be taken into account:

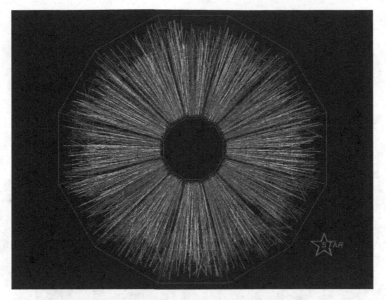

Figure 10.15 A gold–gold collision recorded with the STAR TPC (STAR). Picture courtesy Brookhaven National Laboratory (USA).

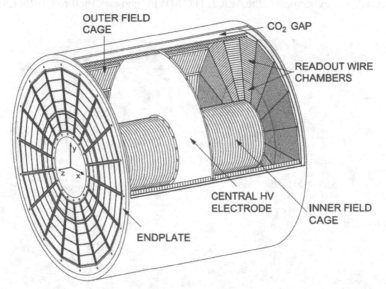

Figure 10.16 Artist's view of the ALICE TPC (Alme *et al.*, 2010). By kind permission of Elsevier.

- diffusion of electrons in their drift to the end-cap detector;
- spread due to the angle of the tracks with respect to the anode wire;
- distortions in the drift path when electrons approach the anode in the presence of a magnetic field, the so-called $E \times B$ effect;
- asymmetry in the energy loss distribution due to clusters.

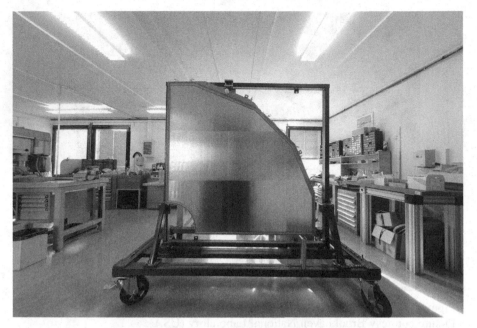

Figure 10.17   A segment of the ALICE TPC MWPC end-cap. Picture CERN (2002).

Figure 10.18   Front view of the ALICE TPC end-plate, with the technical coordinator, Peter Glassel. Picture CERN (2006).

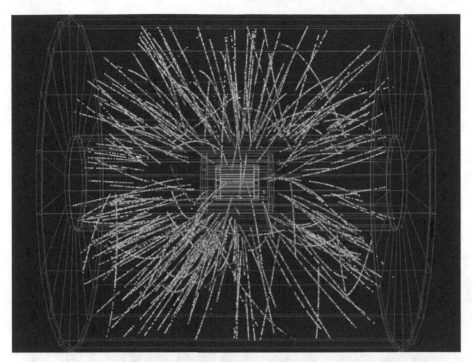

Figure 10.19 A proton–lead interaction recorded with the ALICE TPC at LHC. Picture CERN (2012).

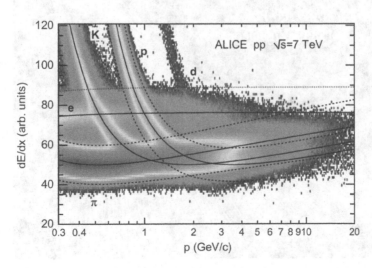

Figure 10.20 Particle identification with the ALICE TPC (You, 2012). By kind permission of Elsevier.

Figure 10.21   Inner view of one of the NA49 TPC modules. Picture CERN (1994).

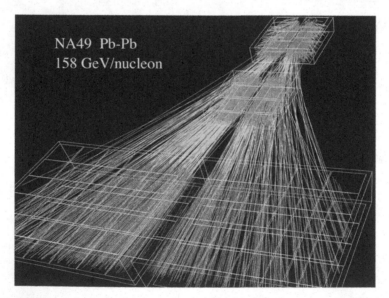

Figure 10.22   A lead–lead interaction recorded with the NA49 TPC complex.
Picture CERN (1996).

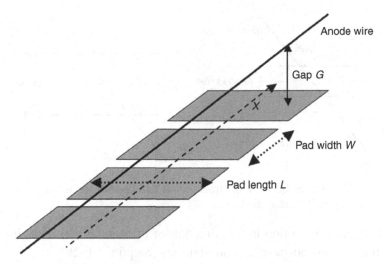

Figure 10.23   Schematics and definitions of the pad row geometry.

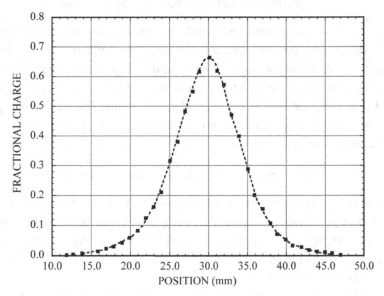

Figure 10.24   Measured (dots) and computed pad response function (curve) (Hargrove *et al.*, 1984). By kind permission of Elsevier.

Many authors have analysed the various dispersive effects and their consequences on the localization properties of the detector (Hargrove *et al.*, 1984; Blum *et al.*, 1986; Amendolia *et al.*, 1989; Saquin, 1992; Anderson *et al.*, 2003); an exhaustive discussion on the subject can be found in the textbook by Blum and Rolandi (1993).

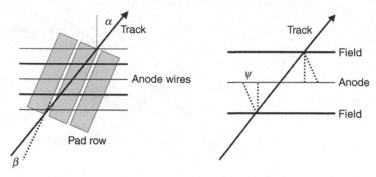

Figure 10.25 Angles between the track, wires and pads (left), and the deflection angle in a magnetic field (right).

The TPC space resolution in the direction of the anodes can be written as the squared sum of four independent contributions (Saquin, 1992):

$$\sigma^2 = \sigma_0^2 + \sigma_D^2(1 + \tan^2\alpha)\cos^2(\alpha - \beta)L$$
$$+ \sigma_A^2(\tan\alpha - \tan\psi)^2\cos^2(\alpha - \beta) + \sigma_P^2\tan^2\beta; \qquad (10.4)$$

the dispersive terms in the sum correspond, respectively, to the point-like PRF, the electron diffusion over the drift length L, the magnetic deflection due to the E × B effect, and to the angle of the tracks. The quantities α, β are the angles of the track with the normal to the anodes and to the pad row direction and ψ is the angle due to the E × B effect (Figure 10.25); σ_D, σ_A and σ_P are ad-hoc constants corresponding to the respective dispersive effects. For the special case of a pad row parallel to the anodes, the expression simplifies as:

$$\sigma^2 = \sigma_0^2 + \sigma_D^2(1 + \tan^2\alpha)L + \sigma_A^2(\tan\alpha - \tan\psi)^2. \qquad (10.5)$$

The various constants appearing in the expressions can be deduced from simulations, or from a fit to measured data. As examples, Figure 10.26 and Figure 10.27 show the dependence of the azimuthal resolution on the drift length L and pad crossing angle β, the second and fourth terms in expression (10.4), measured with the ALEPH TPC at 1.5 T (Atwood *et al.* 1991).

The combined effects of geometry and magnetic deflection angle (third term in the expression) are seen in the example of Figure 10.28 measured with the DELPHI TPC (Brand *et al.*, 1986). As a consequence of the E × B distortion, the minimum is observed for tracks crossing the detector at an angle corresponding to the average deflection of the ionized trail due to the Lorentz force. In the absence of a magnetic field, the best localization accuracy is obtained for tracks perpendicular to the anodes, minimizing the angular dispersion factors.

In the previous considerations, the ionization density along the track was considered uniform. As discussed in Section 2.4, the presence of asymmetries in

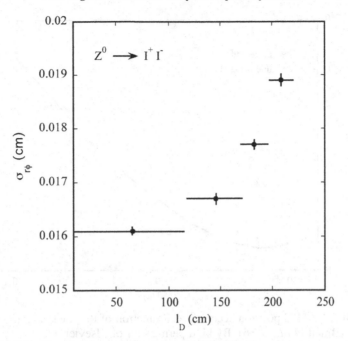

Figure 10.26   Dependence of the azimuthal localization accuracy on the drift length (Atwood *et al.*, 1991). By kind permission of Elsevier.

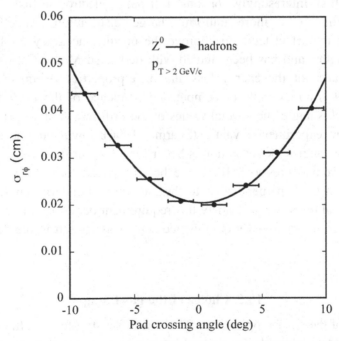

Figure 10.27   Dependence of the localization accuracy on the angle between tracks and pad row (Atwood *et al.*, 1991). By kind permission of Elsevier.

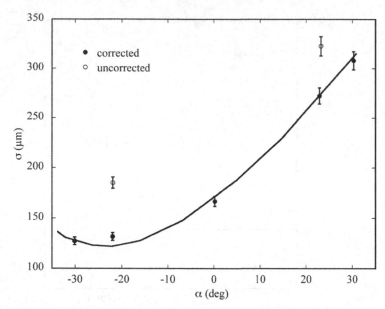

Figure 10.28   TPC position accuracy as a function of the track angle, measured at 1.2 T (Brand *et al.*, 1986). By kind permission of Elsevier.

the energy loss due to delta electrons (clusters) contributes to the worsening of the space resolution. Interestingly, for long drift paths, electron diffusion smears the trails, resulting in a more uniform charge distribution, an effect named declustering; the effect tends to improve the position accuracy for large angles and drift lengths, and has been studied with dedicated Monte Carlo simulations taking into account the energy loss and drift properties (Blum *et al.*, 1986). Figure 10.29 is an example of computed dependence of the position accuracy from the track's angle and several values of the drift length at 1.5 T; points with error bars are experimental values (Sharma, 1996). Owing to the declustering effect, at large angles the resolution is better for a long drift.

As seen from expression (10.4), the best projected localization accuracy is obtained for a track perpendicular to the pad rows. For large end-cap MWPC modules, radial rows of pads satisfy this requirement better for a tracks emerging from a centred interaction vertex; this geometry was chosen in the design of the ALEPH TPC.

10.5 Choice of the gas filling

The choice of the gas for operating a TPC has to satisfy often conflicting requirements, depending on the detector's application and desired performances. For large volume devices it is convenient, if not mandatory, to select an operating point

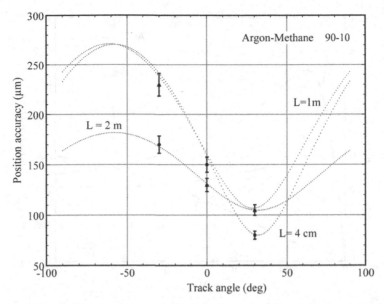

Figure 10.29 Position accuracy as a function of track angle for several drift lengths. Points with error bars are measurements from the ALEPH TPC. (Sharma, 1996). By kind permission of A. Sharma.

requiring moderate electric fields to avoid too high values of voltage on the cathodes. In many mixtures this results in low, non-saturated drift velocities and large electron diffusions, affecting the rate capability and localization accuracy of the detector. As shown in Figure 10.30, a standard argon–methane 90–10 mixture (P10) satisfies the requirements of a high drift velocity and low diffusion at moderate drift field values, and has been the preferred choice for the first generation of TPCs, with a peaking drift velocity around 150 V/cm and a strong reduction of the traverse diffusion at high magnetic fields.

Requiring increasingly stringent safety measures because of their flammability, the use of hydrocarbon-rich gases has become less and less popular. A replacement of methane with carbon dioxide, although satisfying the diffusion requirements despite a lower drift velocity, is discouraged due to the instabilities and secondary processes observed in the MWPC operation. Safer mixtures with reduced methane content have been studied in the framework of the International Linear Collider detector with the so-called TDR[1] gas; reaching a saturated mode of operation requires fields larger than for P10, as seen in Figure 10.31; the transverse diffusion suppression is, however, comparable (Behnke *et al.*, 2001).

For detectors operating without a magnetic field, and if one of the major requirements is a good two-track resolution, neon–carbon dioxide mixtures are

---

[1] ILC Technical Design Report.

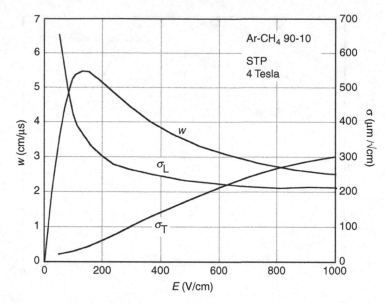

Figure 10.30   Electron drift velocity, longitudinal and transverse diffusion coefficient in 4 T in argon–methane 90–10.

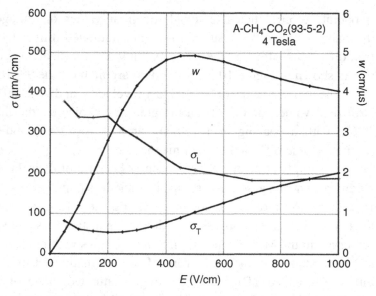

Figure 10.31   Electron drift velocity, longitudinal and transverse diffusion coefficient in 4 T in argon–methane–carbon dioxide.

preferred; in Figure 10.32, measured values of the charge distribution on reaching the anodes after 60 cm drift in the transverse (a) and drift direction (b) are compared (Afanasiev *et al.*, 1999). The neon–CO_2 mixture has been chosen for the vertex TPCs in the NA49 experiment (Fuchs, 1995).

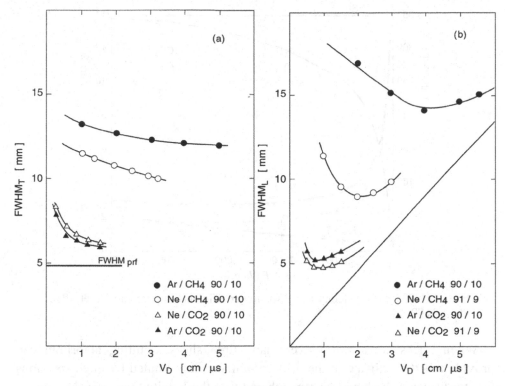

Figure 10.32 Transverse (a) and longitudinal width (b) of the charge distribution after 60 cm drift in several gas mixtures (Afanasiev *et al.*, 1999). By kind permission of Elsevier.

Pure carbon tetrafluoride (CF_4) has intrinsically many desirable properties: the highest known drift velocity, favouring its use in fast detectors, a very low transverse diffusion, even in absence of a magnetic field (Figure 10.33), and reduced sensitivity to thermal neutron background as compared to hydrogenated molecules; these advantages are balanced by the need to apply very high voltages to achieve charge multiplication. Promising results obtained with a prototype TPC with MWPC readout have been reported (Isobe *et al.*, 2006). Coupled to gas electron multiplier sensors, CF_4-filled TPCs have been demonstrated to provide excellent space resolutions (Oda *et al.*, 2006), see Chapter 13.

10.6 Coordinate in the drift direction and multi-track resolution

As already mentioned, most TPCs have been optimized for the measurement of the projected azimuthal coordinate, in the plane of the wires and pad rows; the readout electronic, designed to record the charge distribution of induced signals on pads with good amplitude resolution, usually has a modest time resolution, corresponding to the separation between successive time slices.

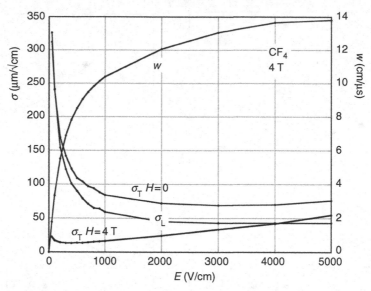

Figure 10.33   Drift velocity and diffusion coefficients in pure carbon tetrafluoride at NTP.

When tracks are sampled over many time slices, a fitting algorithm can
provide a better estimate of the drift coordinate, only limited by the dispersions
due to diffusion. It should be remembered that the longitudinal diffusion coefficient is not reduced by magnetic field, so even using a fast time recording
system on the anode wires, the drift time measurement provides a modest
resolution for long detectors. All considerations discussed in Chapter 9 apply
for this case. Figure 10.34 is an example of longitudinal space resolution as a
function of drift length, measured with the ALICE TPC for increasing values of
the angle between the tracks and the normal to the anode wires (Alme *et al.*,
2010).

The multi-track separation is limited in the projection plane by the width of
the pad response, and in the longitudinal direction by the time width of the signals,
and is therefore geometry and electronics dependent. As an example, Figure 10.35
shows the two-track resolution measured with the STAR TPC (Anderson *et al.*,
2003).

Essential for ensuring the required extreme multi-track capability of the detector
is the use of custom-made fast amplifier-ADC ASIC, performing on-board, digital
pedestal correction and zero-suppression; Figure 10.36 is an example of detected
signal for a high-multiplicity event, before and after the digital shaping performed
by the circuit (Mota *et al.*, 2004).

A detailed study of signal shape in TPCs for various geometry and operating
conditions, aiming at improving the multi-track resolution, is given in Rossegger
and Riegler (2010).

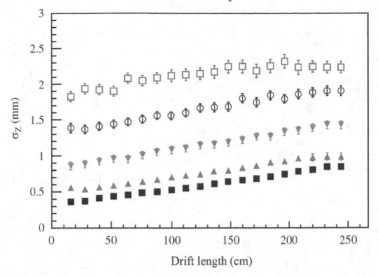

Figure 10.34 Longitudinal resolution as a function of drift length. Sets of points corresponds to tan (α) = 0 (lowest values) up to tan (α) = 0.92 in steps of 0.23, where α is the angle between the tracks and the perpendicular to the anodes (Alme *et al.*, 2010). By kind permission of Elsevier.

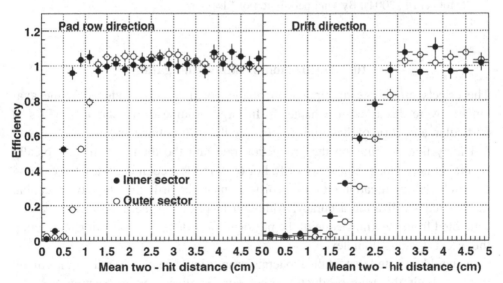

Figure 10.35 Two-track resolution in the pad and drift directions measured with the STAR TPC (Anderson *et al.*, 2003). By kind permission of Elsevier.

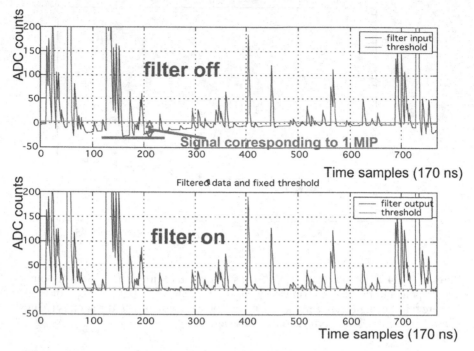

Figure 10.36   Digital filtering of a pad signal to improve multi-track resolution (Mota *et al.*, 2004). By kind permission of Elsevier.

10.7 Positive ion backflow and gating

The effects on charge multiplication of the positive ions produced in the avalanches were discussed in Chapter 8. In large volume drift chambers, TPCs in particular, a fraction of these ions recede into the drift volume and accumulate with a density that depends on the production rate and the total drift time, typically several hundred ms per metre of drift. For even modest counting rates, the ensuing electric field modification introduces distortions in the drift path, largely exceeding the desired localization accuracy. Although in principle these distortions can be estimated from the knowledge of the physical track density distribution, correcting the effect is rather complex.

It should be noted that, at the moderate proportional gains, around 10^4, normally used to limit the ions production rate, and as discussed in Section 8.9, the avalanches develop preferentially on the drift side of the anode, with the consequence that the majority of ions recede in the direction of the incoming tracks; they then transfer into the drift volume in a percentage corresponding to the ratio between the drift and the MWPC gap fields, typically 10–20% of the total.

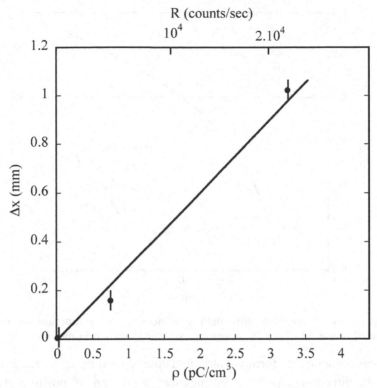

Figure 10.37   Lateral displacement of the recorded position of a collimated ^{55}Fe source after 10 cm of drift close to the positive charge cloud produced by a strong source, 10 mm apart, with the intensity given in the top scale (Friedrich *et al.*, 1979).

For a radiation flux releasing R electrons per second, the positive ions density is given by (Friedrich *et al.*, 1979):

$$\rho^+ = \frac{e\,R\,L\,\varepsilon\,M}{w^+},$$

where w^+ and L are respectively the ion's drift velocity and drift length, M is the proportional gain and ε is the fractional ion feedback. The ensuing field modification can be computed using standard electrostatic methods; for a uniform space charge distribution in the drift volume, the field strength is decreased in the region close to the end-cap and increased towards the drift electrode. Although in this case the electron drift lines are not changed, the variation of field results in a non-constant drift velocity.[2]

---

[2] The integral of the field over the gap has of course to remain constant and equal to the difference of potential divided by the gap.

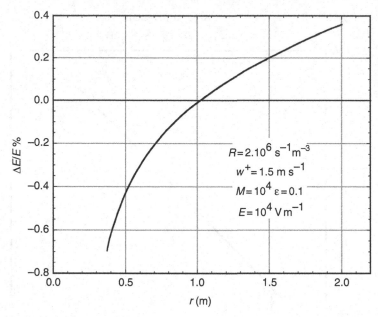

Figure 10.38   Percentage drift field modification as a function of radius, computed in the ALEPH TPC operating conditions shown in the inset.

In the case of a non-uniformly distributed charge accumulation, the field is also modified in direction; electrons drifting near a column of positive charge are attracted inwards. Measured with a small TPC-like drift chamber, Figure 10.37 shows the lateral displacement of the recorded position after 10 cm of drift of a collimated source, 1 cm away from the positive charge sheet produced by a second source with the rate indicated on the upper scale (Friedrich *et al.*, 1979).

For a cylindrical TPC with inner and outer radii r_1, r_2, and assuming a uniform charge density $\rho^+(r) = \rho_0$, the radial perturbation of the field is (Blum and Rolandi, 1993):

$$E(r) = \frac{\rho_0}{2\,\varepsilon_0}\left(r - \frac{r_2^2 - r_1^2}{2r\,\ln(r_2/r_1)}\right).$$

Figure 10.38, computed from the expression above, shows the percentage field modification as a function of radius for the standard operating conditions of the ALEPH TPC, as given in the inset.

An effective way to substantially reduce the ion feedback is to insert between the MWPC cathode and the drift volume a wire mesh, or gating grid, opened or closed to the charge flow under control of an external voltage. While a simple mesh at an appropriate potential cancelling the drift field would be effective, a grid of parallel field wires, alternately connected to positive and negative voltages, is a more convenient geometry, minimizing the transient's pickup on signal electrodes

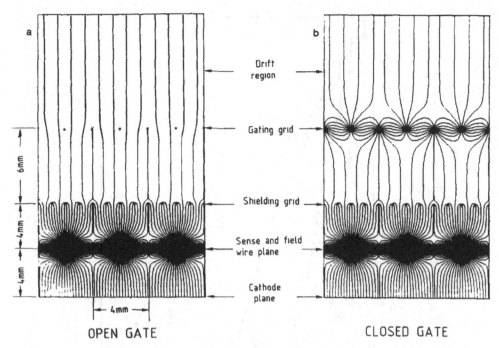

Figure 10.39 Principle of operation of the gating grid; open to the drifting charges when wires are kept at the same potential, the grid is closed with the application of a voltage difference to alternate wires (Amendolia *et al.*, 1985a).

at the time of application of the gating pulse. In the open gate mode, the grid wires are connected to the same voltage V_G, corresponding to the local drift field equipotential; to close the gate, a symmetric bipolar field $\pm \Delta V_G$ is added to alternate wires, as shown schematically in Figure 10.39 (Amendolia *et al.*, 1985a).

The gating grid can be operated in two modes: always open and closed to the ions' backflow on detection of an event, or always closed and open to ionization electrons in coincidence with a trigger and for a time equal to the total drift time. In the second case, due to the time taken to generate and distribute the transparency restoring voltage pulses, a layer of drift a few cm thick close to the end-cap is lost to detection. In either mode, the goal is to reach an ion suppression factor equal to or better than the detector gain.[3]

Aside from small losses due to the different diffusion coefficient, an electrical gate open to electrons in one direction transmits all ions in the opposite one. This is not the case in the presence of a magnetic field, due to the different drift line modification due to the Lorentz force affecting only electrons; for a proper choice of the polarization of the gating grid, all ions are stopped, independently of the value

---

[3] Based upon the consideration that at this value the ions' feedback equals the unavoidable contribution of primary ionization.

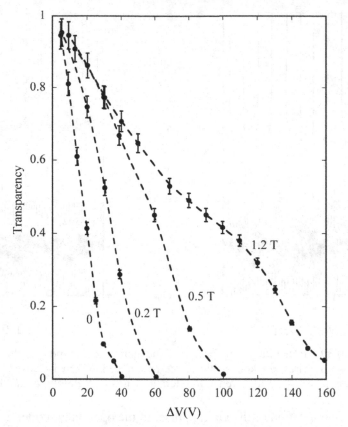

Figure 10.40   Electron transparency of a gating grid as a function of polarization voltage, for increasing values of the magnetic field (Amendolia *et al.*, 1985b).

of the magnetic field, while the electron transparency is partly restored at high field values, as shown in Figure 10.40 (Amendolia *et al.*, 1985b). While very attractive since not requiring the application of pulsed potentials, the 'ion diode' has not often been used because of the large distortions introduced due to the E × B effect.

The problem of the reduction of the positive ions' backflow has been analysed in great detail during the development of the new generation of TPCs with micro-pattern gas detector readout, covered in Chapter 13.

Gating is a very effective way to reduce or eliminate the field distortions generated by the accumulation of positive ions in the drift volume; however, it can only be used for relatively small event rates, with the average distance between events comparable with the maximum electron drift time, as for the LEP detectors. When gating is not possible, operation of the detector at a smaller proportional gain and an optimization of geometry aiming at reducing the ions' backflow, coupled with suitable analytical calculations to estimate the space charge distribution, may allow one to correct the distortions (Rossegger *et al.*, 2010; Rossegger *et al.*, 2011).

The new generation of micro-pattern devices, aside from providing an improved track resolution, permit a strong reduction of the positive ions' backflow when used as end-cap TPC detectors.

10.8 TPC calibration

While a maximum effort is put into the design and construction of large TPCs to keep strict tolerances on all sources of distortion that can affect the final performance, a thorough monitoring of operating and ambient parameters is needed, together with an extensive calibration procedure. The proportional gain can be monitored with the use of external radioactive sources; placed in known positions on the drift electrode, they permit the control of gain and/ or capture losses as well as the presence of distortions of the electrons drift lines.

A very powerful tool to perform tracking calibrations is the use of pulsed laser beams ionizing the TPC gas and emulating straight tracks at strategic positions and angles; advantages of using laser-generated tracks are the favourable ionization statistics (Poisson-like, without long-range δ electrons), controlled ionization density and insensitivity to magnetic field.

Pulsed nitrogen and frequency-doubled Nd:YAG lasers, emitting at 337 nm (3.68 eV) and 226 nm (4.68 eV) respectively, are commonly used, with sub-ns timing resolutions; two-stage N_2 lasers have been developed, offering very high intensity and low divergence beams.

For the gases normally used in detectors, ionization requires two- or three-photon excitation and therefore high power density. The ionization density can be largely increased by adding a controlled amount of a low ionization potential vapour in the gas mixture; the presence of pollutants is often sufficient by itself to enhance the ionization rate. Many compounds have been tested; for a review see for example Hilke (1986a). However, in view of the severe ageing problems that have been encountered by the voluntary or unwanted addition of trace organic compounds, the use of additives should be considered with care (see Chapter 16) and in general the solution requiring higher power density has been preferred, although it is often recognized that a good fraction of the ionization is produced by the presence of unknown impurities; the laser power is adjusted to obtain the desired ionization density.

Various methods have been implemented to generate multiple laser tracks covering the sensitive volume, with one or more high-power lasers and arrangements of fixed beam splitters and mirrors, as used for ALEPH (Decamp *et al.*, 1990), DELPHI (Brand *et al.*, 1989) and ALICE (Alme *et al.*, 2010), or systems of remotely controlled rotating mirrors (Miśkowiec and Braun-Munzinger, 2008).

The straight track reconstruction permits one to correct for mechanical tolerances, drift path distortions, temperature and gas variations; the large ionization yield resulting from a laser beam hitting the cathode can be used for continuous drift velocity monitoring.

10.9 Liquid noble gas TPC

Very powerful in the detection and tracking of complex events, standard TPCs are unsuitable in the search for rare events such as neutrino interactions, proton decays and other exotic particles for which a gas has a very low cross section. Proposed in the seventies by Carlo Rubbia and collaborators, the liquid argon TPC has a density several orders of magnitude larger, and a correspondingly higher inter-action probability.

While observed in liquid noble gases, charge multiplication is hard to exploit as it requires the use of very thin anode wires (Derenzo *et al.*, 1974). The devices operate in the ionization mode, collecting the electrons released in the sensitive volume; the higher ionization density of the charged prongs compensate for the lack of gain. The charge is collected and detected on two perpendicular wire grids at a close distance; in some designs, a screening grid separates the sensitive volume from the sensors electrostatically. Proper choice of the electric field strength ensures that most of the electrons pass the screening and the first induction grid before being collected by the last mesh.

To avoid electron losses by capture, paramount for the operation of the LAr-TPC is the degree of purity of the liquid. A substantial effort has been devoted to the identification and removal of sources of electro-negative pollutants, increasing the electron lifetime to several ms, well above the maximum drift time of large detectors (Buckley *et al.*, 1989; Bettini *et al.*, 1991).

In the framework of the ICARUS experiment, implemented in the underground Italian Gran Sasso National Laboratory (LNGS), increasingly large LAr-TPC prototypes have been built and operated, demonstrating lossless charged particle tracking up to 140 cm of drift length (Arneodo *et al.*, 2000). Commissioned in 2004, the ICARUS T600 detector has 500 tons of sensitive volume (Amerio *et al.*, 2004); its modular conception permits to increase the volume further. Figure 10.41 is an example of a cosmic ray shower detected with the device and demonstrates its multi-track resolution.

Simultaneous detection of the argon scintillation with photomultipliers can be exploited to provide the time of the event, and serve as trigger for the data acquisition (Cennini *et al.*, 1999).

Other developments of cryogenic and dual-phase detectors are described in Section 15.2.

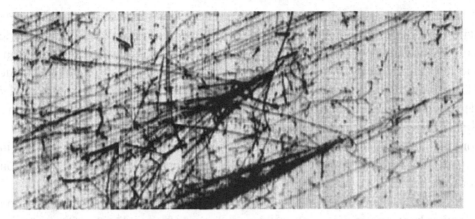

Figure 10.41   Cosmic ray shower recorded with the ICARUS T600 LAr-TPC (Amerio *et al.*, 2004).

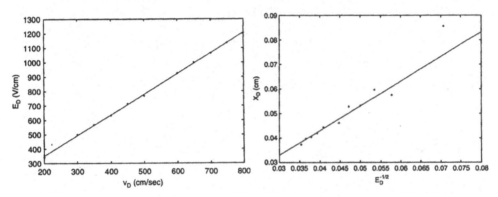

Figure 10.42   Drift velocity (left) and longitudinal diffusion (right) for an 80 cm drift for CS_2^- ions as a function of electric field (Martoff *et al.*, 2005).

10.10 Negative ion TPC

As discussed in the previous sections, the space resolution of drift chambers, and of TPCs in particular, is limited by the diffusion of electrons during their drift. While the transverse diffusion is reduced by a strong magnetic field parallel to the electric field, the longitudinal diffusion, determining the localization accuracy in the drift direction, is not affected. It has been proposed (Martoff *et al.*, 2000) to exploit the electron capture properties of some molecules to transform the drifting species from electrons to negative ions, having intrinsically much smaller diffusion (see Chapter 4). Detection relies on the observation that on arrival in the high field of the multiplying end, negative ions can be stripped of their captured electron and give origin to a charge-amplifying avalanche.

The drift properties of negative ions have been investigated with a small TPC with pure carbon disulfide and Ar-CS$_2$ gas fillings, both at low and atmospheric pressures (Ohnuki *et al.*, 2001; Martoff *et al.*, 2005). Electrons are extracted from the drift electrode by an intense flash of UV light and quickly captured by the electro-negative gas to form negative ions; the arrival time and shape of the delayed pulse at the anodes are analysed to deduce the drift velocity and longitudinal diffusion coefficient. Results of the measurements of drift velocity and longitudinal diffusion for an 8 cm drift in a He-CS$_2$ mixture are shown in Figure 10.42 (Martoff *et al.*, 2005); values are consistent with the expectation for ions.

Potentially interesting to achieve very good space resolutions, a negative ion chamber has a major drawback in the very long drift time, limiting its use to the detection of rare events as, for example, in dark matter and WIMPs research (Alner *et al.*, 2004).

11

Multi-tube arrays

11.1 Limited streamer tubes

Despite their successful use in many experimental setups, gaseous detectors with multi-wire geometry potentially suffer from a reliability problem, since the accidental rupture of a single wire may cause the malfunctioning of a large part if not all of the system. Modular designs, as used for jet chambers, may limit the extent of the damage; nevertheless, the failure of one sector affects the whole experiment. Various technologies have been devised to repair a damaged detector; in general they require the dismounting of the setup, which is particularly difficult if the device is embedded in other components. This led to the development of detectors with each anode enclosed in a box or cylindrical tube protecting the surrounding elements from a local failure; one of them, the drift tube, is described in Chapter 9.

Originally developed as a way to limit the photon- and electron-mediated propagation of avalanches between adjacent wires in electromagnetic calorimeters, the cube lattice MWPC had each anode wire enclosed in an array of cathodes of rectangular cross section (Battistoni *et al.*, 1979b); the coordinates along the wire are measured by recording the induced charge profile on strips perpendicular to the anodes (Battistoni *et al.*, 1978). The early design made use of U-shaped aluminium profiles for the cathodes, with the open side covered by a resistive electrode to permit detection of the induced signals. Further development led to the use of extruded plastic profiles with controlled resistivity, with readouts of both coordinates on external sensing strips.

In systematic studies of performances with single-wire tubes, very large signals were observed using thick anode wires and large fractions of hydrocarbons as gas fillings; their origin was attributed to a transition from proportional avalanche to a streamer, limited in extension towards the cathode by the decreasing field in a process analogous to the self-quenching streamer mechanism described in Section 7.8 (Battistoni *et al.*, 1979a). A comparison of counting characteristics

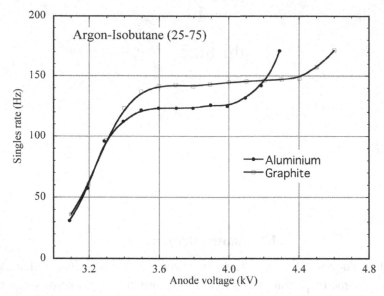

Figure 11.1   Comparison of singles counting rates for streamer tubes with graphite and aluminium cathodes (Battistoni *et al.*, 1983). By kind permission of Elsevier.

measured with metallic and resistive cathodes provides longer efficiency plateaux with the second option, probably due to the suppression of photon-mediated electron extraction from the cathodes, Figure 11.1 (Battistoni *et al.*, 1983).

Named plastic streamer tubes (PST) and built using large plates of extruded plastic profiles with resistive coatings, the detectors have been massively used in low- and medium-rate experiments requiring very large detection areas, as the NUSEX proton decay experiment under Mont Blanc (Battistoni *et al.*, 1986), MACRO at Gran Sasso (Ambrosio *et al.*, 2002), UA1 (Bauers *et al.*, 1987) and DELPHI experiments at CERN (Golovatyuk *et al.*, 1985).

Figure 11.2 shows schematically the construction of a PST module for the Mont Blanc experiment, with two-dimensional readout on external strips, and Figure 11.3 is a detailed view of the detector (Iarocci, 1983). A recording of charge profiles induced on external strips or pads is exploited to achieve localization; requiring rather inexpensive readout electronics, owing to the large signals obtained in the limited streamer operation, the tubes provide sub-mm localization accuracies in the direction of the anode wires. The amount and space distribution of the signals depend on detector geometry, cathode resistivity, distributed capacitance and amplifiers time constants; Figure 11.4 is a measurement of the fractional induced charge, or cathode transparency, as a function of the electrode surface resistivity (Battistoni *et al.*, 1982).

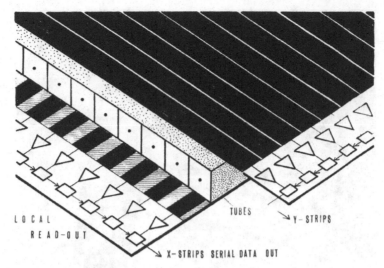

LOCAL
READ-OUT

TUBES

Y-STRIPS

X-STRIPS SERIAL DATA OUT

Figure 11.2   Schematics of a plastic streamer module (Iarocci, 1983). By kind permission of Elsevier.

While the transparency improves with the increase in resistivity, the rate capability of the detector can be affected by the accumulation of surface charges, modifying the field. Figure 11.5 is an example of charge measured as a function of rate, given in counts per cm of anode wire, for a detector with a cathode resistivity of about 50 kΩ/square (Fujii *et al.*, 1984); the gain drops above a particle flux of a few KHz per cm.

Many authors have analysed the process of charge induction through resistive electrodes (Battistoni *et al.*, 1978; Golovatyuk *et al.*, 1985; Fujimoto *et al.*, 1986); the subject is be covered in more details in Chapter 12, which describes the development of resistive plate chambers.

11.2 Drift tubes

Named drift tubes, or straws in their lighter version, arrays of individual cylindrical proportional counters with metallic or low-resistivity cathodes solve the problem of both reliability encountered with multi-wire systems, and the rate limitations of resistive streamer tubes. Assembled in several staggered layers, as shown schematically in Figure 11.6, and with the measurement of drift time on each tube, they provide ambiguity-free reconstruction of tracks with good accuracy.

An early prototype, consisting of two planes of staggered proportional counter tubes operated in the proportional amplification mode, was built in the early eighties using 0.5 mm thick aluminium tubes, 1 m long and 3 mm in diameter, and tested up to a flux above 10^6 particles/s and mm of wire (Hammarström *et al.*,

Figure 11.3   Detailed view of one end of the PST (Iarocci, 1983). By kind permission of Elsevier.

1980b). The technology evolved over the years, with the construction of detectors deploying thousand of tubes, usually assembled in self-supporting modules. For applications as muon detectors, where multiple scattering is not a major concern, both the tube walls and the supporting structure can be rather sturdy; this permits one to operate the tubes at pressures higher than atmospheric, improving the localization accuracy.

To exploit the intrinsic good localization properties of drift tubes, particularly for large systems, thorough position, gain and space–time correlation monitoring are essential, suggesting the name monitored drift tubes (MDT) for the detector (Biscossa *et al.*, 1999). Built in large quantity for CERN's ATLAS barrel and end-cap muon spectrometers, the tubes have cathodes made of 400 μm thick aluminium tubes, 3 cm in diameter, and 50 μm anode wires held in place by

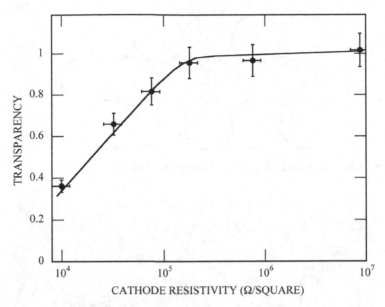

Figure 11.4 Cathode electrical transparency as a function of surface resistivity for 1 cm wide pickup strips (Battistoni *et al.*, 1982). By kind permission of Elsevier.

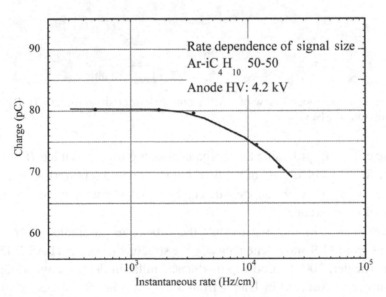

Figure 11.5 PST signal charge dependence from particles rate (Fujii *et al.*, 1984). By kind permission of Elsevier.

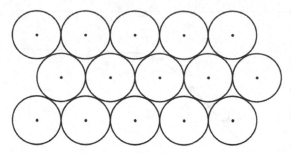

Figure 11.6   Schematic of a stack of staggered drift tubes.

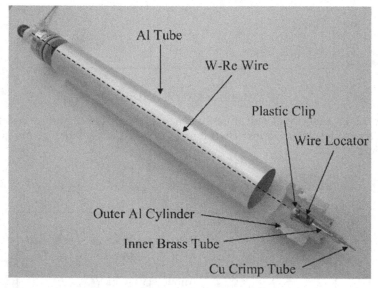

Figure 11.7   Exploded view of a MDT counter (Adorisio *et al.*, 2007). By kind permission of Elsevier.

insulating end-plugs providing the mechanical strength and gas inlets (Figure 11.7); Figure 11.8 is a detailed view of the wire-holding end-plug (Adorisio *et al.*, 2007). In the event of a failure, the anode wire can be replaced with a procedure described in the previous reference.

For installation in the experiment, the tubes are assembled in multi-layer modules; Figure 11.9 show schematically the structure of one ATLAS MDT barrel chamber (Riegler, 2002a). End-cap modules, similar in design, have a trapezoidal shape with tubes varying in length from 1.3 to 6.3 m (Bensinger *et al.*, 2002; Borisov *et al.*, 2002).

During construction, the absolute position of each anode wire is certified with 20 μm accuracy with respect to reference marks by a system of X-ray tomography

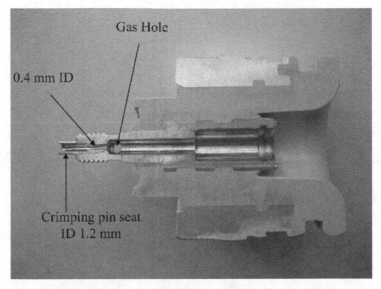

Figure 11.8   Detailed view of the MDT wire holder end-plug (Adorisio *et al.*, 2007). By kind permission of Elsevier.

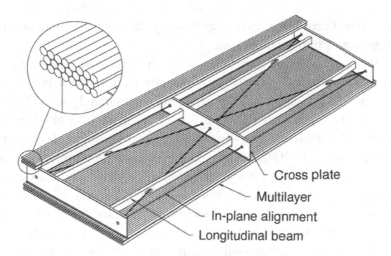

Figure 11.9   MDT module for the ATLAS barrel muon spectrometer (Riegler, 2002a). By kind permission of Elsevier.

(Schuh *et al.*, 2004); optical and capacitive alignment systems, complemented by calibrations on cosmic rays, permit one to reduce the system position errors to a few tens of μm (Cerutti, 2004; Dubbert *et al.*, 2007a). The picture in Figure 11.10 shows a module and monitoring system during assembly. About 1200 modules, for

Figure 11.10 Construction and alignment of a MDT module. Picture CERN (2004).

a total of 370 000 drift tubes, have been built by different collaborating institutes and are operational in the experiment.

While the best position accuracies could be obtained with a mixture of $Ar/N_2/CH_4$, owing to its saturated drift velocity over most of the drift length (Riegler *et al.*, 2000), due to ageing problems encountered the ATLAS MDTs are operated with argon-CO_2 (93–7) at 3 bars, resulting in a more pronounced non-linearity of the space–time correlation and a moderate loss in position accuracy (Aleksa *et al.*, 2002). However, the space–time correlation is affected by the counting rate, due to the accumulation of positive ions and the resulting space–charge field distortion, a process discussed in Section 7.7. The measured position accuracy of the MDT as a function of distance from the anode, both for low and high rates (about 1.4 kHz/cm of wire), are compared in Figure 11.11; the lines correspond to a fit with a model calculation (Aleksa *et al.*, 2002).

Mounted on the coils of the toroidal magnets of the spectrometer, the MDTs operate in a stray magnetic field up to 0.4 T; the position-dependent distortions introduced in the space–time correlation have been thoroughly modelled to achieve the required tracking accuracy (Dubbert *et al.*, 2007b).

An improved design of the MDT, with a tube diameter reduced from 30 to 15 mm, has been developed to cope with the higher counting rates expected after the upgrade of CERN's LHC (Bittner *et al.*, 2011).

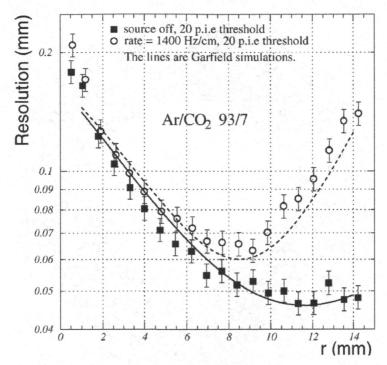

Figure 11.11   MDT space resolution as a function of distance from the wire for two source rates (Aleksa *et al.*, 2002). By kind permission of Elsevier.

11.3 Straw tubes

A limitation for the use of the drift tubes described in the previous section lies in the thickness of the cathodes; acceptable for high-energy muon detection, it causes a deterioration of tracking accuracy for lower momentum particles. The development of techniques capable of manufacturing single-wire counters with thin plastic walls, the so-called straws, permits the realization of large arrays of light detectors, particularly suited for the construction of cylindrical trackers in colliding beam setups (Baringer *et al.*, 1987). Figure 11.12 is a schematic cross section of one of the first large arrays of straw drift chambers for the MARK II detector, including 552 single-wire tubes; made with thin polymer cathodes, metal-coated on the inner side, the straws typically have a thickness of 50 to 100 μm, corresponding to a few tens of a per cent of radiation length (Ford *et al.*, 1987). Planar assemblies of straw arrays, usually glued together in modules to improve mechanical stability, have been used in many experiments (Arai *et al.*, 1996; Armstrong *et al.*, 1999; Ogren, 1995; Fourletov, 2004; Bychkov *et al.*, 2006; Bachmann *et al.*, 2004).

Several technologies have been developed to permit the industrial manufacturing of very large quantities of detectors. Figure 11.13 shows the method used for

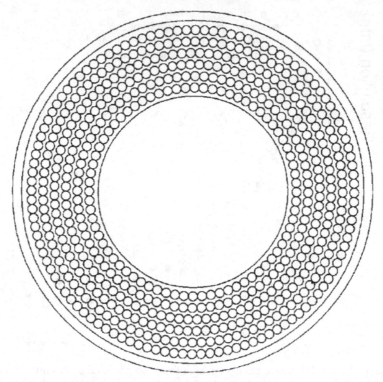

Figure 11.12   Schematic cross section of the MARK II cylindrical straw detector at PEP (Ford *et al.*, 1987). By kind permission of Elsevier.

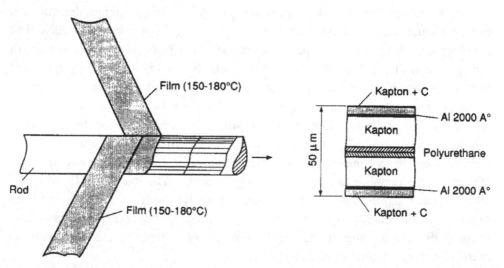

Figure 11.13   Straw manufacturing method, winding two polymer films around a core (Akesson *et al.*, 2004a). By kind permission of Elsevier.

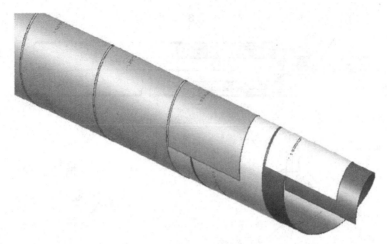

Figure 11.14   Two-layer structure of the straws for COMPASS (Bychkov *et al.*, 2006). By kind permission of Elsevier.

the straws of the ATLAS Transition Radiation Tracker (Akesson *et al.*, 2004a): two thin polymer films are wound in spirals on a precisely tooled mandrel and bonded together at high temperature with a thermoplastic polyurethane layer. On the inner and outer sides, the Kapton sheets have a thin aluminium coating and a thicker carbon-loaded polymer foil. This provides a certain degree of protection against discharges, and avoids the spark damages that may result from the use of a thin conductor as cathode; the added conductor improves the signal transmission along the tube. Illustrated in Figure 11.14, the straws of the forward tracker of the COMPASS spectrometer at CERN are produced with a similar technology, gluing together a 40 μm thick inner sheet of carbon-loaded polymer and a 12 μm thick aluminized polymer. At the two ends of the straw, insulating plugs, shown in Figure 11.15, provide the electrical connections and the gas distribution (Bychkov *et al.*, 2006). More than 12 000 straws, up to 320 cm in length and arranged in planar multi-layer modules, have been built and operated.

Owing to their light construction, straws are rather sensitive to humidity and temperature variations; for long tubes, regularly spaced internal supports depicted in the figure stabilize the anode wire against displacements due to mechanical or thermal deformations (Bychkov *et al.*, 2006).

The ATLAS straw-based transition radiation tracker (TRT) at CERN's LHC consists in three modules, a cylindrical barrel surrounding the central region and two forward wheels. The barrel TRT has three concentric rings, each with 32 identical straw tubes 144 cm in length and 4 mm in diameter; the two end-cap detectors are assembled in 20 wheels with short (37 cm) radially-oriented straws (Akesson *et al.*, 2004a; Martin, 2007). Embedded in foil or fibre radiators and filled with a xenon-rich gas mixture, the TRT serves the purpose of both high-resolution tracker

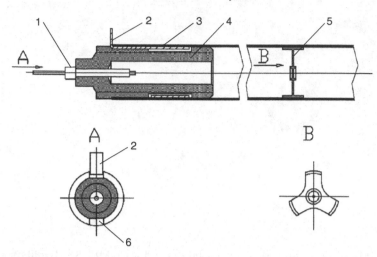

Figure 11.15   Schematics of a straw's end-plug (Bychkov *et al.*, 2006). By kind permission of Elsevier.

Figure 11.16   The ATLAS barrel TRT. Picture CERN (2005).

and particle identification through the detection of soft X-rays generated in the radiator by fast electrons. To improve the rate capability, anode wires are split in the middle and held together with a glass joint and are read out on both sides. Figure 11.16 and Figure 11.17 show respectively the completed barrel and end-cap TRT detectors before installation in the experiment.

Owing to the faster ion collection, due to the small diameter and higher field, straw tubes have a high intrinsic rate capability; Figure 11.18 is a measurement of relative gain on soft X-rays for a 4 mm diameter counter, and drift time resolution

Figure 11.17 The two assembled end-cap ATLAS TRT detectors. Picture CERN (2005).

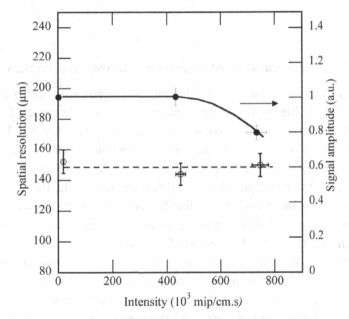

Figure 11.18 Relative gain and position accuracy as a function of particle flux (Akesson *et al.*, 1995). By kind permission of Elsevier.

for charged particles as a function of flux; the performances are unaffected up to a rate around 5.10^5 Hz/cm of wire (Akesson *et al.*, 1995).

As most proportional counters, straw tubes can be operated in a wide choice of filling gases; hydrocarbon mixtures are generally avoided for safety reasons and to avoid the formation of polymers. While the straws' rate capability is limited to a few tens of kHz/mm of wire by the basic space charge processes discussed in Section 7.7, the higher granularity favours their use in high rate experiments if particular care is taken to avoid ageing problems.

Due to the manufacturing process, straws are particularly prone to suffer from the adverse effects of pollutants released by oily residues; addition of small amounts of carbon tetrafluoride can considerably extend the lifetime of the detector, thanks to its etching properties. In laboratory tests, integrated charges up to several C/cm have been reached in argon-CO_2-CF_4 without deterioration of performances (Bachmann *et al.*, 2004). However, attempts to exploit the cleaning effect of carbon tetrafluoride in the ATLAS TRT have been frustrated by the chemical aggressiveness on materials of the fluorinated compound released in the avalanches, particularly in the presence of moisture, and the use of CF_4 has been discontinued in favour of the milder cleaning action obtained with the addition of a few per cent of oxygen (Akesson *et al.*, 2002; Akesson *et al.*, 2004b). The subject of ageing and radiation damage is covered extensively in Chapter 16.

11.4 Mechanical construction and electrostatic stability

In an ideal cylindrical counter, the anode wire is in equilibrium under the effect of electrostatic forces due to the applied voltage. However, because of positioning errors, straw distortions and gravitational sag, the wire may not be perfectly centred and can move to a new equilibrium position, under the effect of the asymmetric electrostatic forces contrasted by its mechanical tension, with a maximum deflection towards the cathode in the centre. Aside from affecting the localization, the displacements reduce the difference between the voltage required for operation and the breakdown point, as shown by the measurements with several gas fillings in Figure 11.19 (Akesson *et al.*, 2004b).

Many authors have computed expressions for the conditions of wire stability under electrostatic forces, similarly to those developed for multi-wire chambers. The electrostatic force per unit length due to an offset of the wire from the centre by a quantity δ is given in MKS units by the expression (Carr and Kagan, 1986; Oh *et al.*, 1991; Akesson *et al.*, 1995):

$$F = \frac{2\pi\varepsilon_0 V^2 \delta}{R^2 (\ln(R/r))^2},$$
(11.1)

where V, R and r are the applied voltage, cathode and wire radius respectively. For equilibrium, the electrostatic force must be equal to or smaller than the

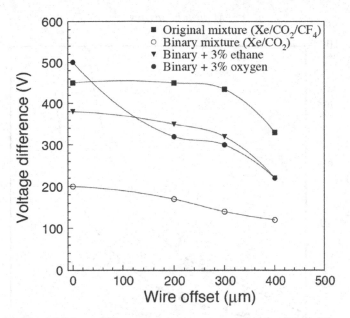

Figure 11.19 Voltage difference between operation and breakdowns as a function of wire offset in the ATLAS TRT (Akesson *et al.*, 2004b). By kind permission of Elsevier.

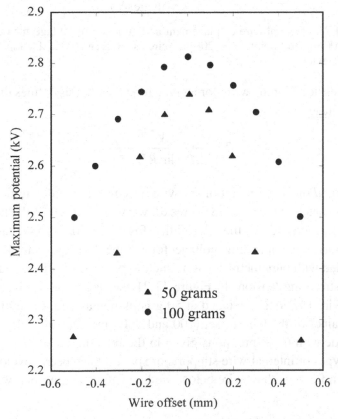

Figure 11.20 Maximum operating voltage as a function of wire offset and two values of tension (Akesson *et al.*, 1995). By kind permission of Elsevier.

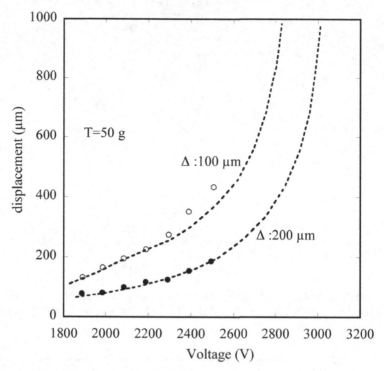

Figure 11.21 Central wire displacement as a function of voltage for two initial offsets, 100 μm (full points) and 200 μm (circles) (Ogren, 1995). By kind permission of Elsevier.

mechanical tension T of the wire; for a given tube length, this defines the maximum operating voltage:

$$T \geq \frac{2\varepsilon_0 V^2 L^2}{\pi R^2 (\ln(R/r))^2}.$$
(11.2)

The deflection d of the midpoint of the wire of length L from its initial position is given approximately by $d = L^2 F/8T$; breakdown will occur when the field increase due to the deflection exceeds the gas rigidity. Figure 11.20, from the last reference, shows the maximum operating voltage for a tube 1 metre long and 4 mm in diameter, filled with pure methane, as a function of the wire offset, and two values of the wire stretching tension. In Figure 11.21 the measured displacement in the centre of a wire 1 m long, stretched at 50 g is plotted as a function of voltage for two initial values of the wire offset (100 and 200 μm) (Ogren, 1995). A fit to the points is made with the expressions given in the reference.

Various types of internal wire supports or spacers have been developed to keep the maximum free length of wire below the limit of instability; one was shown in Figure 11.15.

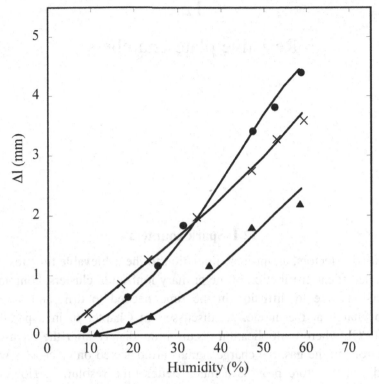

Figure 11.22   Length variations of three 3.2 m long straws as a function of humidity (Bychkov *et al.*, 2006). By kind permission of Elsevier.

The effects on the straws' mechanical properties on ambient variations, temperature and humidity have been analysed in detail; Figure 11.22 shows the elongation of 3.2 m straws made with different polymer materials as a function of relative humidity; the elongation of course affects the mechanical tension applied to the wire (Bychkov *et al.*, 2006). Similar distortions are generated by differences in the thermal expansion coefficients of the straws' components and supporting frames; this shows the importance of controlling the ambient conditions.

12

Resistive plate chambers

12.1 Spark counters

In wire-based detectors, an intrinsic limitation to the achievable time resolution is set by the statistical distribution of the primary ionization clusters, combined with the dispersions due to diffusion in the time needed to drift and amplify the ionization charge at the anode. As discussed in Chapter 6, in a parallel plate counter (PPC), where multiplication occurs immediately after the release of primary electrons in the gas, the charge signals are detected on electrodes without a delay and can therefore provide a much better time resolution. However, this requires detecting the induced signals at the very beginning of the avalanche development; good results can only be obtained with large multiplication factors, obtainable at very high fields. In a standard PPC with conductive electrodes this is a critical operation, seriously hindered by discharges and their propagation throughout the whole counter.

A structure capable of achieving the required high gains while limiting the discharge propagation, named the spark counter, employs one or two high-resistivity electrodes; the large current surge generated by an avalanche then results in a localized field drop, thus dumping the discharge (Pestov, 1982). Operation near to or in the discharge mode is then possible, and provides fast and large signals. Localization can be conveniently performed with external readout strips or pads in contact with the outer face of the resistive electrodes.

Use of a gas mixture designed to absorb photons in a wide energy range avoids propagation and ensures the locality of the discharge; Figure 12.1 is an example of the combined photon absorption coefficients for the mixture with the components and proportions indicated in the inset (Pestov *et al.*, 2000).

A narrow gap (100 to 300 μm), which requires high fields for multiplication, reduces the statistical fluctuations in the avalanche development, and provides the best time resolution; it imposes, however, constraints to the quality and uniformity

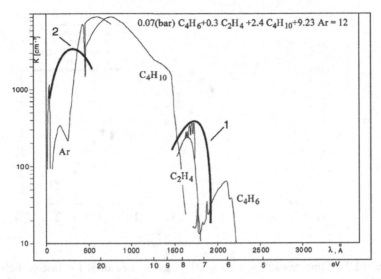

Figure 12.1 Combined photon absorption coefficient/cm of the gas mixture shown in the inset as a function of photon energy and wavelength (Pestov *et al.*, 2000). By kind permission of Elsevier.

of the electrodes. For detection of minimum ionizing particles, to compensate for the reduced efficiency due to the energy loss statistics in a narrow gap, the counter can be operated at high pressures.

By using as cathode a custom-made electron conducting glass with resistivity in the range 10^9 to $10^{10}\Omega$ cm (see Section 13.1), a 100 μm gap and a complex mixture of hydrocarbons at 12 bars, time resolutions down to 30 ps have been demonstrated in the early 1980s (Pestov, 1982); Figure 12.2 is the outcome of systematic time resolution measurements achieved with different gas mixtures; the horizontal scale is the ratio of the operating to the multiplication threshold voltages.

Electrostatic properties of spark counters as well as the signal propagation in transmission lines on resistive supports have been studied in detail, aiming at the construction of large size detectors (Steinhaeuser *et al.*, 1997). Reviews of the technology and its applications are given in Schmidt (1999) and Pestov (2002); the second reference includes a critical comparison of performances of resistive plate detectors built with a range of electrode materials and in different modes of operation, from avalanche to spark.

The difficult procurement of the electron-conducting glass and the high pressures of operation, requiring heavy gas containment vessels, have, however, discouraged the use of the technology; nevertheless the main findings of the development constituted the basis for further developments of parallel plate counters. The electron-conducting glass (also named Pestov glass) has been also used successfully in the development of high-rate microstrip chambers, described in Section 13.1.

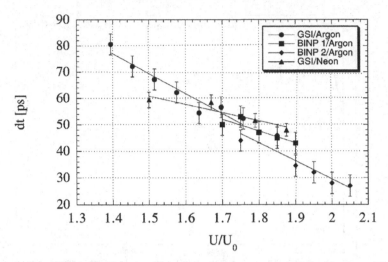

Figure 12.2   Time resolution of a spark counter measured for fast particles in several beam runs and gas mixtures, as a function of relative over-voltage (Pestov *et al.*, 2000). By kind permission of Elsevier.

12.2 Resistive plate counters (RPCs)

Aiming at the realization of very large detection areas, the resistive plate counters (RPC) employ high-resistivity electrodes made of phenolic polymer laminates, which are cheap and easily available, and have a volume resistivity between 10^9 and $10^{10} \Omega$ cm.[1] With sensitive gas gaps a few mm thick, RPCs achieve full detection efficiency and nanosecond time resolutions for fast particles (Santonico and Cardarelli, 1981; Cardarelli *et al.*, 1988). Further developments have permitted one to consolidate the technology and build the very large systems, hundred of square meters in area, needed for muon detection in high-energy physics experiments.

Figure 12.3 is a schematic cross section of a standard single-gap RPC. The high resistivity electrode plates are assembled with insulating support frames, providing tightness and gas distribution; for large detector sizes, regularly spaced insulating pillars between the electrodes ensure the uniformity of the gap. To provide the operating voltage over large detection areas and avoid the potential drops caused by the signal currents, the plates are coated on the outer side with a thin graphite layer having a moderate surface resistivity, around 200–300 kΩ/square[2] that

---

[1] Common trade names: Bakelite, Melamine; the value of the resistivity can be controlled with additives.

[2] The resistance of a uniformly conductive slab of thickness L between two faces of surface A is given by the expression $R = \rho L/A$, where ρ is the bulk resistivity (in Ω cm). For thin conductive sheets of thickness s, the sheet resistance is defined as $R_s = \rho/s$, usually expressed in Ω/square since its value is independent of the size of the square.

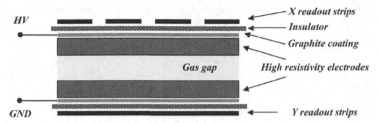

Figure 12.3 Schematics of a resistive plate chamber.

distributes the potentials while permitting detection of the signals induced by the avalanche multiplication on sets of external strips, insulated from the electrodes (see also the discussion on electrode transparency in Section 11.1).

On application of a voltage to the coatings, a migration of charges in the bulk of each electrode cancels the external field, making them equi-potential, so that the full field gradient is applied across the gas layer. As discussed in detail in Chapter 5, following the passage of an ionizing particle and under the effect of the electric field, starting from each primary ionization cluster charge-amplifying avalanches develop in the gas, and add up to build the entire signal; for a few mm thick gap, the process takes 10 ns or less. Due to the exponential avalanche growth, the largest fraction of the detected charge originates from the ionization electrons closer to the cathode; for a discussion of the signal formation process in parallel plate counters, see Section 6.4.

When the avalanche size exceeds 10^6–10^7 electrons, the exponential growth is damped due to the field reduction induced by the growing space charge; this leads to a saturated avalanche regime. On increasing the voltage further, a transition to a streamer may occur; this operating mode was exploited in the early developments, owing to the conveniently large signals obtained, but implies a reduction in the rate capability of the detector caused by the large local voltage drop, only slowly recovering through the high resistivity of the electrodes. A comparison of detection efficiency in the avalanche and streamer modes as a function of particle flux is shown in Figure 12.4 (Arnaldi *et al.*, 1999); in the streamer mode the maximum flux for efficient detection is reduced by several orders of magnitude.

The rate dependence of detection efficiency can be described by a simple model that takes into account only the local drop of effective potential due to the avalanche currents; Figure 12.5 compares the results of the simulation with experimental measurements realized on a standard bakelite RPC with 2 mm gap and bulk resistivity $8 \times 10^{10} \Omega$ cm (Abbrescia, 2004).

To achieve high and stable gains, RPCs are operated with heavily quenched, photon-absorbing gases, as freon (CF_3Br) or mixtures of tetrafluoroethane ($C_2H_2F_4$) and isobutane (i-C_4H_{10}) (Cardarelli *et al.*, 1993; Cardarelli *et al.*, 1996); heavily

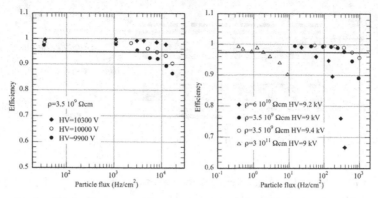

Figure 12.4   RPC efficiency as a function of particle flux and voltage in the avalanche (left) and streamer modes (right) for several values of bulk resistivity (Arnaldi *et al.*, 1999). By kind permission of Elsevier.

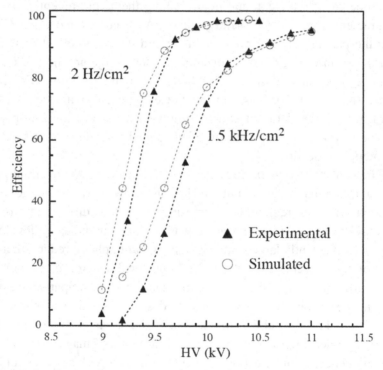

Figure 12.5   Simulated and measured efficiency for minimum ionizing particles at two values of flux (Abbrescia, 2004). By kind permission of Elsevier.

quenched argon-based mixtures are used to reduce the voltage needed for operation (Wang, 2003; Bergnoli *et al.*, 2009; Zhang *et al.*, 2010).

While the streamer mode offers a substantial advantage in terms of signal charge, for many applications an operation in the saturated avalanche mode is

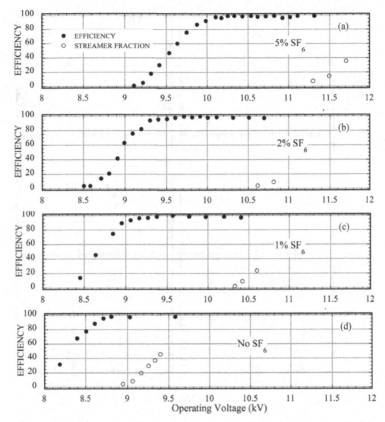

Figure 12.6 Efficiency plateau and streamer onset probability as a function of voltage for several percentages of SF$_6$ addition to the gas (Camarri *et al.*, 1998). By kind permission of Elsevier.

preferred to avoid efficiency losses in the high flux areas of the detector. In this case, owing to the inevitable differences in local gain due to mechanical tolerances, it is mandatory to ensure that the onset of the streamer mode appears at a voltage sufficiently above the one required for normal operation.

The influence of the gas filling on the length of the plateau in the avalanche mode before the streamer has been studied extensively. Addition of small percentage of an electro-negative gas to the mixture considerably increases the gap between the avalanche and streamer modes, as shown in Figure 12.6 for several percentages of sulphur esafluoride (SF$_6$) added to a C$_2$H$_2$F$_4$-C$_4$H$_{10}$ mixture (Camarri *et al.*, 1998).

The streamer suppression effect of the addition of small percentages of SF$_6$ has been confirmed by many systematic studies, motivated by the development of reliable RPCs for massive use as muon detectors in large experiments. Figure 12.7

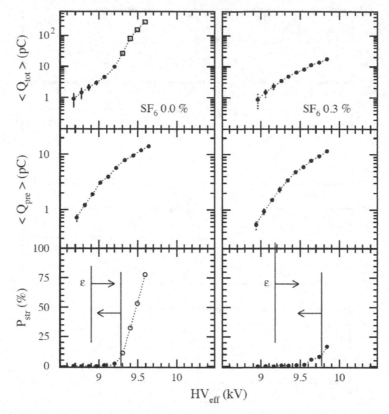

Figure 12.7   Average (top) and streamer precursors total charge (middle) as a function of voltage without and with SF_6 in the gas mixture. The bottom plots show the streamer onset probability (Ahn *et al.*, 2004). By kind permission of Elsevier.

is a representative example (Ahn *et al.*, 2004); this study includes a comparison of the experimental results with a simple avalanche multiplication model. As already mentioned, addition of the quenchers to argon in various proportions permits one to lower the operating voltage, with obvious practical advantages.

   The use of cheap phenolic resin electrodes, while permitting the realization of very large detectors, is hindered by the modest surface quality of the material, thus inducing local spontaneous discharges and leading to high background noise. While several methods of surface treatment have been investigated, a simple solution to the problem was found at the early stage of development by wetting the electrodes with linseed oil dissolved in alcohol, having the property to form a thin, high resistivity polymer coating when drying, covering the surface imperfections (Santonico and Cardarelli, 1981). The considerable reduction in noise rate that can be obtained with the linseed oil conditioning is shown by the comparison

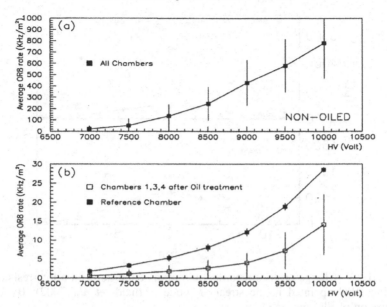

Figure 12.8   Noise rate comparison between non-oiled (a) and two linseed oil treated RPCs (Abbrescia *et al.*, 1997). By kind permission of Elsevier.

in Figure 12.8 (Abbrescia *et al.*, 1997). Despite serious problems encountered in BaBar at SLAC, one of the first experiments making large use of linseed-coated bakelite RPCs, and related to incomplete polymerization, formation of droplets on the surface and on the internal pillars used as gap-restoring spacers (Anulli *et al.*, 2002), with various improvements the technology has been successfully adopted in a majority of RPC-based experimental setups. A long-term decrease in rate capability during operation, imputed to a gradual increase of the electrodes' resistivity with exposure to radiation and/or moisture has also been observed. More detailed studies of long-term degradation (ageing) of RPCs are discussed in Chapter 16.

For a comparison of performances between detectors manufactured with several types of bakelite laminates, with and without the linseed oil conditioning, see for example Biswas *et al.* (2009). Essential to guarantee stable long-term operation of the detectors, the temperature and humidity dependences of the materials' bulk resistivity have been studied systematically by many authors (Ahn *et al.*, 2000; Aielli *et al.*, 2004; Aielli *et al.*, 2006; Doroud *et al.*, 2009b; Moshaii and Doroud, 2009).

RPCs have been produced industrially in massive quantities for use as muon detectors in many large particle physics experiments and are amply described in the literature: BaBar at SLAC (Ferroni, 2009), OPERA (Bertolin *et al.*, 2009), CMS (Colaleo *et al.*, 2009), ATLAS and ALICE at CERN (Bindi, 2012; Arnaldi *et al.*, 2009), BESIII at the Beijing Electron-Positron Collider (Zhang *et al.*, 2010)

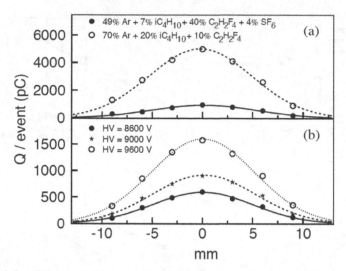

Figure 12.9   Charge profiles recorded in various conditions on a low-resistivity bakelite RPC operated in the streamer mode (Arnaldi *et al.*, 2002). By kind permission of Elsevier.

and many others. The ARGO-YBJ RPC array in Tibet, aimed at the detection of extended cosmic air showers, deploys about 2000 RPC modules covering a sensitive area of 6700 m^2 (Camarri, 2009). Performances and long-term operating stability are described in a vast number of works; good sources of information are the proceedings of regular International Resistive Plate Chambers International Workshops (RPC2005, 2006; RPC2007, 2009) and numerous reviews (Santonico, 2009; Sharma, 2009; Santonico, 2012; Sharma, 2012).

The space distribution of the signals induced on the external electrodes depends on the detector geometry, thickness of the electrodes and operating conditions. Since the induction process is very fast compared to the local RC of the detectors, the resistivity of the electrode has only a small effect on the width, as it can be inferred by simple electrostatics consideration and confirmed by measurements (Arnaldi *et al.*, 2000). Figure 12.9 is an example of charge profiles measured with a low resistivity ($\sim 10^9 \Omega$ cm) bakelite RPC for various gas mixtures and voltages; a Gaussian fit to the distributions gives a standard deviation of about 5 mm, close to the distance between the avalanche maximum in the detector and the pickup strips (Arnaldi *et al.*, 2002). This result is reminiscent of similar observations with MWPCs (see the discussion on the pad response function in Section 10.4).

Recording the charge induced through the electrodes on readout strips or pads provides the time of the event and the space coordinates of tracks crossing the detector; in large area devices, the strip width is typically a few cm, to reduce the number of readout channels and the sharing of charge between adjacent strips.

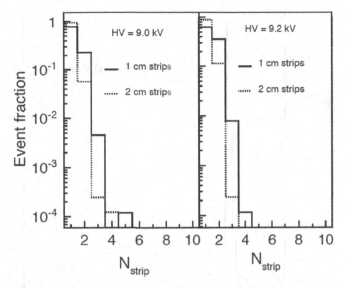

Figure 12.10 Distribution of the number of fired strips for 1 and 2 cm width (Arnaldi *et al.*, 2002). By kind permission of Elsevier.

Figure 12.10, from the previous reference, shows the measured distributions of the number of strips above threshold (or cluster size) for two strip widths and two values of operating voltage; for 1 cm wide strips, the probability of double hit events is about 20%, slightly increasing with voltage.

For a digital readout, the localization accuracy for single strip hit is simply given by $w/\sqrt{12}$, where w is the strip pitch; in the case of charge sharing between adjacent strips, or for the wider clusters induced by inclined tracks, the localization accuracy is improved by making suitable averages (Carrillo, 2012).

The time resolutions achievable with RPCs depend, amongst other factors, on their operating voltage and gap thickness; measured with a low resistivity detector ($\sim 5 \ 10^9$ Ωcm) in the avalanche mode, Figure 12.11 is a representative example, with better than 1 ns standard deviation (Arnaldi *et al.*, 1999). For the large RPC-based systems built for experiments at CERN's LHC, a few ns resolution is sufficient to resolve two beam crossings at 25 ns. Better time resolutions, well below 100 ps and suitable for particle identification, are obtained with the narrow gap RPCs discussed in the next section.

12.3 Glass RPCs

Alternatively to the use of phenolic laminates as resistive electrodes, and close to the original development of spark counters described in Section 12.1, a concurrent approach uses plates of float glass as electrodes, available in a wide range of sizes

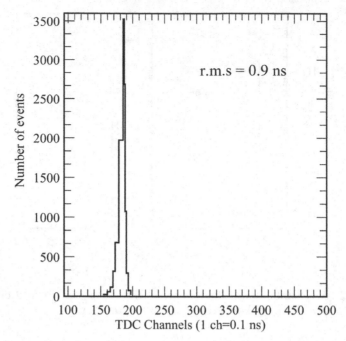

Figure 12.11 Time resolution measured for fast particles of a low resistivity, 2 mm gap RPC (Arnaldi *et al.*, 1999). By kind permission of Elsevier.

and thickness with a resistivity of 10^{11}–10^{12} Ω cm; the operating voltage is also provided by a lower resistivity layer, typically of about 1 MΩ/square deposited on the outer surface of the glass (Anelli *et al.*, 1991; Gustavino *et al.*, 2001; Trinchero *et al.*, 2003). As for the higher conductivity materials, at the application of the external electric field, free charges within the glass (presumably ions) migrate and rearrange, thus creating an internal field opposing the applied field and making the plates equi-potential.

A clear advantage of glass, compared to Bakelite, is its very good surface smoothness, therefore not requiring special conditioning; the drawback, due to the high resistivity, is a limited rate capability. This is, however, compensated by a more stable operation with time (Calcaterra *et al.*, 2004; Candela *et al.*, 2004). High resistivity glass is also used for the electrodes of multi-gap RPCs, discussed in the following section.

In the large glass RPC system built for the BELLE experiment at KEK, commercially available, 2.4 mm thick float glass plates were used, with bulk resistivity in the range 10^{12}–10^{13} Ω cm, coated with a custom-made India ink layer having a surface resistivity of 10^6–10^7 Ω/square to distribute the high voltage (Wang, 2003). Similar detectors have been built for DAΦNE at the Frascati INFN

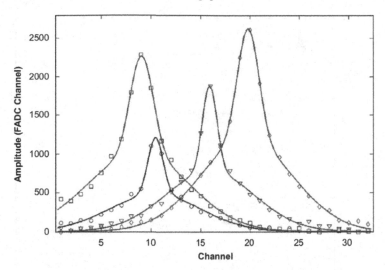

Figure 12.12   Charge profiles for several events recorded on 1 mm wide strips (Ye *et al.*, 2008). By kind permission of Elsevier.

laboratory (Calcaterra *et al.*, 2004), for a digital calorimeter of the International Linear Collider (Bedjidian *et al.*, 2010).

Sub-mm localization accuracies can be achieved with the use of thin electrodes and small pitch readout strips; Figure 12.12 is an example of charge profiles recorded on a 1 mm wide adjacent strips on a detector made with 700 μm thick glass and 2 mm gas gap. The characteristic double Gaussian shape of the distributions is attributed to a narrow signal induced by the avalanches, adding to a wider dispersion due to the low-resistivity external carbon layer used to distribute the voltage; a fit to the narrower distribution provides a position accuracy around 500 μm (Ye *et al.*, 2008).

The use of a custom-made lower resistivity glass ($\sim10^{10}$ Ω cm) permits one to preserve detection efficiency at the higher rates foreseen for the LHC upgrade; Figure 12.13 is a comparison of measured detection efficiency as a function of particle flux for standard and low resistivity glass detectors (Haddad *et al.*, 2013).

Surface damage to the electrodes due to a reaction of water with fluorinated compounds released in the avalanche processes have been experienced in the BELLE experiment at KEK (T. Kubo *et al.*, 2003); more recent observations of the damaging effect of carbon tetrafluoride on glass are discussed in Chapter 16.

12.4 Multi-gap RPCs

In RPCs operated at atmospheric pressure, a few mm thick gas gaps are needed to ensure full detection efficiency for minimum ionizing particles. With an average

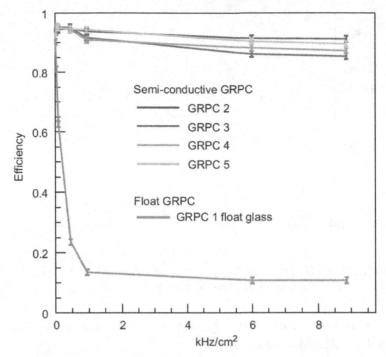

Figure 12.13 Comparative efficiency measured as a function of flux for standard float glass and several lower resistivity devices (Haddad *et al.*, 2013). By kind permission of Elsevier.

distance between primary ionizing collisions of around 300 μm, and since the avalanche size depends on the distance of the ionization clusters from the anode, the ensuing fluctuations limit the intrinsic time resolution of the detectors to a few ns. A substantial improvement in time resolution is achieved by reducing the gap thickness; however, this goes at the expense of the detection efficiency (Zeballos *et al.*, 1996).

The multi-gap RPCs (MRPCs), combining in a single device a stack of interconnected and narrow gap modules, permits one to achieve at the same time good efficiency and time resolution (Cerron-Zeballos *et al.*, 1996; Fonte *et al.*, 2000). Initially manufactured with individually powered resistive plates, the concept evolved with the discovery that, on application of the high voltage only to the external plates in a multi-gap stack, the intermediate floating electrodes reach the appropriate equilibrium potential dynamically and do not need to be connected, as shown schematically in Figure 12.14; this greatly simplifies the detector construction (Akindinov *et al.*, 2000). For large sizes, thin fishing lines stretched across the gaps ensure gain uniformity and prevent the deformations due to electrostatic forces.

Each gap then operates independently; avalanches develop following the release of ionization, and equal and opposite polarity signals are induced through the stack

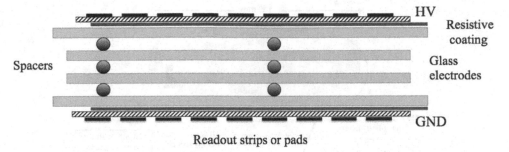

Figure 12.14   Schematics of a five-gap MRPC. The operating voltage is applied to resistive layers applied on the outer plates; signals are detected on external insulated circuits.

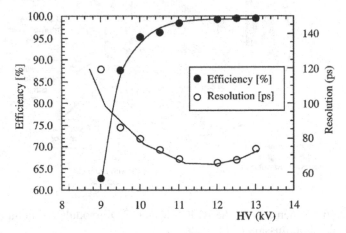

Figure 12.15   Efficiency and time resolution as a function of voltage for a five-gap MRPC (Akindinov *et al.*, 2000). By kind permission of Elsevier.

on the external electrodes; the larger and faster signals determine the time resolution of the structure. Figure 12.15 shows the efficiency and time resolution measured for fast particles as a function of voltage with a five 220 μm thick gaps MRPC built on thin high-resistivity glass plates; the gas filling is a mixture of 90% $C_2F_4H_2$, 5% i-C_4H_{10} and 5% CF_4 (Akindinov *et al.*, 2000).[3]

While limited in rate due to the very high resistivity of the glass electrodes, the structure reacts to the creation of regions of high space charge with a local reduction of gain, making the operation stable.

Thanks to the excellent time resolution achieved, glass MRPCs are used to perform particle identification through a measurement of time-of-flight by the

---

[3] Later studies with the same structures have demonstrated that similar performances can be obtained using only $C_2F_4H_2$ as gas filling.

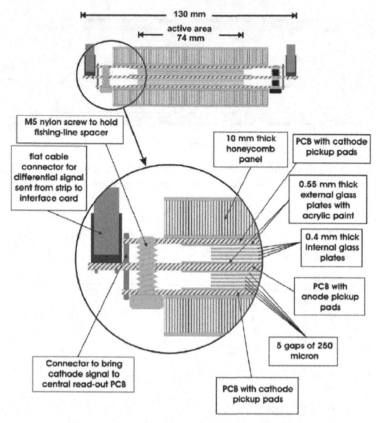

Figure 12.16 Schematics of the ALICE MRPC TOF module (Akindinov *et al.*, 2004). By kind permission of Elsevier.

ALICE experiment at LHC. The resistive plates are made with soda-lime glass plates, 400 μm thick; spacers made with nylon fishing lines 250 μm in diameter ensure the gap uniformity. Figure 12.16 shows a general and detailed schematic of a dual ALICE TOF module (Akindinov *et al.*, 2004); assembled in modules 120×7.4 cm^2 with two stacks of five gaps each, the detectors are read out by arrays of rectangular pads placed on the central and outer surface of the stack.

Measured for fast particles with a set of fast scintillators as time reference, Figure 12.17 shows the efficiency and the standard deviation of the time resolution of the module as a function of voltage.

Crucial to achieve sub-ns time resolutions is the use of very fast amplifiers-discriminators; due to the wide charge spectrum of signals generated in parallel plate detectors, a time walk correction of the recorded time, taking into account the signal amplitude, is needed for best results. To achieve this goal, a dedicated ASIC chip developed for the needs of the ALICE TOF project, named NINO, has eight channels of low-noise, fast amplifiers-discriminators with adjustable threshold

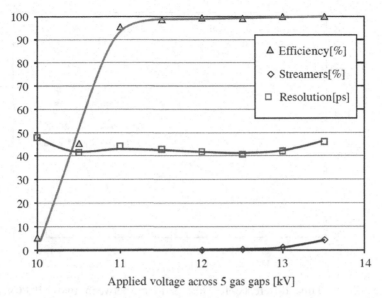

Figure 12.17   Efficiency and time resolution of the ALICE MRPC dual five-gap module (Akindinov *et al.*, 2004). By kind permission of Elsevier.

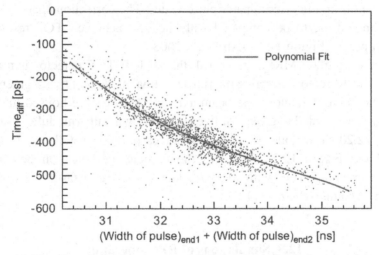

Figure 12.18   Correlation between recorded time and discriminator pulse width, used for time slewing corrections (An *et al.*, 2008). By kind permission of Elsevier.

(Anghinolfi *et al.*, 2004). To improve the signal over noise ratio, the circuit has a differential input, connected to two sets of parallel readout strips on each side of the detector. The recorded width of the discriminated output, proportional to the signal charge, is used for slewing correction. Measured with a 24-gap MRPC, Figure 12.18 is a scatter plot of the measured dependence of the relative time

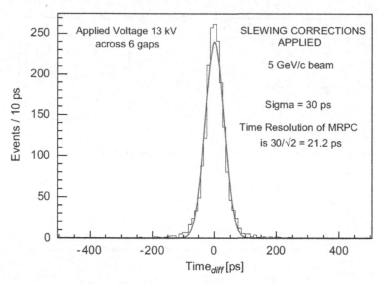

Figure 12.19   Time difference for fast particles between two MRPCs after slewing correction (An *et al.*, 2008). By kind permission of Elsevier.

measurement from the discriminator pulse width. The correction spans over about several hundred ps, to be compared with the best achieved TOF resolution of around 30 ps, see Figure 12.19 (An *et al.*, 2008).

Based on the glass MRPC described, the ALICE TOF detector is a modular structure with 18 sectors covering the full azimuthal angle and five segments in the longitudinal coordinate along the beam axis; the MRPCs are housed in gas tight modules and assembled in groups of five constituting a 'supermodule'. The picture in Figure 12.20 shows one of the supermodules prior to installation (Akindinov *et al.*, 2009); Figure 12.21 illustrates the particle identification power of the detector, combining TOF with the particle momentum measured by magnetic deflection (Akindinov *et al.*, 2012).

12.5 Simulations of RPC operation

Aimed at improving the time and localization properties of the detectors, detailed studies of signal formation and induction process on external electrodes, reproducing the observed signal spectra, have been described. Many authors have investigated the physics of the avalanche development mechanisms in detectors with resistive electrodes, with both analytic calculations and simulation models, analysing the effects of ionization statistics, gap thickness and field modifications due to space charge and negative ion formation due to attachment to electro-negative gas molecules.

Figure 12.20 One of the MRPC supermodules of the ALICE TOF system (Akindinov *et al.*, 2009). By kind permission of Elsevier.

The calculation of the avalanche development and propagation can be done analytically (Ammosov *et al.*, 1997; Mangiarotti *et al.*, 2004) or with Monte Carlo simulation programs, taking into account the strong effect of the space charge at the very high values of fields, typically above 100 kV/cm, met in RPC (Abbrescia *et al.*, 1999a; Abbrescia *et al.*, 1999b; Nappi and Seguinot, 2005; Lippmann and Riegler, 2004; Riegler, and Lippmann, 2004; Khorashad *et al.*, 2011). As an example, Figure 12.22 shows the good agreement between the measured charge spectra with the results of a numerical calculation (Lippmann and Riegler, 2004).

The avalanche growth as a function of time in the MRPC has been simulated using the program MAGBOLTZ (Biagi and Veenhof, 1995b) and with the algorithms developed by Lippmann and Riegler (2004), taking into account the creation of negative ions in the gas containing electro-negative molecules. An example of the results is shown in Figure 12.23, the evolution of the charge density and of the electric field in the gap at successive time intervals following the ionization process (Doroud *et al.*, 2009a). The high density of negative ions and consequent recombination with the positive ions generated in the avalanches

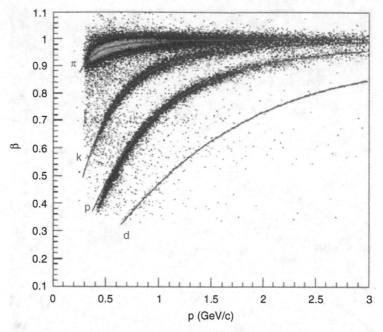

Figure 12.21   Particle identification power of the ALICE MRPC time-of-flight detector (Akindinov *et al.*, 2012). By kind permission of Elsevier.

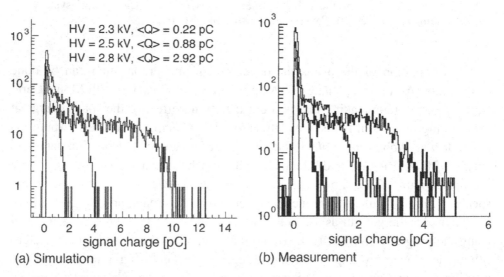

(a) Simulation

(b) Measurement

Figure 12.22   Simulated and measured total detected charge for a narrow gap RPC at three values of applied voltage (Lippmann and Riegler, 2004). By kind permission of Elsevier.

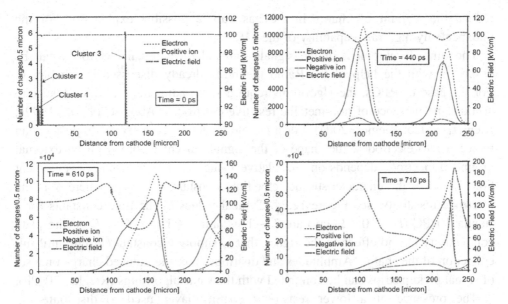

Figure 12.23   Simulation of the charge density evolution with time in an MRPC gap (Doroud *et al.*, 2009a). By kind permission of Elsevier.

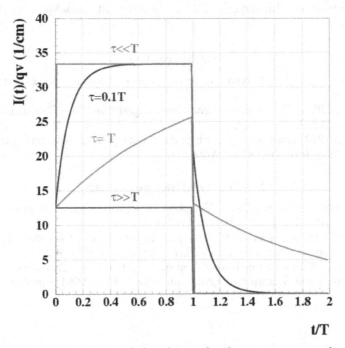

Figure 12.24   Current induced by the avalanche on an external electrode for several values of the time constants (see text) (Riegler, 2002b). By kind permission of Elsevier.

reduce the amount of charge in the gas gap, a possible explanation for the unexpectedly high rate capability of the MRPC.

The time development of the signals induced on external electrodes can be described with the help of the Ramo theorem, already discussed in Chapter 6. For the case of RPCs, the algorithms require some adjustments, which take into account the presence of non-metallic resistive electrodes (Abbrescia *et al.*, 1999a; Riegler, 2002b; Riegler, 2004). For a simple detector geometry, a single gas gap between two electrodes, the shape of the signal current recorded on an external grounded electrode depends on the relative values of the avalanche development time T and of the characteristic detector time constant $\tau = \varepsilon_0 R$, where R is the electrode resistivity, see Figure 12.24. Typical values of the time constants for a standard RPC ($R \sim 10^{12}$ Ω cm) are $T \sim 1$ ns and $\tau \sim 1$ ms ($\tau \gg T$) so the plate conductivity has no effect on the signal, that is simply a constant current until full collection of the charge. Amplitudes and distribution of the induced charge on a set of signal pickup strips can be computed with the same algorithms (Riegler, 2002b).

The presence of a lower resistivity coating layer used to distribute the high voltage affects both the signal shape and distribution; the effect has been studied in detail in the development of a structure aiming at reducing the number of strips needed for the readout of micro-pattern gas detectors, exploiting the charge dispersion on thin-layer resistive anodes (Litt, and Meunier, 1973; Dixit *et al.*, 2004).

Further reading

Riegler, W. (2002) Induced signals in resistive plate chambers. *Nucl. Instr. and Meth.* **A491**, 258.

Abbrescia, M. (2004) The dynamic behaviour of Resistive Plate Chambers. *Nucl. Instr. and Meth.* **A533**, 7.

Sharma, A. (2012) Muon tracking and triggering with gaseous detectors and some applications. *Nucl. Instr. and Meth.* **A666**, 98.

Selected conference proceedings

RPC2005 (2006) VIII International Workshop on Resistive Plate Chambers, S. P. Park, R. Santonico and S. Ratti (eds.) *Nucl. Phys. B (Proc. Suppl.)* **158**.

RPC2008 (2009) IX International Workshop on Resistive Plate Chambers, N. K. Mondal, S. P. Ratti and R. Santonico (eds.) *Nucl. Instr. and Meth.* **A602**.

RPC2011 (2012) X International Workshop on Resistive Plate Chambers, N. Herrmann, R. Santonico and S. Ratti (eds.) *Nucl. Instr. and Meth.* **A661 Suppl. 1**.

13

Micro-pattern gaseous detectors

13.1 The micro-strip gas counter

The localization and rate limits of wire-based detectors were discussed in previous chapters. An innovative device named the micro-strip gas counter (MSGC) (Oed, 1988) appeared to fulfil the increasingly demanding requirements of particle physics experiments: improved position and multi-track resolution, and larger particle flux capability.

The MSGC structure consists of thin parallel metal strips, alternately narrow and wide, engraved on a thin insulating support, and connected respectively as anodes and cathodes; the detector is completed by an upper electrode delimiting the sensitive gas-filled drift gap. Using standard photolithographic technologies, a distance between strips, or pitch, of one hundred microns or below can be obtained, an order of magnitude improvement in granularity over wire chambers; Figure 13.1 is a close view of the anode ends in one of the first MSGCs, with 10 μm wide anode strips at 200 μm pitch, alternating with wider cathodes. The rear side of the support plate, or backplane, can also have a field-defining electrode that may be segmented to perform two-dimensional localization.

With appropriate potentials applied to the electrodes, electrons released in the drift gap move towards the strips and multiply in the high field region close to the anodes. In most cases, for convenience of readout, the anode strips are at ground potential, with the cathodes connected individually or in groups to the negative potential through high value protection resistors. Figure 13.2 shows the electric field in the vicinity of the strips, computed with anodes and backplane at equal potentials. All field lines from the drift volume terminate on the anodes, providing full electron collection efficiency; due, however, to the transverse dispersion of the avalanche during multiplication, a large fraction of the positive ions generated near the anode spreads into the field lines connecting to cathodes, and is quickly

Figure 13.1   Close view of an MSGC plate, with narrow anodes alternating with wider cathode strips (Oed, 1988). By kind permission of Elsevier.

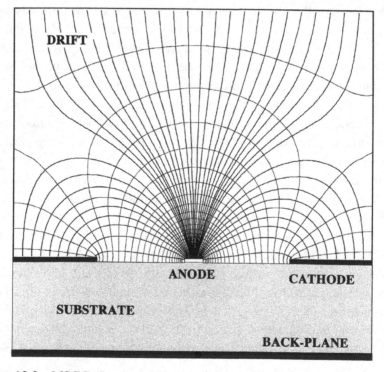

Figure 13.2   MSGC electric field in the vicinity of the strips.

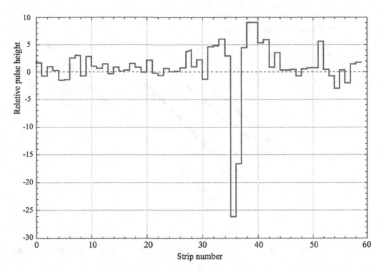

Figure 13.3   Anodic pulse height distribution for a charged particle track perpendicular to the detector. The strip pitch is 200 μm (Bouclier *et al.*, 1995a). By kind permission of Elsevier.

collected; this substantially reduces the positive ion backflow in the drift gap, and results in much smaller field distortions as compared to wire-based devices.

At the occurrence of an avalanche, the fast electron collection and the retrograde motion of ions generate negative signals on the anodes; signals of opposite polarity are induced on the near cathodes, backplane and drift electrode. Due to stray capacitance between strips and to the drift electrode, a fraction of the signal induced on one set of strips is injected into the other, with amplitudes that depend on geometry, giving the typical charge distribution shown in Figure 13.3, sum of signals and noise (Bouclier *et al.*, 1995a). A calculation of the centre of gravity of the recorded distribution provides the position of localized avalanches; the width of the profile, 400 μm in this example, determines the two-track resolution.

Microstrip detectors can be operated in a wide range of gases and pressures, and reach gains up to 10^4 with good proportionality. A compilation of measured gains in several gases at 1 bar, as a function of anode potential for a 200 μm pitch, 10 μm (100 μm) anode (cathode) strips width is given in Figure 13.4 (Beckers *et al.*, 1994). Particularly interesting choices are mixtures containing dimethyl ether (DME), thanks to its non-polymerizing properties (see Chapter 16). Detailed performance studies permitted one to optimize the geometry and operating conditions of the devices in view of their use for fast particle detection and localization (Florent *et al.*, 1993; Beckers *et al.*, 1994).

MSGC plates have been manufactured by photolithographic processing on a variety of supports and with several metals: chromium, aluminium, silver and gold.

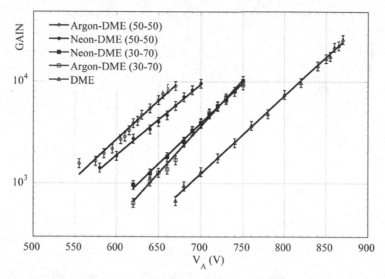

Figure 13.4   Gain as a function of voltage for several gas mixtures at STP (Beckers *et al.*, 1994). By kind permission of Elsevier.

The mechanical sturdiness of chromium is compensated by a higher resistivity, resulting in signal attenuation for long strips; gold plating is used to improve the performances and ease the connection to the readout electronics. The quality and uniformity of the artwork, and in particular smooth and defect-free edges of the strips, are paramount for a correct operation of the detectors. For a summary of manufacturing methods see Sauli and Sharma (1999).

Industrially produced up to sizes of 30×30 cm^2, the rigid MSGC plates can be easily mounted as detector, inserted in a gas containment box or in framed assemblies, suitable for many applications. Figure 13.5 is a schematic view of the components of a light MSGC detector, used in extensive laboratory and beam tests (Bohm *et al.*, 1995). The active plate and a thin glass roof with an internal drift electrode are glued to an insulating frame with gas inlet and outlet; the readout electronics is connected on the anode strip side, and the high voltage provided to cathode strips and drift electrode through protection resistors. For fast beam tracking, the drift gap is usually around 3 mm thick; other values can be used for more efficient detection of soft X-rays.

Thanks to their high granularity and fast ion collection by neighbouring cathode strips, MSGCs were expected to tolerate high radiation fluxes. From the very beginning, however, various operating instabilities have been observed: time-dependent gain shifts, attributed to substrate polarization and charging up; permanent deterioration during sustained irradiation; tendency to discharge. The physical parameters used to manufacture and operate the detectors (substrate material, metal of the strips, type and purity of the gas mixture) play dominant

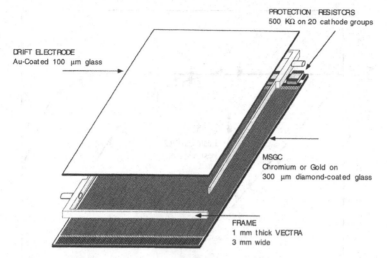

PROTECTION RESISTORS
500 KΩ on 20 cathode groups

DRIFT ELECTRODE
Au-Coated 100 μm glass

MSGC
Chromium or Gold on
300 μm diamond-coated glass

FRAME
1 mm thick VECTRA
3 mm wide

Figure 13.5   Schematics of a light MSGC assembly (Bohm *et al.*, 1995). By kind permission of Elsevier.

roles in determining the medium- and long-term stability (Bouclier *et al.*, 1992; Florent *et al.*, 1993). A research effort was undertaken to understand the MSGC operation, to improve their performance and lifetime, as well as reducing manufacturing costs, an essential goal in view of intensive use in large systems; at the peak of its activity, a collaboration for the development of MSGCs included more than 40 laboratories worldwide[1] (Sauli, 1998).

Use of supports with controlled resistivity permits one to neutralize the surface charge, and extends considerably the rate capability of the MSGC. Commercial, low resistivity glasses have been employed for this purpose, but appeared to rely on ionic conductivity, changing with time from the application of the electric field, an effect referred to as polarization of the dielectric. Better stability of performance has been obtained using electron-conducting substrates, originally developed for spark counters and named Pestov glass from the main developer (Frolov *et al.*, 1991); manufactured in a wide range of resistivity, they exhibit stable operation up to very high radiation fluxes, as shown in Figure 13.6, comparing the normalized gain as a function of rate, measured for soft X-rays, of MSGCs manufactured on commercial borosilicate glass 2 DESAG D-263, SHOTT speciality glass, Germany and on Pestov glass of two values of resistivity. For 10^9 Ohms cm, the proportional gain is unaffected up to a flux of a MHz mm^{-2} (Bouclier *et al.*, 1995b).

An alternative solution is to cover a standard glass with a thin resistive coating before metallization and photolithographic processing; a diamond-like layer, produced by carbon vapour deposition (CVD) and with a surface resistivity around

---

[1] CERN RD28, F. Sauli, spokesman (1993–1996).

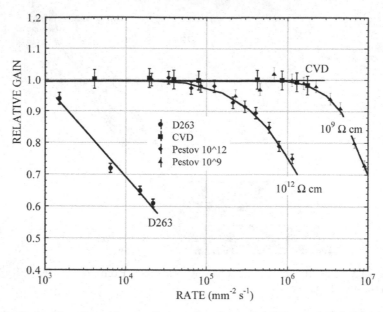

Figure 13.6   Rate dependence of normalized gain for different MSGC substrates (Bouclier *et al.*, 1995b). By kind permission of Elsevier.

10^{14} Ω/square provides equivalent or better rate capability, as shown in the figure (Bouclier *et al.*, 1996b). As the production of large plates of electron conducting glass turned out to be problematic, the CVD coating remains a good option, and has been used for detectors used in systematic measurements of performances (Barr *et al.*, 1996; Angelini *et al.*, 1996; Abbaneo *et al.*, 1998). As an example, Figure 13.7 shows the residual distribution recorded with a set of MSGCs in a minimum ionizing particle beam perpendicular to the detectors; the position accuracy is better than 40 μm rms (Bouclier *et al.*, 1995a).

Introduced originally for applications as detectors in neutron spectrometry, MSGCs are successfully used in this field, using ^{3}He as sensitive gas filling or internal thin-foil converters. A large one-dimensional neutron diffractometer making use of 48 MSGC plates in a semi-cylindrical array, four metres long, shown in Figure 13.8, has been operating for many years at Grenoble's Institute Laue Langevin (ILL), despite some observed deterioration of the anode strips (Clergeau *et al.*, 2001; Hansen *et al.*, 2008).

Two-dimensional localization can be performed with a segmented backplane readout, or various schemes of pad rows replacing the cathode strips on the active side of the plates, see for example Vellettaz *et al.* (2004); Masaoka *et al.* (2003); Takahashi *et al.* (2004); Fujita *et al.* (2007); Bateman *et al.* (2010); Bateman *et al.* (2012). An overview and comparison of neutron detectors is given in Oed (2004) and Buffet *et al.* (2005).

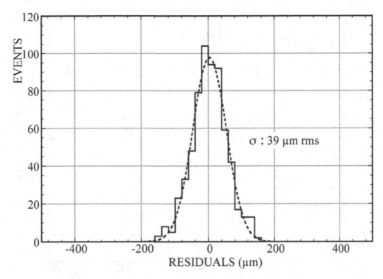

Figure 13.7 MSGC space accuracy for minimum ionizing articles (Bouclier *et al.*, 1995a). By kind permission of Elsevier.

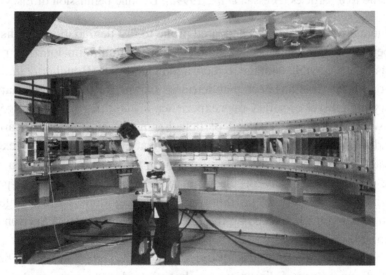

Figure 13.8 MSGC array for the ILL neutron diffractometer (Clergeau *et al.*, 2001). By kind permission of Elsevier.

The use of MSGCs, requiring a large number of readout channels, has been facilitated by the coincidental compatibility with the performances of circuits mass-produced for silicon strip detectors: indeed, the smaller ionization signal is compensated by the gain of the gaseous devices. Several particle physics experiments have been designed to use large arrays of MSGCs (Angelini *et al.*, 1995; Zeuner, 2000; Ackerstaff *et al.*, 1998). Disappointingly, and despite large

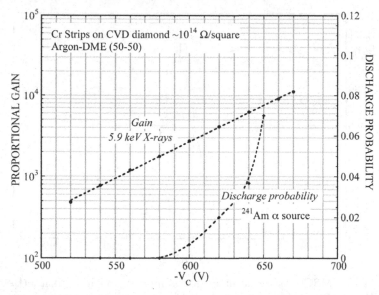

Figure 13.9   MSGC gain measured for soft X-rays, and discharge probability on exposure to α particles (Bressan *et al.*, 1999a). By kind permission of Elsevier.

development efforts, the devices appeared prone to fast degradation and discharges when operated at the gains needed for detection of minimum ionizing particles, often causing irreversible damage to the fragile electrodes (Bouclier *et al.*, 1995c). A common observation has been that detectors that perform well in the laboratory under very high fluxes of soft X-rays experience instabilities and discharge-related damage when exposed to charged particle beams (Schmidt, 1998; Barr *et al.*, 1998). Indeed, albeit with small probability, ionization densities largely exceeding the normal can be deposited by interactions with the detector gas or materials of neutrons, nuclear fragments or electromagnetic showers; a neutron-produced ~MeV proton releases in the gas three orders of magnitude more charge than a minimum ionizing track, exceeding the Raether limit discussed in Section 8.8 and leading to a discharge.

Since beam conditions are difficult to reproduce, a standard test for defining the threshold of appearance of discharges has been introduced, exposing the detectors to heavily ionizing sources; in a 3 mm sensitive gap, the average number of ionization electrons released by ~MeV α particles is around 5000. Figure 13.9 is an example of the gain measured with a soft X-ray source, and of the discharge probability on exposure to α particles as a function of voltage[2] (Bressan, A. *et al.* (1999a)). Capable of reaching gains up to 10^4 in a benign environment, when

---

[2] The definition of discharge probability is somewhat arbitrary, as it depends on the source rate and the observation time. In view of its rapid exponential growth, however, it helps in defining a threshold voltage for breakdown.

exposed to highly ionizing radiation the onset of discharges limits the operation to gains below those needed for the detection of fast particles (around 5000 using standard available electronics).

A convenient α-particle source is ^{220}Rn, introduced in the chamber by inserting a natural thorium cartridge in the gas flow. Many innovative MPGD detectors, described in the following sections, have been systematically tested for resilience to discharges using this method, with comparable outcomes, a demonstration of the fundamental nature of the Raether limit (Bressan *et al.*, 1999a). A noticeable exception is observed by cascading several elements of amplification, each operated at reduced gain, with the use of multiple gas electron multiplier electrodes (GEM), discussed in Section 13.4. A GEM pre-amplifier electrode, added to the MSGC detector in the HERA-B experiment, permitted one to solve the discharge problem and operate the tracker for several years (Schmidt, 1998).

Aside from manufacturing defects, a source of discharges even in the absence of radiation has been identified in a process related to the double amplification discussed in Section 8.8 for wire counters. The electric field near the edges of the cathode strips can be high enough to impart a pre-amplification to ionization electrons released in their proximity, or by spontaneous field emission. A calculation of avalanche multiplication for electrons released in the vicinity of the cathode edges provides the equal-gain lines shown in Figure 13.10 (Beckers *et al.*, 1994); for a normal anodic gain around 2000, electrons released near the cathode edge are pre-amplified and experience a much larger total gain, easily leading to charge densities satisfying the Raether condition and provoking a discharge.

Coating the cathode strip edges with a thin layer of polyimide insulator, using a technique named advanced passivation, prevents or delays the spontaneous field emission at the cost of an increased complexity of manufacturing (Bellazzini *et al.*, 1998b).

Due to the small surface of the anodes MSGCs are particularly sensitive to the presence of thin insulating layers, as those created by gas polymerization or pollutants; great care in the gas purity and choice of construction materials is needed to prevent a fast deterioration (Bouclier *et al.*, 1994b). The subject of MSGC ageing will be discussed in detail in Chapter 16.

13.2 Novel micro-pattern devices

The problems encountered with the MSGCs spawned the development of alternative structures, collectively named micro-pattern gas detectors (MPGDs), capable of achieving comparable performances but more resistant to radiation and damages: micro-gap, micro-wire, micro-dot, field gradient lattice, 'compteur à trous' and others. For an overview of these devices and related references see Sauli and Sharma (1999).

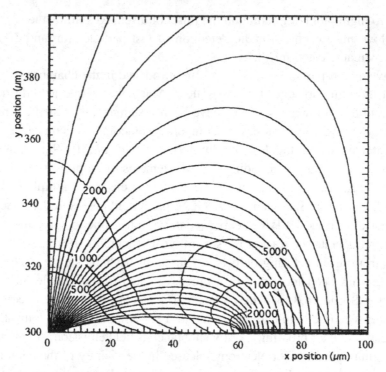

Figure 13.10 Computed equal gain lines in an MSGC. Charge pre-amplification can occur for electrons released in the vicinity of the cathode strip edges (Beckers *et al.*, 1994). By kind permission of Elsevier.

The technology of combining photolithography and thin-layer polyimide depositions, used to coat the cathode edges as described in the previous section, led to the development of structures designed to improve the insulation between anode and cathode strips, as the micro-gap (Angelini *et al.*, 1993) and the micro-groove chambers (Bellazzini *et al.*, 1998a), shown in Figure 13.11.

As many of the gain instabilities and discharge problems encountered in MPGDs are related to the presence of insulators, attempts have been made to build structures almost substrate-free: successive steps of photolithographic metal engraving and polymer wet etching result in devices with the insulator reduced to narrow pillars or bridges between metal strips, minimizing charging-up processes. Examples are the micro-wire chamber (Adeva *et al.*, 1999) and the field gradient lattice detector (FGLD) (Dick *et al.*, 2004), shown in Figure 13.12.

Alternative structures have been developed by exploiting the refined technologies used for solid-state circuit manufacturing on silicon supports. The micro-dot, Figure 13.13 left (Biagi and Jones, 1995), is a matrix of planar proportional counters, each consisting of a central round electrode (the anode) surrounded by one or more concentric electrodes; with suitable potentials applied to the

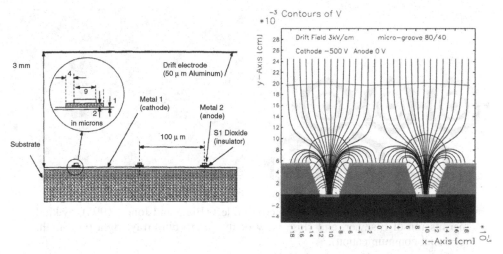

Figure 13.11 Schematics of the microgap, left (Angelini *et al.*, 1993) and micro-groove chambers, right (Bellazzini *et al.*, 1998a). By kind permission of Elsevier.

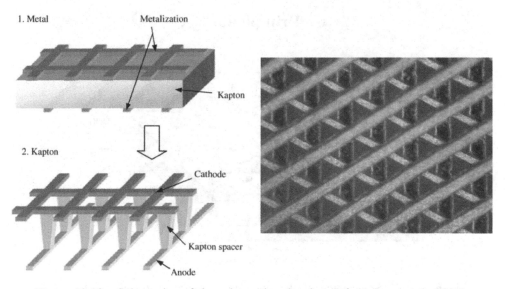

Figure 13.12 Schematics of the micro-wire chamber, left (Adeva *et al.*, 2001) and view of the FGLD, right (Dick *et al.*, 2004). By kind permission of Elsevier.

electrodes, each dot acts as an individual proportional counter, read out individually or interconnected in lines. The micro-pin array (MIPA), right, is a three-dimensional extension of the same structure, with metal points acting as anodes (Rehak *et al.*, 2000).

Similar in concept to the previous devices, but on a coarser scale, the micro-pixel chamber (μPIC), Figure 13.14, (Ochi *et al.*, 2002), features a matrix of

Figure 13.13   Schematics of the micro-dot, left (Biagi and Jones, 1995), by kind permission of Elsevier, and close view of the micro-pin array, right (G. Smith, personal communication).

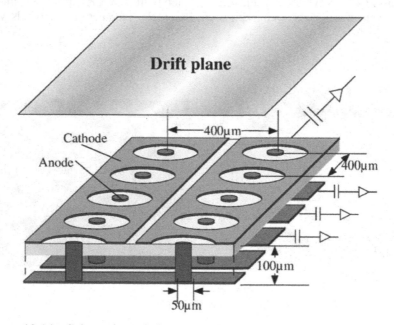

Figure 13.14   Schematics of the micro-pixel chamber (Ochi *et al.*, 2002). By kind permission of Elsevier.

individual circular anodes ensuring charge multiplication, interconnected on the back plane to provide one coordinate; a second coordinate is given by orthogonal cathode strips, as shown in the figure. While suffering, as most other MPGDs, with an initial gain shift at power on due to charging-up of the dielectric substrate, the μPIC chamber has been successfully used as end-plane detector for TPC readout (H. Kubo *et al.*, 2003) and time-resolved neutron imaging (Parker *et al.*, 2013).

In the second application, a selection of the detected charge and a detailed analysis of the conversion tracks permit an efficient neutron identification and localization in the presence of a gamma background.

All MPGD structures described have been tested successfully; in many cases, however, they appeared to be difficult to manufacture in reasonable sizes and quantities, and several have been discontinued.

Although not belonging to the MPGD family, the so-called 'compteur à trous' (CAT), shown in Figure 13.15, is an interesting addition to the field of gaseous detectors (Bartol *et al.*, 1996). Consisting of a single hole drilled through a thick printed circuit board, the counter collects and amplifies the charge released in the upper drift gap; large area devices can be implemented with a matrix of adjacent holes. Despite the expected dependence of gain from the field integral along the line of approach, the device exhibits surprisingly good energy resolution

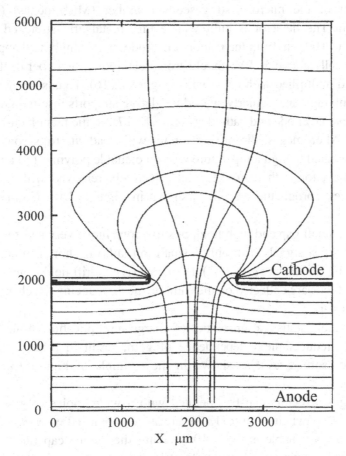

Figure 13.15   Schematics of the 'compteur à trous' (CAT) (Bartol *et al.*, 1996). By kind permission of EDP, les Editions de Physique.

(Chaplier *et al.*, 1999). Built to a reduced scale, the gas electron multiplier (GEM), described in Section 13.4, was inspired by this work.

13.3 Micro-mesh gaseous structure (Micromegas)

The successful development of thin-anode multi-wire and micro-strip structures sidestepped the research on gas detectors by exploiting multiplication in uniform fields. Mechanically sturdier, parallel plate counters have intrinsically better energy resolution and higher rate capability; however, the exponential dependence of gain on the gap and the sensitivity to defects causing field emission result in unstable operation and discouraged their use.

It was observed, however, that in thin, sub-mm gaps, due to saturation of the Townsend coefficient at very high fields, large gains can be attained, with a decreased sensitivity to gap variations and imperfections; this has led to the development of the micro-mesh gaseous chamber (Micromegas) (Giomataris *et al.*, 1996). The detector is built with a thin metal grid stretched at a small distance, 50 to 100 μm, from the readout electrode; with a high field applied across the gap, typically above 30 kV/cm, electrons released in the upper drift region are collected and multiplied with large gains (Figure 13.16). To ensure the uniformity of the multiplying gap, regularly spaced insulating supports inserted during manufacturing separate anode and cathode (Figure 13.17). In the manufacturing process named bulk Micromegas, the printed circuit with readout strips constituting the anode is laminated at high temperature with an etchable polymer foil and the metal mesh; the insulator is then masked and selectively removed with a photolithographic process, producing the pillars shown in Figure 13.18 (Giomataris *et al.*, 2006).

Due to the small gap and high field, positive ions move very swiftly and induce on the anodes fast signals with only a small ion tail, the shorter the narrower the gap (Figure 13.19) (Bay *et al.*, 2002); as the overlying drift field is generally much smaller, most ions are collected on the cathode mesh, reducing the charge backflow into the drift gap.

The favourable statistical properties of charge multiplication in uniform fields, added to the reduced field dependence of gain, result in a very good energy resolution for soft X-rays, as shown in the example of Figure 13.20 (Delbart *et al.*, 2001).

Micromegas detectors can be operated with a wide choice of gases; the gain measurements shown in Figure 13.21 are an example (Bay *et al.*, 2002). The maximum gain attainable exceeds 10^5 making the device capable of efficiently detecting and localizing small amounts of ionization, at least in the absence of a heavily ionizing background.

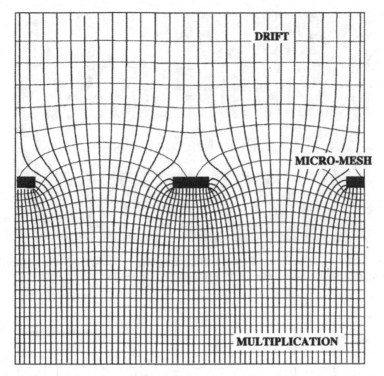

Figure 13.16   Electric field in the Micromegas (Giomataris *et al.*, 1996). By kind permission of Elsevier.

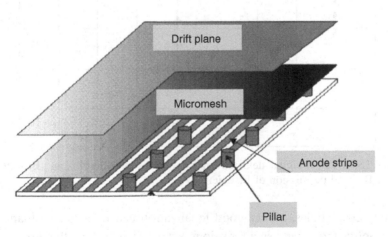

Figure 13.17   Schematics of the Micromegas construction, with insulating spacers across the multiplying gap (Giomataris *et al.*, 1996). By kind permission of Elsevier.

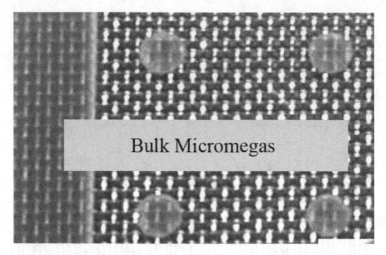

Figure 13.18   Close view of the insulating spacer pillars, 300 μm in diameter (Giomataris *et al.*, 2006). By kind permission of Elsevier.

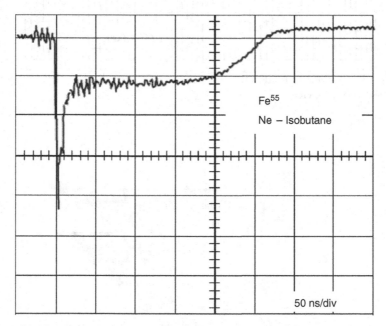

Figure 13.19   Fast signal detected on the anodes of Micromegas (Bay *et al.*, 2002). By kind permission of Elsevier.

For charged particles, a very good localization can be achieved thanks to the narrow amplification gap, small readout strip distance and the use of a low-diffusion gas. In a small system, a position accuracy of 12 μm rms for a CF_4-isobutane filling and 100 μm strips pitch has been obtained for fast particles

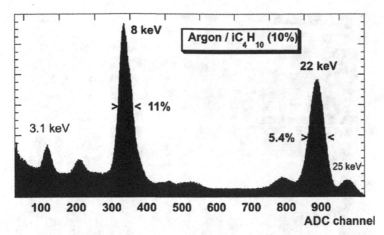

Figure 13.20 Micromegas energy resolution for a ^{109}Cd source (Delbart *et al.*, 2001). By kind permission of Elsevier.

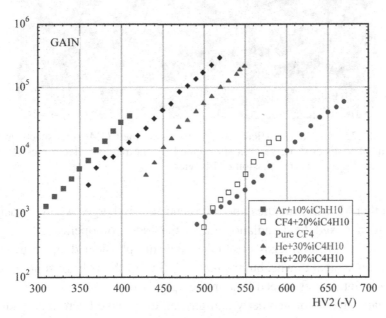

Figure 13.21 Micromegas gain as a function of voltage for several gas fillings at STP (Bay *et al.*, 2002). By kind permission of Elsevier.

perpendicular to the detector (Derré *et al.*, 2000). For larger detector sizes, practical considerations on the electronics channels density bring the space resolution to ~70 µm, as in the Micromegas tracker for CERN's COMPASS experiment, using 360 µm readout strips pitch and operated with a Ne-C_2H_6-CF_4 gas mixture, Figure 13.22 (Bernet *et al.*, 2005).

Figure 13.22   One of the Micromegas detectors in the COMPASS spectrometer, with a 40×40 cm^2 active area surrounded by the readout electronics (Bernet *et al.*, 2005). By kind permission of Elsevier.

As in other MPGD structures, the insurgence of discharges when the detector is exposed to a mixed field radiation limits the range of operating voltage. The sturdiness of the Micromegas electrodes prevents physical damage even in the case of repeated breakdowns, but the recovery time of the voltage supply, typically of a few ms, introduces a dead time in the operation.

Systematic studies of geometry and gas mixtures have been undertaken to try and limit the discharge probability (Bay *et al.*, 2002); Figure 13.23 is a comparison of discharge probability as a function of gain for two gas fillings, measured in a high intensity hadron beam[3] (Delbart *et al.*, 2002). Higher gains can be reached in helium-based mixtures, probably as a consequence of the lower ionization density of background tracks, but this is balanced by a decrease in the charge release of the tracks.

---

[3] For beam measurements, the discharge probability is normalized to the beam intensity.

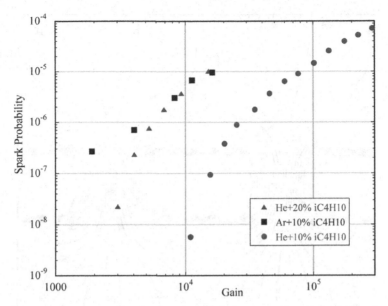

Figure 13.23   Micromegas discharge probability as a function of gain for two gas mixtures (Delbart *et al.*, 2002). By kind permission of Elsevier.

As for the MSGCs, higher gains before breakdown can be reached by adding a GEM pre-amplification stage, an advantage balanced by an increased complexity of construction of the detector (Charles *et al.*, 2011; Procureur *et al.*, 2011). Use of a high resistivity layer insulating the multiplication gap from the anode helps in dumping the discharges before full breakdown, similarly to the approach used for RPCs, see Chapter 12; this implies, however, a noticeable reduction in rate capability and can lead to gain instabilities due to charging-up (Alexopulos *et al.*, 2011).

13.4 Gas electron multiplier (GEM)

The gas electron multiplier (GEM) consists of a thin, metal-clad polymer foil chemically perforated by a high density of holes, typically 100 per square mm (Sauli, 1997). Inserted between a drift and a collection electrode, with a suitable choice of the voltage on electrodes all electrons released by ionization in the overlaying gas gap drift into the holes, where charge multiplication occurs (Figure 13.24). Most of the electrons generated in the avalanches transfer into the lower region; the GEM foil acts then as a charge pre-amplifier, to a large extent preserving the original ionization pattern. Each hole acts as an independent proportional counter, screened from the neighbours; due to the high density of holes, the gain is not affected by space charge up to very high radiation fluxes.

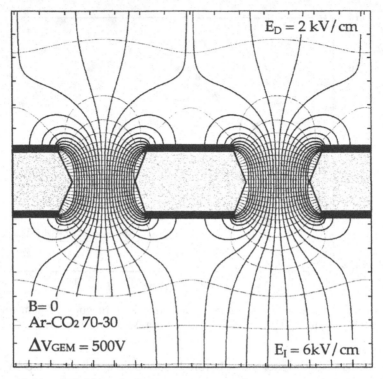

Figure 13.24   GEM electric field near the holes in typical operating conditions
(Bachmann *et al.*, 1999). By kind permission of Elsevier.

As avalanche multiplication occurs almost entirely in the high dipole field within
the holes, the gain is only mildly affected by external fields,[4] and insensitive to the
foil shape, considerably relaxing the mechanical requirements. Separated from the
multiplying electrode, the charge collection and readout plane can be patterned at
will with strips, pads or a combination of the two.

GEM electrodes can be easily shaped to match the experimental requirements;
with proper supports, they can be used to manufacture non-planar detectors. The
so-called 'standard GEM', produced in large quantities and a wide range of shapes
and size, has 70 μm holes 140 μm apart in a triangular pattern, etched on 50 μm
copper-clad kapton, as seen in Figure 13.25 (Altunbas *et al.*, 2002).

The GEM manufacturing method is a refinement of the double-sided printed
circuit technology.[5] The metal-clad polymer is engraved on both sides with the
desired hole pattern; a controlled immersion in a polymer-specific solvent opens the
channels in the insulator. As a result of the process, the holes in the insulator tend to

---

[4] The fraction of electrons in the avalanche leaking into the lower gap and to the following electrode depends
instead on the transfer field; the detected charge defines the effective gain, smaller than the total.

[5] Developed at CERN by Rui de Oliveira.

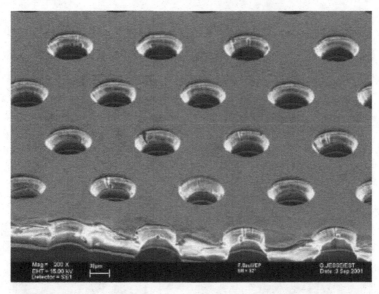

Figure 13.25 Microscope view of a GEM electrode. The holes' diameter and pitch are 70 and 140 μm, respectively (Altunbas *et al.*, 2002). By kind permission of Elsevier.

have a double-conical shape; a thorough control of the etching conditions permits one to approach a near-cylindrical shape, demonstrated to reduce the insulator charging up and consequent gain modifications (Bachmann *et al.*, 1999). A detailed description of the manufacturing procedure is given in Walz (2010).

The described manufacturing process requires the use of two identical masks with the holes' pattern, which have to be aligned with a tolerance of a few μm to avoid creating slanted holes. For large areas, this becomes exceedingly difficult; a single-mask process has been developed permitting the realization of GEM foils close to a square metre (Alfonsi *et al.*, 2010; Villa *et al.*, 2011).

While the majority of GEM-based detectors have a rectangular geometry, the electrode geometry can be shaped according to the experimental needs; Figure 13.26 is an example of an electrode developed for the forward tracker of the TOTEM experiment at CERN (Lami *et al.*, 2006; Antchev *et al.*, 2010).

Alternative manufacturing methods making use of mechanical or laser drilling of thicker supports, generally with a coarser holes' pitch, have been developed for applications requiring moderate localization accuracies (Periale *et al.*, 2002; Chechik *et al.*, 2004; Badertscher *et al.*, 2010); they are generally referred to as thick GEM (THGEM). Figure 13.27 shows a large THGEM electrode, developed for the COMPASS RICH upgrade, having 400 μm diameter holes on a 600 μm thick printed circuit board (Alexeev *et al.*, 2010). For a review see Breskin *et al.* (2009).

Figure 13.26   A semi-circular GEM electrode, 30 cm in diameter, for the TOTEM experiment. Picture CERN (2004).

Figure 13.27   A prototype thick-GEM electrode developed for the COMPASS RICH upgrade (Alexeev *et al.*, 2010). By kind permission of Elsevier.

GEM-based detectors have been tested in a variety of gas fillings and operating conditions, including low and high pressures; for a comprehensive review see Sauli (2014). Gains above one thousand can be reached in the detection of fast charged particles and soft X-rays; in the presence of heavily ionizing tracks, however, the discharge probability is comparable to the one observed in other micro-pattern devices, confirming the fundamental nature of the Raether limit (Bressan et $al.$, 1999a).

A unique feature of the GEM concept is that several amplifiers can be cascaded within the same detector, separated by low field transfer gaps (Bouclier et $al.$, 1997; Büttner et $al.$, 1998; Bressan et $al.$, 1999b). The overall gain of a multiple structure corresponds to the product of the gains of each element, once the transfer efficiency is taken into account (the so-called effective gain); as shown in Figure 13.28 for a double-GEM detector, operated with an argon-CO_2 gas mixture, high gains can be obtained with each element at a much lower voltage than for a single device, largely improving the reliability of the detector (Bachmann et $al.$, 1999). Extensive tests demonstrate that a cascade of three electrodes, named triple-GEM (Figure 13.29) guarantees the reliability and breakdown suppression needed in harsh beam conditions (Ziegler et $al.$, 2001; Ketzer et $al.$, 2004); the required potentials can be applied to the electrodes through simple resistor chains or dedicated power supply systems.

Measurements of discharge probability on exposure to a heavily ionizing background demonstrate that in multiple structures discharges occur at higher

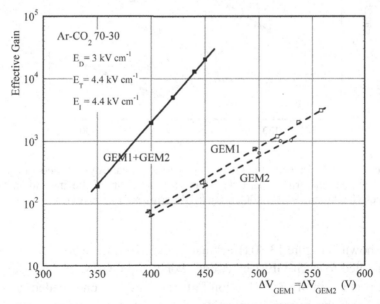

Figure 13.28   Effective gains of two single GEM electrodes, and combined gain of a double structure (Bachmann et $al.$, 1999). By kind permission of Elsevier.

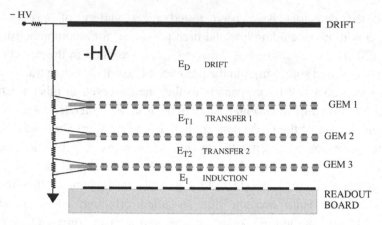

Figure 13.29 Schematics of a triple-GEM chamber and resistor chain used to distribute the voltage to the electrodes (Ketzer *et al.*, 2004). By kind permission of Elsevier.

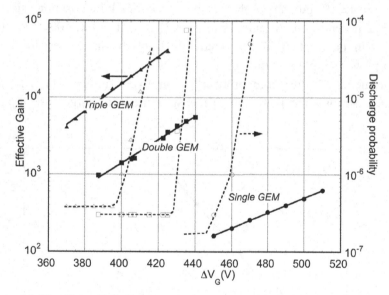

Figure 13.30 Gain (full lines) and discharge probability (dashed curves) of single, double and triple-GEM detectors as a function of the individual GEM voltages (Bachmann *et al.*, 2002). By kind permission of Elsevier.

gains, as shown in Figure 13.30 (Bachmann *et al.*, 2002); the horizontal scale is the voltage applied to each GEM electrode. For a triple-GEM device, the onset of discharge is at a gain of $\sim 3 \times 10^4$, comfortably above the one needed to detect fast particles. Since the average ionization charge of the α particles used for the measurement is around 5000 (see Section 13.1) this seems to violate the Raether

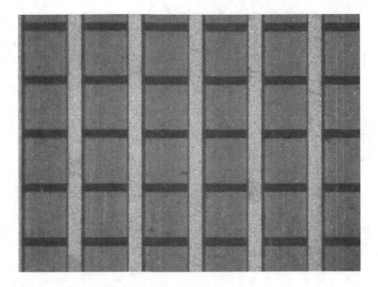

Figure 13.31   Two-dimensional readout circuit with strips at 400 pitch (Bressan *et al.*, 1999c). By kind permission of Elsevier.

limit, but is probably a consequence of the added lateral electron diffusion in the cascaded structures, spreading the avalanche charge into many multiplying holes, kept below the discharge limit in each hole.

Separated from the multiplying elements, the signal collection electrode can be patterned with strips or pads of arbitrary shape. A circuit widely used for two-dimensional projective readout is shown in Figure 13.31; realized on a 50 μm polymer foil with the same technology used for GEM manufacturing, it has 80 μm wide strips on the side facing the multiplier and, insulated from them, wider strips on the lower coordinate, to ensure equal sharing of the collected charge along the two directions (Bressan *et al.*, 1999c). The circuit can be realized with individual readout pads in the part of the detector exposed to higher radiation flux, and projective strips to cover the remaining area (Krämer *et al.*, 2008).

Depending on the experimental needs, the detectors can be operated with a wide choice of gas fillings; Figure 13.32 gives examples of effective gain measured with a triple-GEM detector in several gas mixtures (Breskin *et al.*, 2002). Very high gains can be reached in pure carbon tetrafluoride, permitting the detection of single photoelectrons as for Cherenkov ring imaging applications (see Chapter 14).

With non-flammable argon-CO_2 mixtures, GEM chambers provide routinely for fast particles detection efficiencies close to 100%, localization accuracies around 70 μm rms and 10 ns resolution (Ketzer *et al.*, 2004). Use of a faster gas mixture containing CF_4 permits one to achieve a time resolution better than 5 ns rms, an essential advantage when operating around fast cycling colliders (Alfonsi *et al.*, 2004).

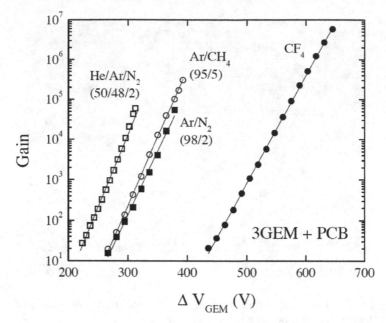

Figure 13.32   Triple-GEM gain as a function of voltage in several gases (Breskin *et al.*, 2002). By kind permission of Elsevier.

A set of medium size (30×30 cm^2 active) triple-GEM chambers with projective cartesian coordinates readout has been used for many years in the harsh running conditions of the COMPASS spectrometer at CERN. One of the detectors, instrumented by a high-density charge-recording readout electronics, is shown in Figure 13.33 (Altunbas *et al.*, 2002; Ketzer *et al.*, 2004). The TOTEM detector at CERN, shown in Figure 13.34, deploys two sets of triple-GEM chambers built with the semi-circular elements shown in Figure 13.26 (Bagliesi *et al.*, 2010).

GEM detector foils can be assembled in non-planar geometry with the use of appropriate edge supports; Figure 13.35 is a cylindrical triple-GEM detector in construction for the KLOE-2 inner tracker at the Italian Laboratori Nazionali di Frascati (LNF) (Bencivenni and Domenici, 2007; Balla *et al.*, 2011).

Owing to the absence of thin anodes, GEM detectors have a very high rate capability; on exposure to a soft X-ray generator, the gain remains constant up to and above a flux of 1 MHz mm^{-2}, as shown in Figure 13.36, measured with a single GEM chamber operated at a gain around 10^3 (Benlloch *et al.*, 1998).

Thanks to the absence of thin anodes, GEM detectors are insensitive to radiation damage induced by polymerization, up to very high integrated fluxes, as discussed in Section 16.5.

Figure 13.33 One of the triple-GEM detectors installed in the COMPASS spectrometer at CERN (Ketzer *et al.*, 2004). By kind permission of Elsevier.

Figure 13.34 A module of the TOTEM tracker, assembly of five triple-GEM detectors. A symmetric segment closes in from the left and completes the detector (Bagliesi *et al.*, 2010). By kind permission of Elsevier.

Figure 13.35   Prototype cylindrical GEM detector for KLOE-2 (Bencivenni and Domenici, 2007). By kind permission of Elsevier.

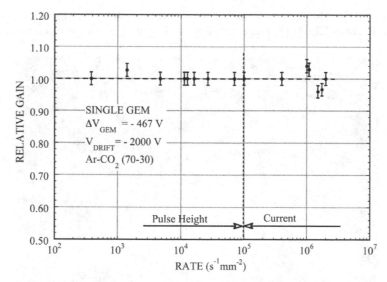

Figure 13.36   Normalized gain of a single GEM detector as a function of rate (Benlloch *et al.*, 1998). By kind permission of Elsevier.

13.5 MPGD readout of time projection chambers

Used for time projection chambers end-cap readout, MPGDs have many advantages compared to the standard MWPC: simpler and more robust construction, better space and multi-track resolution, absence of distortions due to non-parallel

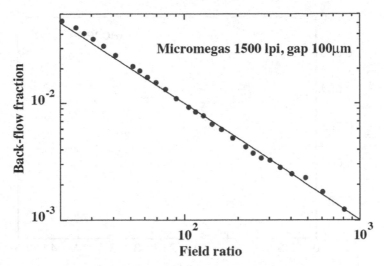

Figure 13.37   Positive ion backflow in Micromegas as a function of the ratio of multiplication and drift fields (Colas, 2004). By kind permission of Elsevier.

electric and magnetic fields close to the anode wires, and a substantial reduction of the positive ion backflow into the drift volume, thanks to their electric field structure. With Micromegas, using a narrow pitch cathode mesh and a gas filling chosen to enhance the transverse spread of the avalanches, the ion feedback is simply equal to the ratio between the drift and multiplying fields, typically around 1%, see Figure 13.37 (Colas, 2004).[6] With a thorough adjustment of the various fields in a multi-GEM structure, the backflow can be reduced below 1% (Bondar *et al.*, 2003); in the presence of a longitudinal magnetic field, due to the different effects of diffusion on ions and electrons, this ratio is further decreased below a few parts in a thousand, Figure 13.38 (Killenberg *et al.*, 2004).

A system of large volume TPC detectors with MPGD readout operates in the T2K neutrino experiment at KEK surrounding the active target. Each TPC module is instrumented with a matrix of Micromegas modules with pads readout, and records the pattern and differential energy loss of charged particles resulting from the interaction of neutrinos in the target (Abgrall *et al.*, 2011). Fundamental to the particle identification power of the detector, a calibration procedure and ambient parameter control ensures a uniformity of energy loss measurement better than 3% over the whole detector area (Delbart, 2010).

---

[6] The fractional ion backflow is defined as the ratio of the ion's current reaching the drift electrode to the electron current collected at the anodes.

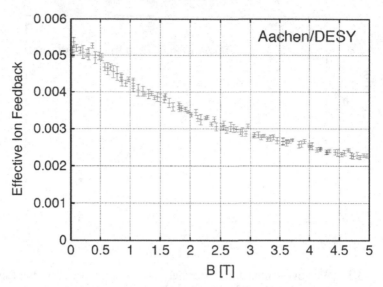

Figure 13.38   Fractional ions backflow in a TPC with GEM readout as a function of magnetic field (Killenberg *et al.*, 2004). By kind permission of Elsevier.

MPGD end-cap systems for TPCs have been the subject of many studies, aiming at optimizing the signal pads shape and reducing the number of readout channels needed (Kaminski *et al.*, 2006; Ledermann *et al.*, 2007; Oda *et al.*, 2006). A way to preserve and even improve the localization properties of detectors, while reducing the number of readout channels, has been devised using a resistive foil between the last multiplication stage and the signal readout electrode (Dixit *et al.*, 2004; Boudjemline *et al.*, 2007). With careful adjustment of the parameters, the collected charge is spread over a larger area, reducing the required density of readout channels. A possible drawback of the method is a reduced rate capability, due to the longer signal integration times.

The choice of the operating gas in TPC-like detectors is dictated by several requirements: reasonably high drift velocities at moderate electric fields, low electron diffusion and non-flammability. In the framework of the study of detectors for the International Linear Collider (ILC), a good compromise has been found with the so-called TDR gas mixture (Ar-CH_4-CO_2 in the volume proportions 93–5–2). Figure 13.39 shows measured values of transverse space resolution as a function of drift distance and magnetic field, for two choices of the readout pads geometry (Janssen *et al.*, 2006).

Mixtures containing carbon tetrafluoride provide even better space resolution, thanks to the lower diffusion, with the disadvantage of requiring higher operating voltages (Oda *et al.*, 2006; Kobayashi *et al.*, 2011).

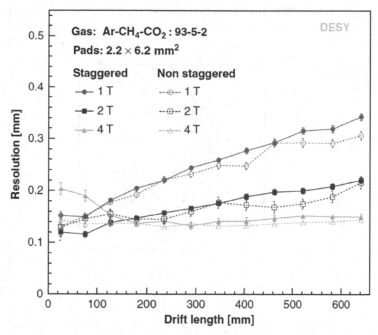

Figure 13.39   Transverse resolution of a TPC with GEM readout as a function of drift distance for several values of magnetic field and two readout pads geometries (Janssen *et al.*, 2006). By kind permission of Elsevier.

13.6 Active pixel readout

The use of anode pad arrays for the signal readout permits one to exploit at best the intrinsic high-granularity properties of MPGDs. The use of conventional systems of external front-end electronics becomes, however, very impractical for small pixel sizes; this has suggested the use of solid-state active pixel electronic circuits for direct collection and recording of the charge generated in the gaseous multiplication process.

In the X-ray polarimeter, a detector developed for astrophysics imaging applications, a custom-made CMOS integrated circuit with ~2000 pixels at 80 μm pitch collects the ionization released in the gas, after amplification with a GEM electrode (Bellazzini *et al.*, 2004). Figure 13.40 is an example of the two-dimensional recording of the ionization trail of a 5.2 keV photoelectron released by soft X-rays interacting in the gas (Bellazzini *et al.*, 2007a); the inset gives the size of the image. Analysis of the differential charge deposition along the track, represented in the figure by the size of the pixels, permits one to identify the higher energy loss at the track's end, and therefore the point and direction of emission, providing information on the X-ray polarization.

Thanks to the very low input capacitance and an electronic noise (typically around 50 electrons rms), even at moderate proportional gains the detector can

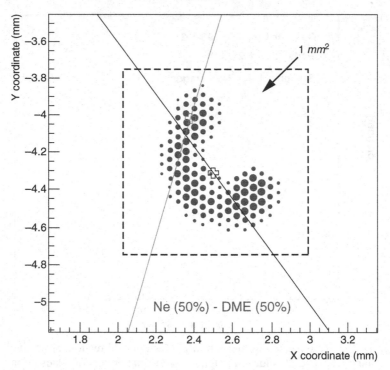

Figure 13.40　A 5.2 keV photoelectron track recorded with the GEM polarimeter (Bellazzini *et al.*, 2007a). By kind permission of Elsevier.

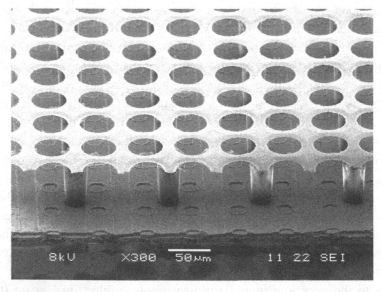

Figure 13.41　The GridPix detector, integrating a Micromegas-like structure over a solid-state pixel readout chip (van der Graaf, 2007). By kind permission of Elsevier.

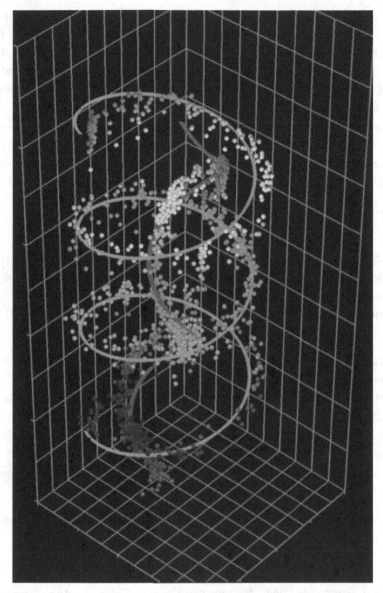

Figure 13.42   Low-energy electron tracks spiralling in a magnetic field, detected by a GridPix detector with TimePix readout (van der Graaf, 2011). By kind permission of Elsevier.

achieve single electron sensitivity and can be used for UV photon imaging (Bellazzini *et al.*, 2007b).

The active gaseous device can be directly built over an existing solid-state pixel circuit using silicon wafer post-processing technologies (Chefdeville *et al.*, 2006). Named InGrid or GridPix, these detectors combine the Medipix (Llopart *et al.*, 2002) or Timepix (Llopart *et al.*, 2007) CMOS pixel readout chips, developed for medical imaging using solid state sensors, with a Micromegas or GEM structure directly built over the solid state sensor (Figure 13.41) (van der Graaf, 2007; Malai *et al.*, 2011). Combined with a drift space in a TPC-like structure, GridPix detectors are capable of imaging sub-mm ionization trails, recording individual ionization clusters, as shown by the example in Figure 13.42 (van der Graaf, 2011).

A drawback of the described approach is the possible damage to the sensitive electronics caused by gas discharges. Various methods of spark protection have been developed, coating the silicon chip with insulating or high-resistivity layers (Bilevych *et al.*, 2011; van der Graaf, 2011).

13.7 MPGD applications

While the main motivation for the development of the new generations of micro-pattern devices has been particle tracking in high-energy physics, many other applications have been developed by exploiting the performances of the detectors. Only a short summary is given here; for detailed reviews see Titov (2007); Titov (2012); Sauli (2014):

High rate imaging of soft X-rays for plasma diagnostics (Pacella *et al.*, 2001; Pacella *et al.*, 2003);

X- and γ-ray astronomy and polarimetry (Bellazzini *et al.*, 2002; Black *et al.*, 2003; Bellazzini and Muleri, 2010; Bernard and Delbart, 2012);

High rate γ-ray imaging for medical portal imaging (Östling *et al.*, 2000; Östling *et al.*, 2003);

UV and visible-range gaseous photomultipliers (Buzulutskov *et al.*, 2000; Breskin *et al.*, 2001; Mörmann *et al.*, 2003; Breskin *et al.*, 2010).

The detection and imaging of UV photons for Cherenkov ring imaging are covered in the next chapter.

Further reading

Sauli, F. and Sharma, A. (1999) Micropattern Gaseous Detectors. *Ann. Rev. Nucl. Part. Sci.* **49**, 341.

Titov, M. (2007) New developments and future perspectives of gaseous detectors. *Nucl. Instr. And Meth.* **A581**, 25.

Sauli, F. (2013) Gas Electron Multiplier (GEM) detectors: principles of operation and applications, in *Comprehensive Biomedical Physics Vol. 6*, M. Danielsson (ed.) (Elsevier).

Selected conference proceedings

MPGD2011 (2013) Second International Conference on Micro-Pattern Gaseous Detectors, *JINST* **8**.

14

Cherenkov ring imaging

14.1 Introduction

Charged particles traversing a medium with a velocity exceeding the velocity of light in the material emit a coherent front of electromagnetic radiation, an effect discovered in the thirties by Pavel Cherenkov, Nobel laureate in physics 1958. Detection of the photons emitted by the Cherenkov effect provides information on the particle's velocity; combined with an independent measurement of total energy or magnetic deflection, it helps in identifying the mass of the particle. Although the emission occurs over the whole spectrum, photons can only be detected within the wavelength interval for which the medium is transparent.

The effect is widely exploited in particle physics for particle identification in specific ranges of mass and velocity, deploying a succession of radiators with different index of refraction; using photomultipliers as photon detectors, threshold Cherenkov counters have a limited angular acceptance, and are used mainly for particle tagging of un-separated high energy beams.

Above the threshold, the angle of emission of photons depends on the particle velocity. In early devices exploiting the measurement of the emission angle, named differential Cherenkov counters, optical means restrict the detection of photons with ring-shaped diaphragms, and are therefore sensitive in a well-defined range of velocity; more sophisticated methods for detecting and localizing individual photons over a wider angular range have been attempted, using various systems of image intensifiers, but are limited in use due to their high cost and narrow angular coverage. For a review of classic Cherenkov counter techniques see Benot *et al.* (1972); Litt and Meunier (1973); and Gilmore (1980).

The development in the early 1970s of large-area position-sensitive detectors opened up new possibilities. Filled with a photosensitive vapour and operated at high gains, MWPCs could be used to detect and localize single photons in the ultra-violet (UV) domain, as proposed in the late seventies by the seminal work of

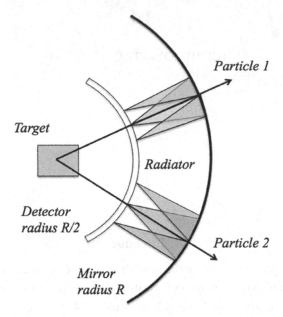

Figure 14.1   Schematics of a ring imaging Cherenkov detector (Seguinot and Ypsilantis 1977. By kind permission of Elsevier.

Jacques Séguinot and Thomas Ypsilantis (Seguinot and Ypsilantis, 1977). The cone of photons emitted in a radiator by charged particles above the Cherenkov threshold can be reflected in a ring pattern on a detector in the focal plane of a spherical mirror, Figure 14.1, or detected directly with the so-called proximity focussing geometry (Figure 14.2). In both cases, a measurement of the ring radius provides information on the particle's velocity; more dispersive, proximity focusing requires the use of thin liquid or solid radiators with large values of the refractive index. The two methods are often combined within the same detector with different radiators, to extend the velocity range of identification, Figure 14.3.

To be exploited, the method requires finding a gas with a photo-ionization threshold lower than the transparency cut-off of the windows separating detector and radiator. In the original proposal, the choice was the use of lithium fluoride windows, expensive and easily damaged by exposure to moisture, with acetone or benzene as additives to the gas. Further work by the same authors identified triethyl amine (TEA), with a photo-ionization threshold around 7.6 eV, as a more convenient choice (Séguinot *et al.*, 1980). Combined with easily available and stable calcium fluoride windows, it permits one to attain quantum efficiency close to 50% at the peak of the window transparency region.

The discovery of a compound, TMAE, with even lower photo-ionization threshold (~5 eV) (Anderson, 1980) and the later development of solid caesium

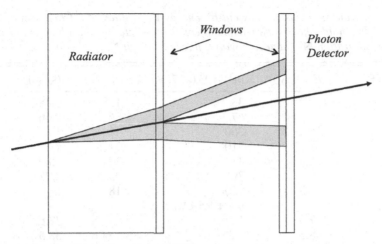

Figure 14.2   Proximity focusing RICH.

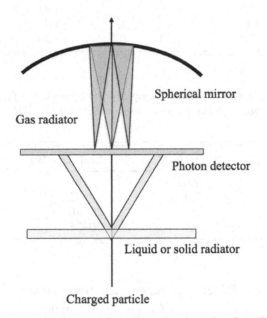

Figure 14.3   Combined gas ring imager and proximity focusing on a single photon detector.

iodide photocathodes that allowed one to use fused silica windows, easily available in large sizes, permitted the conception and construction of the large ring imaging Cherenkov counters (RICH)[1] particle identification systems that are described in the next sections.

---

[1] Also named CRID (Cherenkov ring imaging detectors)

Table 14.1 *Radiators: transparency cut-off energy* E_{TC}, *index of refraction* n *or* (n − 1), *Cherenkov threshold velocity* γ_T, *maximum emission angle* θ_M *and photons per cm of radiator at saturation* N_{PH} *detected with a figure of merit* $N_0 = 80$ *cm*$^{-1}$.

Gases (NTP)	E_{TC} (eV)	$(n-1)10^6$ (at 7eV)	γ_T	θ_M (°)	N_{PH} /cm
He		33	123	0.48	0.005
Ne		67.3	86	0.66	0.01
Ar	16	300	41	1.4	0.05
CH_4	8.6	510	31	1.83	0.08
CF_4	12.5	488	32	1.80	0.08
C_2F_6		793	25	2.28	0.13
i-C_4H_{10}	8.3	1500	18		0.24
Liquids (NTP)		**n (at 6.5 eV)**			
C_5F_{12}	7.5	1.262	1.638	37.6	30
C_6F_{14}	8.3	1.278	1.606	38.5	31
Solids		**n (5.5 eV)**			
LiF	11.6	1.42	1.41	45.2	43
MgF_2	10.8	1.41	1.41	44.8	40
CaF_2	10	1.47	1.36	47.1	43
Fused silica	7.5	1.52	1.33	48.8	45

Table 14.2 *Photo-ionization threshold energy* E_T, *peak quantum efficiency of photosensitive compounds QE and vapour pressure P at NTP (QE values depend on sources).*

	E_T (eV)	Peak QE (%)	P (NTP) (Torr)
Acetone	9.65	31	251
TEA $(C_2H_5)_3N$	7.5	33–42	73.2
TMAE $C_2[(CH_3)_2N]_4$	5.3	29–40	0.50
Caesium iodide CsI	5.9	20–35	$<10^{-5}$

Values of quantum efficiency of vapours and transparency of gases and window materials are covered in detail in Chapter 3. Compiled from different sources, Table 14.1 and Table 14.2 summarize some useful parameters for widely used components in the realization of RICH counters (Gray, 1963; Ekelöf, 1984; Nappi and Seguinot, 2005). It should be noted that often values differ depending on sources, measurement conditions and the presence of trace pollutants. Several works describe the purification systems and methods of use of the fluorocarbon fluids used as radiators (Albrecht *et al.*, 2003; Wiedner *et al.*, 2008; Ullaland, 2005; Hallewell, 2011).

By inspection of the tables, one can see that using TEA as photoionising vapour ($E_T = 7.5$ eV), the choice of radiators is restricted to gases and fluoride crystals; with TMAE or solid CsI the range extends to fluorinated liquids, covering a wider region of refractivity and therefore of threshold velocity.

14.2 Recalls of Cherenkov ring imaging theory

Only some basic expressions are given here; detailed discussions of the theory and practice of the ring imaging technology can be found in many reviews (Ypsilantis, 1981; Arnold *et al.*, 1992; Seguinot and Ypsilantis, 1994; Ypsilantis and Seguinot, 1994; Nappi and Seguinot, 2005; Dalla Torre, 2011).

A charged particle traversing an optical medium with refractive index n emits a Cherenkov radiation when its relativistic velocity β exceeds a threshold value $\beta_T = 1/n$. The angle of emission θ relative to the particle direction is:

$$\cos\theta = 1/n\beta, \tag{14.1}$$

reaching a maximum value for $\beta \to 1$:

$$\theta_{\mathrm{MAX}} = \cos^{-1}\frac{1}{n}. \tag{14.2}$$

Computed from expression (14.1), Figure 14.4 shows the dependence of the Cherenkov angle on the particle velocity and several values of refractive index; large values of n result in low thresholds and large emission angles, and conversely for small values.

The number of photons emitted in the energy interval ΔE is given by the Frank–Tamm expression:

$$\frac{\Delta N}{\Delta E} = \left(\frac{\alpha}{\hbar c}\right)Z^2 L\sin^2\theta = \left(\frac{\alpha}{\hbar c}\right)Z^2 L\left[1 - \left(1 - \frac{n}{\beta}\right)^2\right], \tag{14.3}$$

where α is the fine structure constant, $\hbar$ Plank's reduced constant and Z and L the particle's charge and path length in the radiator.

Over the narrow wavelength interval between the window's transparency cut-off and the detector's photo-ionization threshold, for which n can be considered constant, integration of expression (14.3) gives a simple expression for the number of detected photons:

$$N = N_0 Z^2 L\sin^2\theta, \tag{14.4}$$

where the figure of merit N_0, characteristic of each detector, depends on the detector efficiency ε_D, the reflectivity of the mirror ε_R, and the optical transmission ε_T, all functions of the photon energy; it can be computed by integration over the

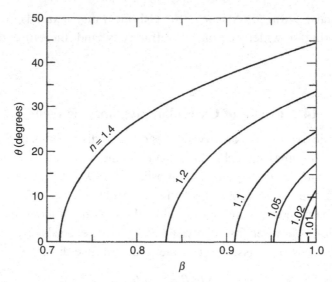

Figure 14.4   Cherenkov emission angle as a function of velocity for several values of the medium's index of refraction.

spectral sensitivity range, between the photo-ionization threshold and the window transparency cut-off:

$$N_0 = \frac{1}{137\hbar c} \int_{E_1}^{E_2} \varepsilon_D(E)\varepsilon_R(E)\varepsilon_T(E)\,dE. \tag{14.5}$$

Typical values of the parameter N_0 for RICH detectors are between 50 and 100 photons/cm.

For an ideal optical system such as the one shown in Figure 14.1, the relative velocity resolution for N detected photons is given by:

$$\frac{\Delta\gamma}{\gamma} = \frac{\gamma^2\beta^2 n}{\sqrt{N_0 L}}\Delta\theta, \gamma = 1/\sqrt{(1-\beta^2)},$$
$$\Delta\theta = \Delta\theta_1/\sqrt{N}, \tag{14.6}$$

where $\Delta\theta_1$ is the single photon angular error, that includes all dispersive effects: the photon and particle direction localization, geometrical and optical aberrations. At high photon energies, an unavoidable contribution is the chromatic aberration caused by the dependence of the index of refraction of the medium on the wavelength of the emitted photons. The wavelength dependence of the refractive index in argon is given in Figure 14.5 (Langhoff and Karplus, 1969); the refractive index of fluoride crystals, used as detector windows or solid radiators, is shown in Figure 14.6 (Gray, 1963; Roessler and Walker, 1967). Insets in the figures provide

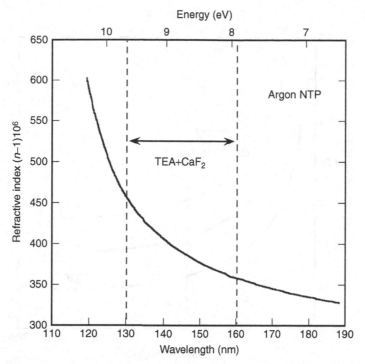

Figure 14.5: Wavelength dependence of the refractive index in argon. Data from Langhoff and Karplus (1969).

the approximate regions of sensitivity for a detector using TEA as photosensitive vapour and a calcium fluoride window, and respectively quartz windows and TMAE or CsI as photosensitive agents. A compilation of refractive index values as a function of wavelength for gases and liquids is given in Ullaland (2005); Figure 14.7 is an example of the dependence for several fluorocarbons used as radiators.

For optimal RICH detector performance, the various dispersive contributions should be minimized. Figure 14.8 is an example of computed dependence of the single photon angular resolution on three dispersive sources for a proximity focusing detector with a CsI photocathode, as a function of the Cherenkov angle and for two values of the particle incidence angle (Nappi and Seguinot, 2005).

The optimization depends on the detector geometry and operating parameters, and determines the particle identification power of the detector. Figure 14.9 gives a representative example of computed resolution of a TEA-operated RICH device with a 50 cm long, 10 atmosphere argon radiator and a 50 cm focal length mirror; the right scale gives the expected number of detected photons per ring (Charpak *et al.*, 1979a).

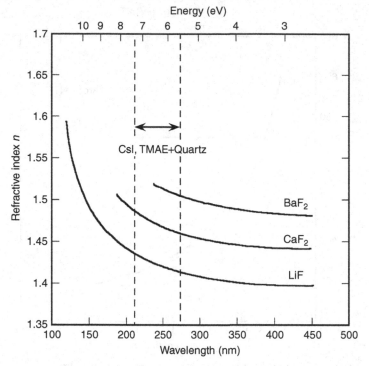

Figure 14.6   Wavelength dependence of the index of refraction of fluorides. Data from Gray (1963) and Roessler and Walker (1967).

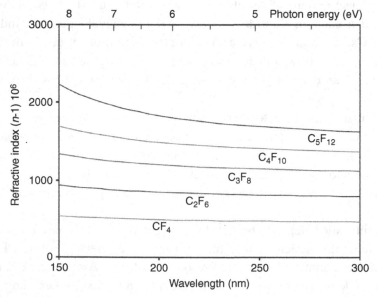

Figure 14.7   Index of refraction of *n*-perfluorocarbons as a function of wavelength (Ullaland, 2005). By kind permission of Elsevier.

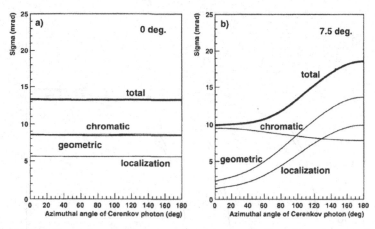

Figure 14.8   Cherenkov angle error due to several dispersive factors, for two values of the particle incidence angle (Nappi and Seguinot, 2005). With kind permission of Springer Science+Business Media.

14.3 First generation RICH detectors

The practical implementation of the RICH technology with gaseous detectors depends on solving two crucial issues: the conversion of photons into electrons by a suitable photo-ionizing agent, and the possibility of reaching gas gains high enough to detect single photoelectrons. As discussed in Section 8.10, due to the copious emission of UV photons by the avalanches, reconverting in the sensitive volume, the two requirements are conflicting. A solution was found in the late 1970s by operating the multi-step avalanche chamber structure with a triethyl amine (TEA)-rich gas mixture (Charpak and Sauli, 1978), see Section 8.10.

Owing to high vapour pressure of TEA, 55 torr at room temperature, Figure 14.10 (Anderson, 1988), the absorption length for UV photons is a few mm; detectors can therefore achieve a few ns time resolutions and operate at high event rates. Sensitive in the VUV region, between 7.5 and 9.5 eV with a calcium fluoride window, and thanks to the suppression of photon-induced feedback between the two amplification elements, the device can attain gains well above 10^5, adequate for fully efficient detection of single photoelectrons. The region of sensitivity restricts the choice of the main filling gas, which has to be transparent for the photons; a mixture of helium with 3% TEA was used in the detector described below.

Recording the avalanche charge on anodes and on cathode wires at an angle to the anodes with the methods described in Chapter 8, the detector can localize unambiguously multiple photoelectrons with sub-millimetre accuracy. Figure 14.11

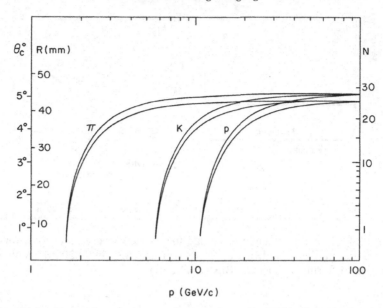

Figure 14.9   Example of computed resolution of a TEA-CaF$_2$ RICH detector with a 50 cm long, 10 atmosphere argon-filled radiator as a function of momentum (Charpak *et al.*, 1979a). By kind permission of Elsevier.

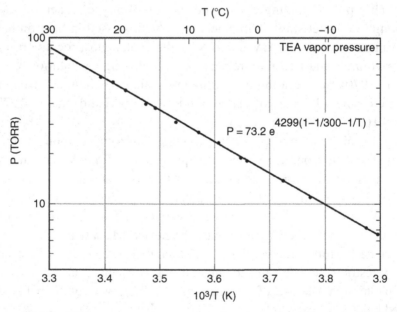

Figure 14.10   TEA vapour pressure as a function of temperature (Anderson, 1988). By kind permission of Elsevier.

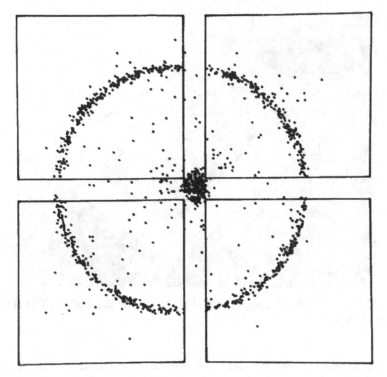

Figure 14.11   Overlapping Cherenkov rings detected with a TEA-filled multi-step chamber; the ring radius is 7 cm (McCarty *et al.*, 1986). By kind permission of Elsevier.

is an example of a ring image obtained by overlapping several hundred events produced by a proton beam in a helium-filled radiator, 8 m long, perpendicular to a multi-step chamber with an active surface of 20×20 cm^2. The squares are the boundaries of the composite four-crystal calcium fluoride window; the central region corresponds to the direct detection of the beam (McCarty *et al.*, 1986).

A large RICH detector of the described design was built and operated for many years for particle identification in the experiment E605 at FERMILAB (Adams *et al.*, 1983; McCarty *et al.*, 1986). Two identical multi-step chambers with 40×80 cm^2 active area, mounted on off-beam flanges of a 15 m long helium radiator, detect the photons reflected by a set of mirrors at the far end of the vessel; the calcium fluoride windows separating detector and radiator, assembled at CERN with a composite matrix of 32 crystals, are probably the largest CaF$_2$ windows ever built (Figure 14.12). Localization is performed by recording the signals on anodes and on groups of cathode wires at $\pm$ 45° to the anodes; Figure 14.13 shows a five-photon event, demonstrating the ambiguity-free reconstruction properties of the detector. Operated for many

Figure 14.12   The large composite CaF_2 window of the E605 RICH. Picture CERN (1982).

years, with an average of 3.5 detected photons for saturated rings, the E605 RICH could identify pions from kaons up to 200 GeV/c; an early example of experimental ring radius distribution is shown in Figure 14.14 (McCarty *et al.*, 1986).

Alternatively to bulky gas radiators, and capable of identifying lower momentum particles, lithium and calcium fluoride can be used as radiators with the proximity focusing geometry. Owing to the large index of refraction of the crystals, however, photons emitted by near perpendicular particles undergo total internal reflection and are not detected. An elegant solution to the problem was developed for the CLEO III LiF-TEA ring imaging system, using an assembly of long, wedge-cut crystals arranged in a saw-tooth geometry; with around ten detected photons per ring, after thorough reconstruction taking into account the radiator geometry, a single photon angular resolutions of 12 mrad could be achieved (Mountain *et al.*, 1999; Artuso *et al.*, 2003).

14.4 TMAE and the second generation of RICH detectors

Despite their successful use in experiments, due to the sensitivity in the far UV domain, TEA-based photon detectors suffer from the described limitations in the choice of windows and radiators; they have also large intrinsic chromatic aberrations, which affect the resolution. Originally introduced in the 1980s to detect the

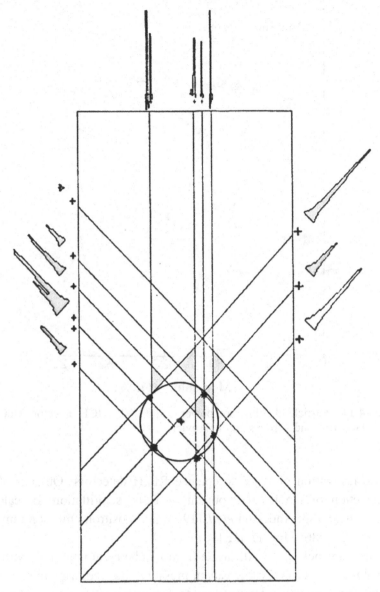

Figure 14.13   A five-photon ring detected in the E605 RICH (McCarty *et al.*, 1986). By kind permission of Elsevier.

photons emitted by the scintillation of xenon (Anderson, 1980), tetrakis dimethyl amino ethylene (TMAE),[2] thanks to its very low photo-ionization threshold, 5.3 eV, which permits the use of fused silica windows, TMAE represents a big step

---

[2] Colloquially pronounced 'Tammy', TMAE is part of a family of compounds spontaneously fluorescing by oxidation in air, developed as trackers for military applications. While this property imposes strict tightness requirements on detectors, it has been exploited to detect leaks in the detector.

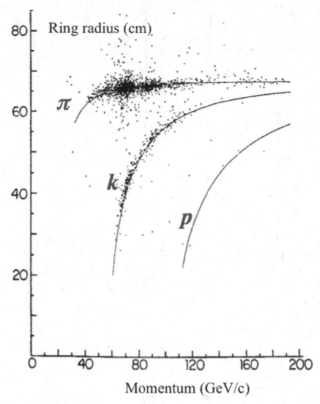

Figure 14.14   Particle identification power of the E605 RICH detector (McCarty *et al.*, 1986). By kind permission of Elsevier.

towards the realization of large acceptance RICH detectors. Other applications include detection of the fast component of BaF_2 scintillation for calorimetry (Anderson *et al.*, 1983; Anderson *et al.*, 1984) and positron emission tomography (Shotanus *et al.*, 1986; Miné *et al.*, 1988).

The main drawback of TMAE, aside from its chemical reactivity with oxygen and many materials, is its very low vapour pressure, thus requiring thick absorption gaps and/or high operating temperature to achieve good photon detection efficiency, Figure 14.15 (Anderson, 1988). Computed for a peak cross section of 30 Mb, Figure 14.16 provides the equivalent absorption length for the saturated vapour as a function of temperature; it is about 1 cm at 30°C. While this sets an intrinsic limit to the achievable time resolution, it was suitable for use in the forthcoming generation of colliding beam experimental systems. For historical surveys of TMAE-based RICH detectors see for example Buys (1996) and Engelfried (2011) as well as the proceedings of regular workshops dedicated to the subject.

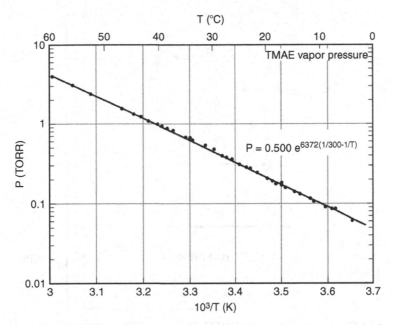

Figure 14.15 TMAE vapour pressure as a function of temperature (Anderson, 1988). By kind permission of Elsevier.

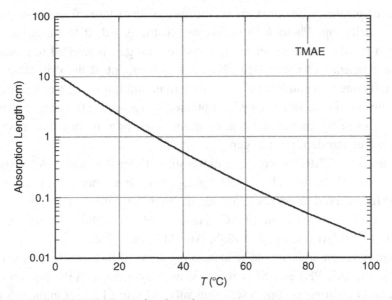

Figure 14.16 TMAE absorption length.

TMAE has been successfully used in detectors based on the multi-step chamber (Baur *et al.*, 1994). However, a two-dimensional time projection (TPC)-like drift chamber of the design shown in Figure 14.17, developed in the early 1980s, appeared more suitable to cover the large detection areas required by accelerator

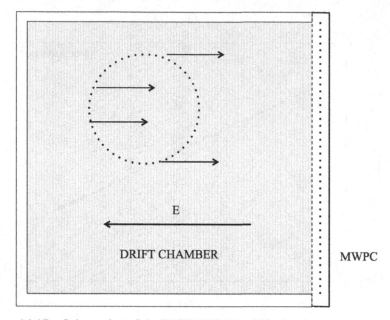

Figure 14.17   Schematics of the RICH TPC-like drift chamber.

experiments with moderate event rates (Ekelöf *et al.*, 1981; Barrelet *et al.*, 1982). Two sets of parallel wires or strips on the inner side of large fused silica windows define the drift gap. Photoelectrons created in the gas drift towards the end-cap detector, a multi-wire chamber; a thick conversion gap is needed to ensure good detection efficiency (5 cm in DELPHI). A measurement of the drift time on each wire permits one to reconstruct the ring pattern unambiguously; a coarse determination of the coordinate along the wire, obtained by recording the charge induced on cathode strips or by current division on the anodes, permits one to correct for the photon conversion depth in the gap.

First used at CERN's proton–antiproton collider for the UA2 experiment (Botner *et al.*, 1990), the TPC RICH concept was developed for the construction of the OMEGA fixed target spectrometer at CERN (Apsimov *et al.*, 1986), and the collider experiments SLD at SLAC (Ashford *et al.*, 1986; Abe *et al.*, 1998), DELPHI at LEP (Arnold *et al.*, 1988b; Arnold *et al.*, 1988a).

Similar in design and capable of identifying charged particles in a wide range of velocities, the DELPHI and SLD modules employ a single TPC photon detector between two radiators, as shown schematically in Figure 14.18. On the gas radiator side, a set of spherical mirrors reflects the photon ring through the upper quartz window of the drift chamber; on the lower side, thin liquid radiator boxes with quartz windows on the detector side generate a ring pattern by proximity focusing. Figure 14.19 shows the computed Cherenkov angle as a function of particle momentum for the gas and liquid radiators indicated in the insets.

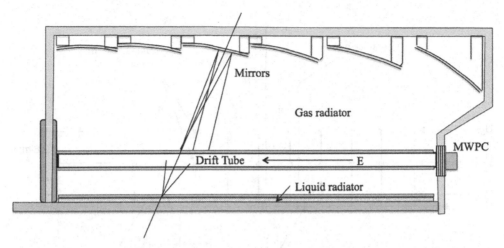

Figure 14.18   Schematics of the DELPHI barrel RICH (Arnold *et al.*, 1988b). By kind permission of Elsevier.

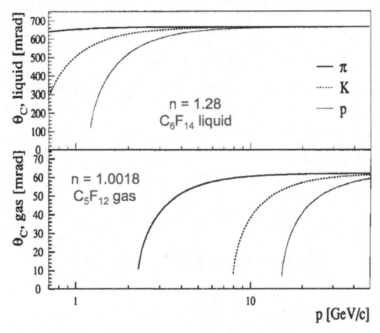

Figure 14.19   Cherenkov emission angles as a function of momentum for liquid and gas radiators (Adam *et al.*, 1996). By kind permission of Elsevier.

Due to the long photon absorption path in the low concentration photosensitive vapour, the gain achievable with the end-cap MWPC is limited by the appearance of secondary processes. Several detector geometries have been developed to limit the avalanche spread; Figure 14.20 shows two solutions, with sectored blinds

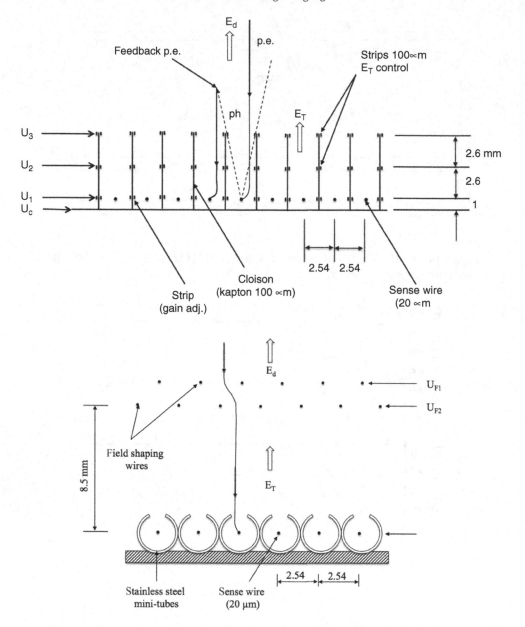

Figure 14.20 Photon feedback limiting structures: sectored blinds (top) and open-slit tubes (bottom) (Arnold *et al.*, 1986). By kind permission of Elsevier.

('cloisons') or sliced proportional tubes (Arnold *et al.*, 1986); geometry and electric field are designed so that all photoelectrons are collected by the anode wires, while the largest fraction of the photons emitted in the avalanches is absorbed by the walls. The first solution has been retained for most experiments.

Figure 14.21   Two DELPHI barrel RICH TPC modules. Picture CERN (1986).

Figure 14.21 shows two 160 cm long modules of the DELPHI barrel RICH during construction, and Figure 14.22 the insertion of the 24-module full assembly into the mirror's barrel. After completion, the detector is installed horizontally between the time projection chamber and the magnet inner coil of the experiment.

Due to the detector geometry, for angled tracks the liquid radiator rings are in fact elongated hyperbola-like patterns, truncated at large angles by the total reflection in the liquid–window interfaces (Figure 14.23); the Cherenkov angle can be reconstructed by measuring the particle trajectory directly in the detector or in external devices (Arnold *et al.*, 1988b; Adam *et al.*, 1996; Erzen *et al.*, 1999). Figure 14.24 shows the single photon and ring radius angular resolutions measured for the liquid and gas radiators (Albrecht *et al.*, 1999); the particle identification capability of the detector is illustrated in Figure 14.25 (Schins, 1996).

A big effort was undertaken to refine the TMAE purification process needed to attain the desired quantum efficiency; its chemical reactivity with many materials has been studied in detail to permit the construction and long-term operation of detectors (Hallewell, 1994). The successful development of solid photocathodes, described in the next section, led to the dismissal of TMAE-based detectors.

14.5 Third generation RICH: solid caesium iodide (CsI) photocathodes

Compared to photo-ionizing gas fillings, a solid photocathode would have the advantage of an intrinsicly better time resolution due to the isochronous emission

Figure 14.22   Insertion of the completed DELPHI RICH into the mirror barrel. Picture CERN (1986).

and collection of photoelectrons. The possibility of using a thin layer of caesium iodide (CsI), which has a 5.9 eV photo-ionization threshold, was discussed in the early works on RICH detectors (Arnold *et al.*, 1992). However, due to the reactivity of CsI with oxygen and water, it took several years to develop the appropriate deposition method, substrate kind and preparation and handling technologies to ensure high and uniform quantum efficiency in large area gaseous detector (Rabus *et al.*, 1999; Schyns, 2002 Braem *et al.*, 2003; Cisbani *et al.*, 2003). As shown by the example in Figure 14.26, the QE for the large areas required for HEP applications has been improved by a factor of two from the early measurements, approaching the values measured for monochromatic UV photons on small laboratory samples (Di Mauro, 2004). The CsI layer is deposited on the photocathodes by vacuum evaporation; although a brief exposure to air is known not to damage the layer, systems of leak-tight transfer boxes in argon are used to store the electrodes and mount them in the final detector assembly. A detailed description of the system developed for the production, handling and quality control of the large CsI photocathodes for the ALICE RICH is given in Hoedlmoser *et al.*, (2006).

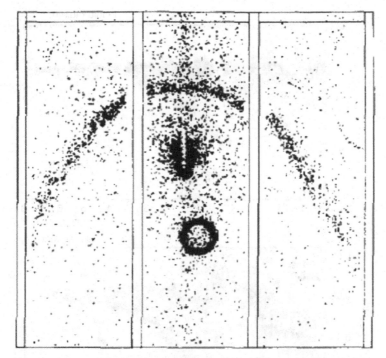

Figure 14.23   Superimposed image of 200 ring images for beam tracks at 18° with a prototype detector plane. The small ring and the elongated hyperbola are due respectively to the gas and the liquid radiators. The central scattered points are due to the beam (Arnold *et al.*, 1988b). By kind permission of Elsevier.

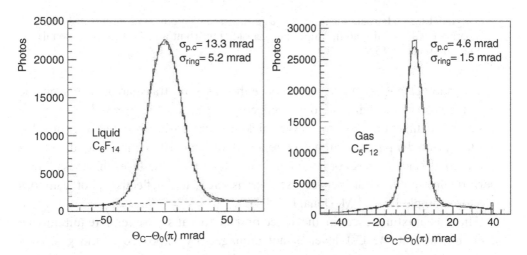

Figure 14.24   Cherenkov single photon and ring resolutions achieved with the DELPHI RICH (Albrecht *et al.*, 1999). By kind permission of Elsevier.

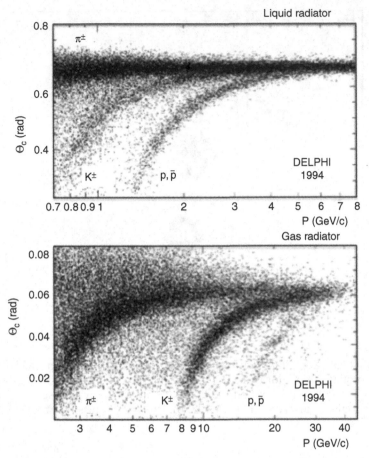

Figure 14.25   Particle identification as a function of momentum in the DELPHI barrel RICH for the liquid (top) and gas radiators (bottom). E. Schins personal communication (Schins, 1996).

CsI photocathodes can be deposited either as semi-transparent layers on the detector side of the window, or as reflective layers on the electrode facing the window. In the first case, the layer thickness has to be thoroughly controlled not to exceed the photoelectron capture length, around 100 Å; the second solution is preferred since it does not require a strict tolerance in the layer thickness. For a performance comparison between semi-transparent and reflective photocathodes see for example Lu and McDonald (1994).

The gas mixture used in the detector has to satisfy several requirements. The first is that the CsI layer is not damaged by exposure to the gas, thus imposing a strict control of the oxygen and water contaminants. For reflective photocathodes, as photons reach the layer through the gas gaps, the transparency should be close to 100% in the energy range between threshold

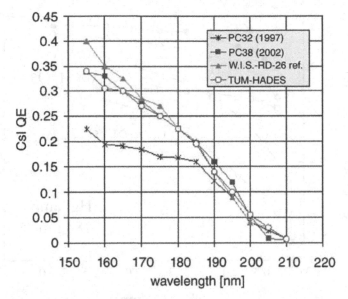

Figure 14.26 Improvement in quantum efficiency of large area CsI photo-cathodes (Di Mauro, 2004). Data are for the first ALICE prototype (PC32), the production modules ALICE (PC38) (Friese *et al.*, 1999) and HADES (TUM) (Fabbietti *et al.*, 2003). The curve labelled RD-26 is the best value obtained for small samples (Breskin *et al.*, 1995). By kind permission of Elsevier.

and window cut-off; photon absorption curves for gases commonly used in counters were given in Section 3.4. Once produced, photoelectrons have to escape from the layer; the extraction efficiency is a function of the external electric field, and depends on the gas, as shown in Figure 14.27, providing the recorded photocurrent in a cell normalized to the value in vacuum (Breskin *et al.*, 2002).

For mixtures including noble gases, due to their large electron backscattering cross sections, losses are considerable even at very high values of the field (Coelho *et al.*, 2007b). Despite the safety concerns caused by its flammability, methane has been adopted as gas filling by most experiments since it provides near-vacuum extraction at moderate values of the field; moreover, it permits one to reach the gas gains required to detect single photoelectrons, and it is fully transparent in the desired region of wavelength. The use of carbon tetrafluoride, which would solve the safety issues, raises concerns about the long-term damages induced by fluorinated compounds released in the avalanches (see Section 16.4).

Possible alternatives using neon mixtures are compared with pure methane in Figure 14.28 (Azevedo *et al.*, 2010).

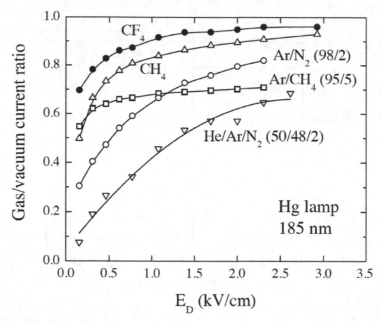

Figure 14.27 Ratio of photocurrent in different gases to that in vacuum as a function of field for CsI (Breskin *et al.*, 2002). By kind permission of Elsevier.

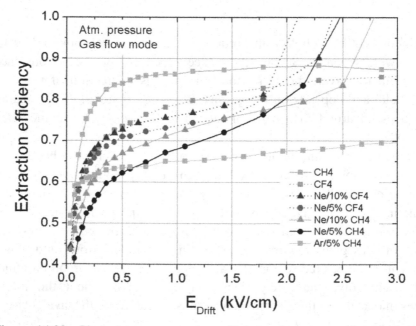

Figure 14.28 Photoelectron extraction efficiency in neon- and argon-based mixtures (Azevedo *et al.*, 2010). © IOP Publishing Ltd and Sissa Medialab srl. Reproduced by kind permission of IOP Publishing.

14.6 CsI-based RICH particle identifiers

CsI-based gaseous photon detectors have been operated in many experimental setups; for comprehensive reviews see Piuz (2003) and Dalla Torre (2011). Differing in the detailed geometry and choice of the radiator, these systems share a common design of the detector, an MWPC with the photosensitive layer deposited on the cathode facing the window (Figure 14.29); a mesh in contact with the window serves the purpose of collecting the direct ionization of the particles. In proximity focusing systems, where the liquid radiator is in contact with the window, the distance between the MWPC and upper mesh is larger, up to 10 cm, to increase the size of the ring. In most cases, pure methane at atmospheric pressure is used as gas filling.

Due to the expected multiplicity of the events (several rings with many photo-electrons), a projective readout is generally unsuitable; the readout plane is then segmented into a matrix of pads individually instrumented. Photoelectrons extracted from the CsI layer drift to the anodes and multiply in avalanche, thus inducing a charge distribution on the cathode pads; the geometry and size of the pads are chosen to permit a centre of gravity interpolation of the recorded charge and improve the coordinate determination.

The COMPASS RICH-1 detector is shown schematically in Figure 14.30. Two sets of CsI-MWPCs are mounted on a large volume gas radiator; the mirror array reflects the ring images on the detector planes, situated outside the high flux region of the primary beam (Albrecht *et al.*, 2003). Figure 14.31 is an example of a multiple ring pattern recorded with the detector, and Figure 14.32 shows the particle identification capability of the instrument (Abbon *et al.*, 2007).

Named the high momentum particle identification detector (HMPID), the ALICE proximity focusing RICH device deploys seven CsI-MWPC modules surrounding the interaction region at CERN's LHC; each module mounts six CsI-coated 64×40 cm^2 photocathode panels, segmented into pads of 8×8.4 mm^2.

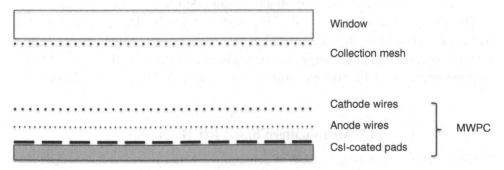

Figure 14.29   Schematics of the photon detector with a CsI photocathode.

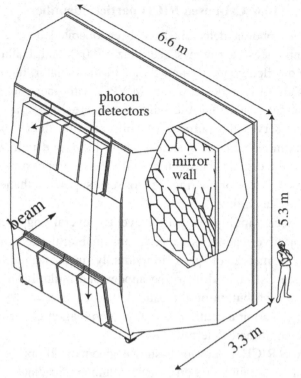

Figure 14.30   Artist's view of the COMPASS RICH-1 detector (Albrecht *et al.*, 2003). By kind permission of Elsevier.

Full-scale prototypes of the ALICE HMPID were tested both at CERN and in the STAR experiment at RHIC in Brookhaven. With a 15 mm thick perfluorhexane (C_6F_{14}) radiator, close to twenty reconstructed photon hits per ring are observed at saturation; Figure 14.33 is a single-track event display recorded in a test beam run (Cozza *et al.*, 2003). Most events share the charge between adjacent pads, as shown by the shades of grey in the picture, permitting interpolation. The resolution of the reconstructed ring radius is about 3 mrad (De Cataldo *et al.*, 2011).

Successfully operating in the ALICE experiment, but limited for particle identification to about 5 GeV by the use of high refractive index radiators, the detector is getting upgraded, adding a mirror-focused gas radiator to extend the useful range of momentum up to 10–30 GeV (Volpe *et al.*, 2008; Di Mauro *et al.*, 2011).

14.7 Micro-pattern based RICH detectors

While they are potentially capable of achieving a better time resolution than gas-phase photon detectors, experience has shown that MWPC-based CsI devices

Upper Photon Detectors

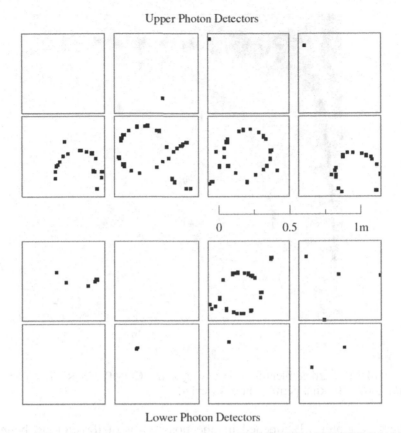

0   0.5   1m

Lower Photon Detectors

Figure 14.31   A multi-ring event recorded with the COMPASS RICH-1 (Abbon *et al.*, 2007). By kind permission of Elsevier.

suffer from a number of limitations affecting their performance. The use of a gas filling transparent to UV photons and the open MWPC geometry do not prevent photons emitted by the avalanches from reaching the photocathode, thus imposing a limit on the maximum gain that can be safely reached; as a consequence, relatively slow, low noise amplifiers have to be used, partly losing the fast intrinsic detector response. As observed in the COMPASS RICH, after a discharge induced by local defects, the detectors have a long-term memory of the event, preventing for days the application of the full voltage (Albrecht *et al.*, 2005). Even in the best operating conditions, pronounced long-term photocathode degradations have been observed and attributed to the bombardment by the ions produced in the avalanches (Hoedlmoser *et al.*, 2007). The cathode-induced signal readout method also sets a limitation on the two-photon separation, typically of several cm. The advent of the new generation of micro-pattern gas detectors, and in particular of the gas electron

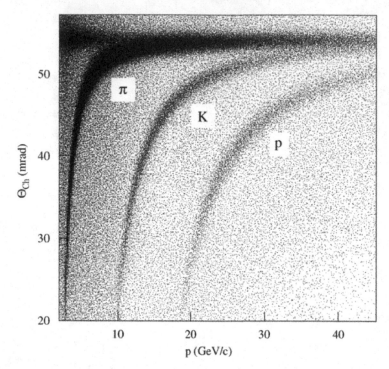

Figure 14.32   Particle identification power of the COMPASS RICH-1 (Abbon *et al.*, 2007). By kind permission of Elsevier.

multipliers (Chapter 13), opened up the possibility of developing better performing RICH devices.

Figure 14.34 shows schematically a GEM detector having the upper surface of the first electrode coated with a CsI photosensitive layer (Meinschad *et al.*, 2004); photoelectrons extracted from the layer are pulled by the applied field into the nearest hole, where a first step of charge amplification occurs. One or more GEM foils in cascade then permit one to attain the gain needed for detection, even in pure noble gases and their mixtures, an essential property in view of the realization of sealed counters (Buzulutskov *et al.*, 2000).

The device has several potential advantages. As most of the multiplication occurs in the last step of a multi-GEM structure, photons emitted by the avalanches cannot reach the photocathode and induce secondary processes. Only a minor fraction of the ions (those produced in the first stage of multiplication) hit the photocathode; using a small reverse electric field in the first gap, most primary ionization goes undetected. As in standard multi-GEM detectors, the amplified electron charge is directly collected on patterned electrodes, and can provide sub-mm position resolution. Due to the constant collection time, the intrinsic time resolution, if not degraded by the readout electronics, is of a few ns (Mörmann *et al.*, 2003).

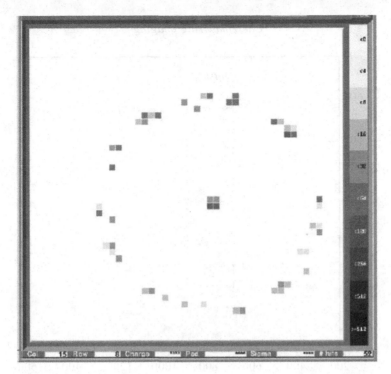

Figure 14.33   High momentum pion Cherenkov ring in the ALICE HMPID (Cozza *et al.*, 2003). By kind permission of Elsevier.

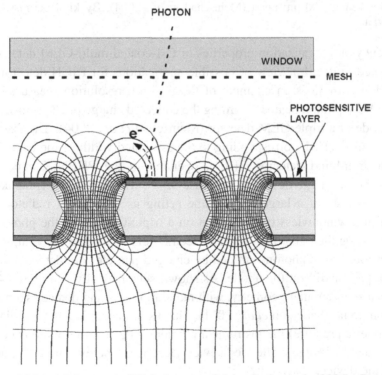

Figure 14.34   CsI-coated GEM detector (Meinschad *et al.*, 2004). By kind permission of Elsevier.

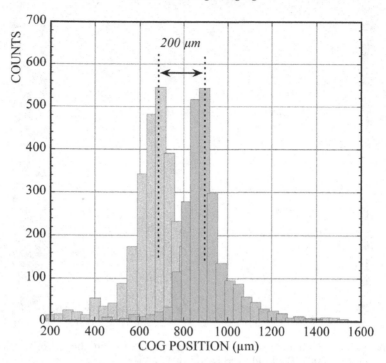

Figure 14.35   Position resolution measured with a CsI-GEM for two collimated photon beams 200 μm apart (Meinschad *et al.*, 2004). By kind permission of Elsevier.

Efficiency and localization properties of CsI-coated multi-GEM detectors have been studied in detail in view of possible applications for RICH detectors (Sauli, 2005). Figure 14.35 is an example of the position resolution measured for two collimated UV photon beams recording the collected charge on 200 μm wide strips on the anode of a triple-GEM detector (Meinschad *et al.*, 2004). The distributions have 200 μm FWHM, corresponding to the estimated width of the beam.

The hadron-blind detector (HBD), operated in the PHENIX experiment at RHIC, is the first large device making use of a CsI-coated GEM (Fraenkel *et al.*, 2005). It consists of a large gas volume acting as Cherenkov radiator, directly mounted in a windowless configuration on a triple-GEM with the photosensitive electrode facing the radiator. A reverse electric field applied to the volume collects the direct ionization; photons emitted by charged particles are detected and localized by a pad matrix on the anode. As shown schematically in Figure 14.36, the HBD has a semi-circular geometry, with radial radiators and detectors aimed at the interaction point (Anderson *et al.*, 2011). Thanks to the choice of the radiator (CF_4 at atmospheric pressure) the instrument efficiently detects the photons emitted by electrons and positrons, while the heavier hadrons are below threshold, hence the name of the device.

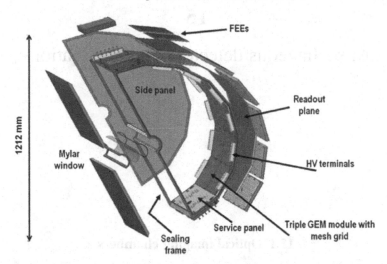

Figure 14.36   Schematic assembly of the HBD (Anderson *et al.*, 2011). By kind permission of Elsevier.

Manufactured on thin printed circuit board, the scaled-up thick GEM (TGEM, see Section 13.4) has the mechanical advantages of a rigid structure in view of the realization of large detector areas; optimization of performances and design show that gains sufficiently large for single photon detection can be obtained with a single multiplier (Alexeev *et al.*, 2011; Alexeev *et al.*, 2012).

Further reading

Ypsilantis, T. (1981) Cherenkov Ring Imaging. *Phys. Scripta* **23**, 371.
Nappi, E. and Seguinot, J. (2005) Ring Imaging Cherenkov Detectors: the state of the art and perspectives. *Riv. Nuovo Cimento* **28**, 1.

Selected conference proceedings

RICH2007 (2008) Sixth International Workshop on Ring Imaging Cherenkov Detectors, S. Dalla Torre, B. Gobbo and F. Tessarotto (eds.) *Nucl. Instr. and Meth.* **A595**, 1.
RICH2010 (2011) Seventh International Workshop on Ring Imaging Cherenkov Detectors, G. Hallewell, R. Forty. W. Hofmann, E. Nappi and B. Ratcliff (eds.) *Nucl. Instr. and Meth.* **A639**, 1.

15

Miscellaneous detectors and applications

15.1 Optical imaging chambers

The process of electron–molecule collisional photon emission, described in Chapter 5, is exploited in the gas scintillation proportional counters; with photomultiplier readout, the devices detect the primary or the field-enhanced secondary gas scintillation, providing energy resolutions approaching the statistical limit for soft X-rays. At high charge gains, the photon emission is copious enough to permit the use of pixelated photon sensors, image intensifiers, solid state cameras and similar, for imaging the position of the radiation through a window and with suitable optics. While limited in rate by the recording hardware, optical imaging has the potential advantage of a high granularity at low cost.

In a TPC-like configuration, ionization released in the gas volume drifts to a charge-multiplying element; a suitable optics focuses the image on the sensor (Figure 15.1). Due to the long integration time of standard cameras, the recorded image corresponds to the projection of the tracks on the plane of the multiplier, integrated over the full drift time; addition of a photomultiplier to record the time structure of the light emission can provide further information on the event.

As the scintillation of most gases is peaked in the far or vacuum ultra-violet, in order to permit the use of cheap windows and sensors, the wavelength of the photon emission has to be shifted close to the visible range. This can be obtained with a thin layer of wavelength shifter (WLS) deposited on the inner side of the window, or more conveniently using an additive to the main gas acting as internal wavelength shifter. Early work in this direction, using acetone as internal WLS and an image intensifier as sensor, demonstrated the feasibility of the technique (Gilmore *et al.*, 1983).

The low-ionization threshold vapours studied in the development of the Cherenkov ring imaging technique, TEA and TMAE (see Chapter 14) are particularly efficient internal shifters, with a peak of emission under avalanche conditions

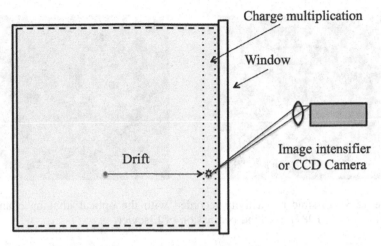

Figure 15.1   Scheme of the optical imaging chamber.

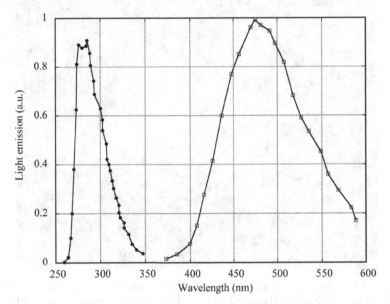

Figure 15.2   Scintillation emission spectra of TEA and TMAE. Data from Suzuki *et al.* (1987) and Charpak *et al.*, (1988). By kind permission of Elsevier.

around 280 and 480 nm, respectively (Figure 15.2) (Suzuki *et al.*, 1987; Charpak *et al.*, 1988). Although the longer wavelength of the TMAE emission is easier to detect and image, the difficult manipulation and chemical reactivity have generally discouraged its use.

A standard MWPC with a gas filling containing 7% triethyl amine (TEA) was used in the early development of the optical imaging chamber; however, a parallel-plate single or multi-step structure appeared to be a better solution because of the

Figure 15.3 Cosmic ray activity recorded with the optical imaging chamber (Charpak *et al.*, 1987). By kind permission of Elsevier.

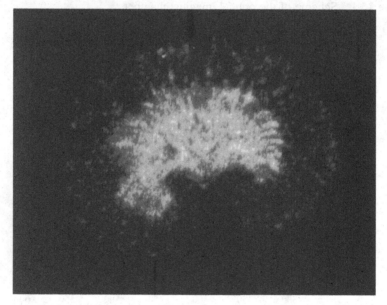

Figure 15.4 Autoradiography of a tritiated rat kidney slice recorded with the optical imaging chamber (Dominik *et al.*, 1989). By kind permission of Elsevier.

larger gains achievable and the uniform response, not modulated by the wire spacing; the photon emission close to the surface of the window reduces parallax errors, particularly when using a thin foil WLS (Charpak *et al.*, 1987; Suzuki *et al.*, 1988). Screen shots of cosmic ray activity recorded with a small optical imaging chamber are shown in Figure 15.3 (Charpak *et al.*, 1987).

The optical imaging chamber has been used for applications in several fields. The image in Figure 15.4 is the optical recording of the activity from a tritium-loaded slice of a rat's kidney, placed inside the detector over the cathode of a parallel plate

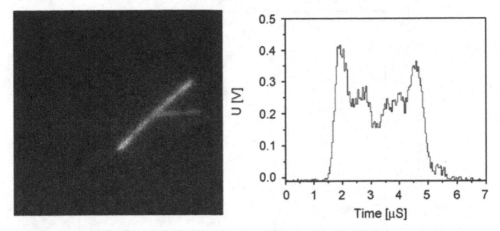

Figure 15.5   Projected image of a nuclear decay (left) and corresponding time distribution of the scintillation (right) (Miernik *et al.*, 2007). By kind permission of Elsevier.

avalanche counter; the exponential avalanche growth in the gap favours the localization of the emission point of the positron (Dominik *et al.*, 1989).

Other applications include the imaging of charged particle tracks for dosimetry (Titt *et al.*, 1998) and nuclear isotope decays (Cwiok *et al.*, 2005). As seen in Figure 15.5, a simultaneous recording of the projected tracks and of the time structure of the scintillation, detected with a photomultiplier operated synchronously with the imaging camera, provides information on the ionization density of the tracks (Miernik *et al.*, 2007).

The introduction in the late nineties of the gas electron multiplier (GEM), described in Section 13.4, opened up new possibilities for the development of optical imagers. Capable of achieving high proportional gains in ^{3}He-CF$_4$ gas mixtures, the devices have been used to image the proton–triton pairs issued by neutron conversions in the gas. While the main reason for adding carbon tetrafluoride to the detectors is to decrease the range of the ionizing prongs, it has been found that CF$_4$ has a strong scintillation component in the visible, between 500 and 700 nm, matching the spectral sensitivity of solid state CCD cameras, Figure 15.6 (Fraga *et al.*, 2002). Figure 15.7 is an image of multiple neutron interactions in the gas, showing several proton–triton pairs (Fraga *et al.*, 2002). Integration of the scintillation yield along the tracks permits the identification of the prongs; in the detection of ions stopping in the gas volume, the time dependence of the yield clearly shows the typical Bragg peak energy loss profile, Figure 15.8 (Margato *et al.*, 2004).

The very good proportionality between the scintillation yield and the energy loss of charged particles is also a promising tool for dose assessment in hadron therapy (Seravalli *et al.*, 2009; Klyancho *et al.*, 2012).

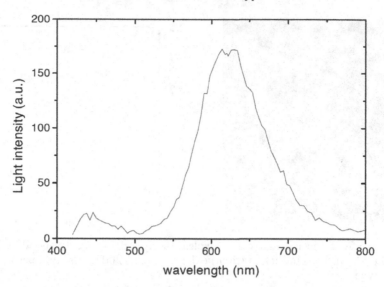

Figure 15.6 Scintillation yield of ^{3}He-CF$_4$, measured at a GEM charge gain of 190 (Fraga *et al.*, 2002). By kind permission of Elsevier.

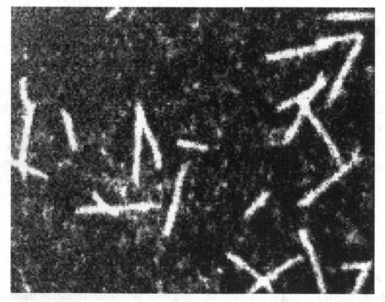

Figure 15.7 Neutron interactions in ^{3}He detected with a GEM-based optical imaging chamber (Fraga *et al.*, 2002). By kind permission of Elsevier.

15.2 Cryogenic and dual-phase detectors

Liquid xenon is used as sensitive volume in a TPC-like configuration for applications requiring the detection of low interaction cross section events, exploiting its higher density and atomic number. Figure 15.9 shows the LXe Gamma-Ray

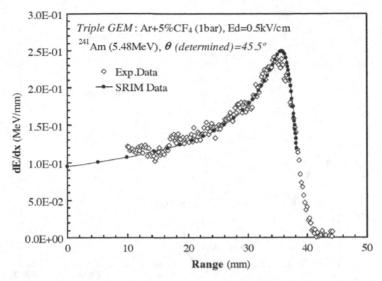

Figure 15.8 Computed (full line) and measured light yield for stopping 5.58 MeV alpha particles (Margato *et al.*, 2004). By kind permission of Elsevier.

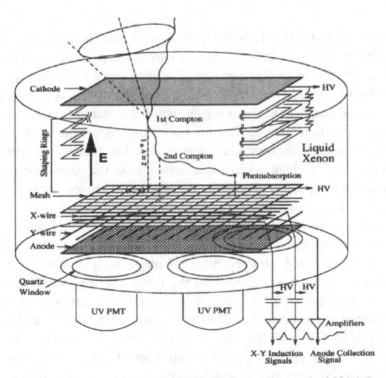

Figure 15.9 Scheme of the LXe GRIT TPC (Aprile *et al.*, 2001). By kind permission of Elsevier.

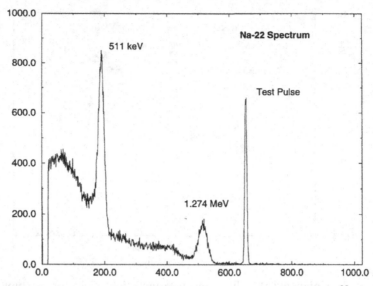

Figure 15.10   Energy resolution of the LXe TPC measured for a ^{22}Na source (Aprile *et al.*, 2002). By kind permission of Elsevier.

Imaging Telescope (GRIT), a cryogenic TPC developed for the imaging of high-energy gamma rays in balloon-borne astrophysics experiments (Aprile *et al.*, 2001). The detector, with 400 cm^3 sensitive area and 7 cm drift gap, is filled with high purity liquid xenon. UV-sensitive photomultipliers detect the primary scintillation and generate the event trigger, while the charge signals collected on two orthogonal wire meshes provide the three-dimensional image of the ionization trails. Systematic measurements demonstrate the energy and spatial resolution of the detector; Figure 15.10 is an example of the scintillation spectrum recorded with a ^{22}Na source (Aprile *et al.*, 2002). The device has been successfully tested for applications as Compton imager of gamma ray sources in astrophysics (Aprile *et al.*, 2008) exploiting the method outlined in Section 3.7. In later developments, the direct charge collection in the liquid has been replaced by the detection, on a second set of photomultipliers, of the secondary scintillation of electrons extracted from the liquid into an overlying gas layer and subject to a high electric field; other dual-phase devices are described below. A prototype detector, XENON10 (Sorensen *et al.*, 2009), and the larger sibling XENON100, with 65 kg of active LXe, operated in the Italian Gran Sasso Laboratory aiming at the detection of dark matter and weakly interacting massive particles (WIMPs); Figure 15.11 is an artist's view of XENON1T, an advanced project with a sensitive volume close to 1 ton (XENON). Similar in conception, the LUX dark matter detector, installed in a

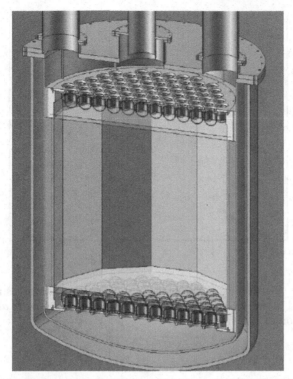

Figure 15.11   Artist's view of XENON1T, showing the cryostat with the LXe sensitive volume and the photomultiplier arrays to detect the primary scintillation in the liquid and the secondary emission from the gas (XENON). E. Aprile personal communication.

mine in South Dakota, has an active volume of 350 kg liquid xenon (LUX). Due to the very low expected event rate, if any, extreme care is devoted to the choice of the components and screenings to reduce the accidental counts due to radioactivity.

Alternatively to the use of external devices for detection of the scintillation, several successful attempts have been described which make use of internal CsI sensitive layers with extraction of the photoelectron in the liquid and successive transport and detection of the charge (Aprile *et al.*, 1994; Periale *et al.*, 2004).

Exploited in the devices described above, the dual-phase concept, where electrons are extracted from the liquid and amplified by charge multiplication in the gas phase, has been studied by many authors, in particular for applications in astrophysics and rare event detection (Bolozdynya, 1999; Aprile *et al.*, 2004).

The gains achievable in gases at cryogenic temperatures have been studied systematically by using as charge amplifying detectors gas electron multipliers

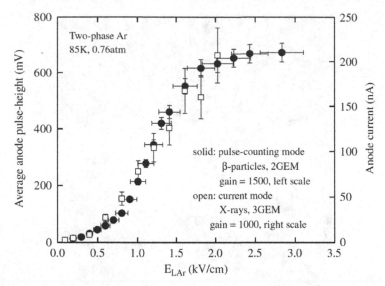

Figure 15.12   Electron extraction probability from liquid argon as a function of field, recorded in the pulse or current detection modes (Bondar *et al.*, 2006). By kind permission of Elsevier.

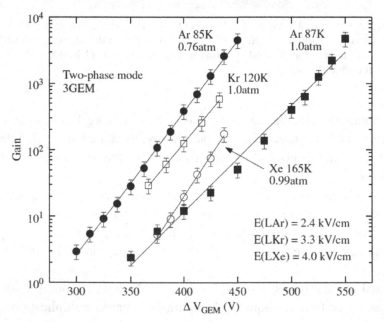

Figure 15.13   Two-phase gains measured with a triple-GEM as a function of voltage applied to each multiplier (Bondar *et al.*, 2006). By kind permission of Elsevier.

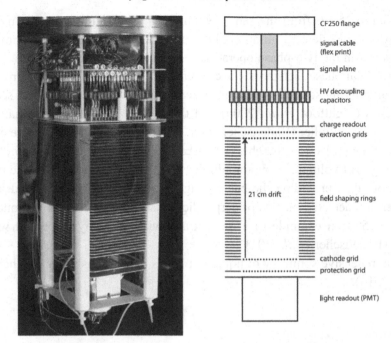

Figure 15.14 The LAr LEM-TPC prototype (Badertscher *et al.*, 2011). By kind permission of Elsevier.

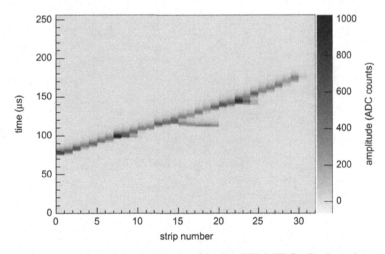

Figure 15.15 Muon track recorded with the LEM-TPC (Badertscher *et al.*, 2011). By kind permission of Elsevier.

(GEMs) in cascade (see Section 13.4), assembled at a short distance above the liquid level; the voltage applied to the lower GEM electrode facing the liquid controls the electron extraction probability. As shown in Figure 15.12, a full extraction from liquid argon is obtained at fields above 2 kV/cm (Bondar *et al.*,

2006). Figure 15.13, from the same reference, is an example of gains measured with a triple-GEM detector in argon, krypton and xenon at different temperatures and pressures in the two-phase operation, as a function of GEM voltage.

Several dual-phase detectors have been built making use of micro-pattern gaseous amplifying structures, aiming at neutrino and dark matter direct searches: Micromegas (Lightfoot *et al.*, 2005) and the coarser GEM structure named the large electron multiplier (Badertscher *et al.*, 2010). The liquid argon large electron multiplier time projection chamber (LAr LEM-TPC), shown in Figure 15.14, has a 21-cm long drift volume. A photomultiplier at the bottom end detects the primary scintillation, thus providing the energy trigger; ionized trails are drifted into the multiplier and detected on sets of perpendicular strips on the LEM readout plane. Figure 15.15 is an example of a muon track with delta rays recorded with the detector (Badertscher *et al.*, 2011).

A comprehensive summary of liquid noble gas detectors is given in Chepel and Araujo (2013).

16

Time degeneracy and ageing

16.1 Early observations

A progressive degradation of performance with the appearance of discharges was observed in the early development of MWPCs, enhanced by the need to operate the detectors at high gains due to the limitations of available electronics. A gaseous detector, in which owing either to improper cleaning or to the creation of deposits on electrodes due to molecular dissociations or polymerization on exposures to radiation, a secondary discharge mechanism is activated, has a very characteristic behaviour (see also Section 7.9). Normally well behaving at low counting rates, a degraded detector shows an increasing background counting rate (or dark current) when exposed even for a short time to higher radiation fluxes; this current may persist after removal of the source.

The appearance and degree of radiation damage can be quantitatively assessed by monitoring the single counting rates at increasing intensities of a radioactive source, as a function of the voltage. Figure 16.1 is a measurement realized by exposing a detector to a ^{55}Fe X-ray source before and after a long-term irradiation of around 10^7 counts/cm^2, operating the counter with an argon-isobutane-freon gas mixture (Charpak et al., 1972). Before irradiation, the efficiency plateau is reached at the same voltage, independently of the rate; the onset of discharges at lower voltages for increasing source intensities is due to spontaneous electron field emission from the cathodes. After a long-term exposure, however, a decrease of the discharge point and a shift to higher voltages of the beginning of the plateau are observed, attributed to an increase of the effective anode wire diameter due to conductive deposits; while hydrocarbon polymers are indeed generally insulating, the anodic layer may be at least partly reduced to elemental carbon by chemical or thermal effects due to energetic electron bombardment in the avalanches.

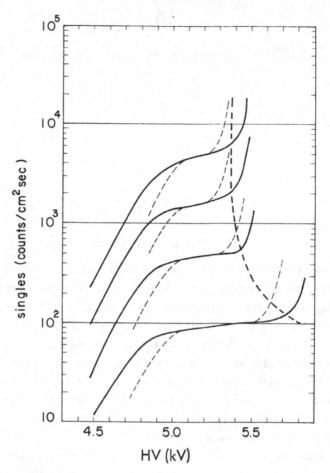

Figure 16.1   Singles counting rates at increasing X-ray fluxes before (full lines) and after a long-term irradiation (dashed lines). 'Magic gas' filling (Charpak *et al.*, 1972). By kind permission of Elsevier.

  The use of non-polymerizing additives having an ionization potential lower than any other constituent in the gas mixture, by a process of charge transfer between species (see Section 4.4), was found to prevent the degradation, or at least to increase by several orders of magnitude the integral flux capability of a proportional chamber (Charpak *et al.*, 1972). In Figure 16.2, the singles counting rate plateaux for increasing X-ray source intensities are measured in a chamber operating with 4% methylal[1] added to magic gas; no change is observed in the main operational parameters. Total exposures up to 10^{12} particles/cm^2 have been reported, without detectable ageing effects.

[1] $(OCH_3)CH_2$

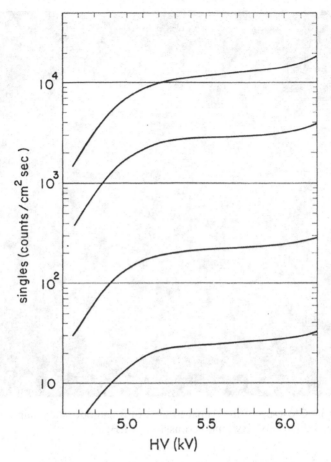

Figure 16.2   Singles counting rates at increasing source intensities measured in a MWPC with 4% methylal added to the 'magic gas' (Charpak *et al.*, 1972). By kind permission of Elsevier.

16.2 Phenomenology of the radiation damages

The radiation-induced damages are local and permanent. Discolourations and thin layers of translucent, whitish or dark materials are often uncovered in the exposed regions, particularly well visible on planar electrodes; Figure 16.3 is a close view of the irradiated area on a chamber with aluminium cathodes (Majewski, 1986); the pattern, possibly due in this case to oxidation by ions released in the avalanches, reflects the pattern of the wires. Damaged wires appear instead darker, suggesting a carbon-like deposit.

Cleaning the damaged regions with various solvents has been attempted, in many cases removing the source of the problems, until the detector is irradiated again. A fast and repeated switching off and on of the operational voltage may

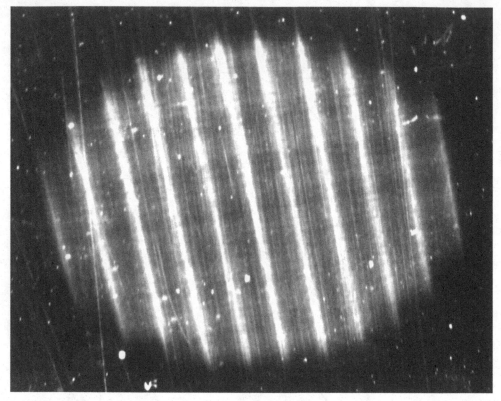

Figure 16.3   Close-up of the damaged area of a cathode after long-term irradiation (Majewski, 1986). By kind permission of LBL.

restore the original low noise operation, probably by capacitive removal of the positive ion deposits on the insulating layers responsible for the secondary sustained discharge, and was actually implemented in early setups to cure the disease, albeit as a temporary solution (Schilly *et al.*, 1970).

The accumulation of positive ions on the thin insulating layers building up on the cathode results in the appearance of a local dipole field, getting stronger the higher the radiation flux; electrons can then be extracted from the layer and injected into the gap by the Malter effect (Malter, 1936). When these electrons reach the anodes they multiply, thus generating more ions; the process easily becomes self-sustaining, and maintains a continuous dark current even after removal of the source of radiation. Temporary suppression of the operating voltage stops the process, which reappears, however, when the detector is exposed to radiation, the more rapidly the higher the flux.

In gas mixtures rich in heavy hydrocarbon quenchers, often used to allow large gains, the conjecture that the degradation is due to a polymerization process occurring in the gas under avalanche conditions comes naturally. These polymers, appearing in the liquid or solid phase, deposit on all electrodes, inducing the described secondary phenomena. A similar phenomenology observed in new chambers can be associated with the presence of greasy layers on the electrodes due to improper cleaning.

An obvious way to prevent ageing is to use a gas not containing hydrocarbon additives. However, good energy resolutions and large gains can only be obtained in mixtures having a high photon-absorption cross section, thus preventing secondary effects due to photoelectric emission at the cathode (photons are copiously emitted in the avalanches); this explains the common use of hydrocarbons quenchers. Moreover, it appears that the use of non-polymerizing quenchers, such as carbon dioxide, does not always prevent the ageing process; chemical analysis of the deposits found on damaged wires shows the presence, sometime as dominant species, of plasticizers, silicone and sulphur compounds that could not have been present in the mixture itself but are introduced into the gas flow by the flux regulators, tubes and gas connections.

The standard way to assess the radiation resistance of a detector is to expose it for a long term to an intense radiation from radioactive sources or X-ray generators, and measure at regular intervals the local gain at low rate. Figure 16.4,

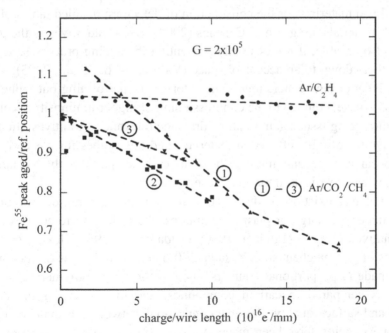

Figure 16.4   Normalized gain as a function of collected charge, measured with a MWPC under sustained irradiation in different gases (Kotthaus, 1986). By kind permission of Elsevier.

providing the normalized gain as a function of collected charge in several gas mixtures, is a typical example (Kotthaus, 1986). Since the measurement can take weeks if not months, care should be taken to correct for gain variations not related to ageing (temperature, pressure); a convenient procedure is to monitor the detector gain in a non-irradiated part of the chamber.

The effect of gain, materials and gas composition on ageing has been studied for decades; summaries of the status of the knowledge in the field can be found in the proceeding of dedicated workshops and many dedicated articles (Kadyk, 1986; Hohlmann *et al.*, 2002; Capeáns, 2003; Titov, 2004). Aged chambers can be sacrificed to inspect the electrodes, subjecting damaged wires to a wide variety of surface analysis, from microscopy to mass spectrometry. Figure 16.5 collects a representative selection of images of damaged anode wires showing all kinds of deposit, from bulbous layers to crystal-like spikes (Juricic and Kadyk, 1986; Hilke, 1986b). The micron-scale filament growth shown in Figure 16.6 has been found to be mainly composed of silicone polymers, probably seeded by residues of lubricants used in the detector and carried in the gas flow (Binkley *et al.*, 2003).

Despite often contradictory results, from these works emerges a set of general guidelines on how to build and operate gaseous detectors capable of withstanding high radiation fluxes. The 'golden rules' include warnings against the use of heavy organic quenchers, care to avoid sources of silicon oil pollution, indications on good and bad materials for the construction of detectors; detailed tables of 'good' and 'bad' materials are given in Capeáns (2003). Use of additives in the gas flow appears also capable, if not of totally preventing the ageing processes, at least of slowing them down to an acceptable pace (Va'vra, 1986b; Sauli, 2003).

The advent of new generations of detectors, very performing but vulnerable to radiation damage, coupled to the commissioning of higher luminosity accelerators, opened the ageing issue again in all its dramatic relevance. Additives often turned out to have undesirable effects on materials, and the recipes for gas and material choice became more and more difficult to enforce, particularly for large area systems, where cost is an important issue.

Abundant data exist on the dissociation and polymerization processes of hydrocarbons under discharge or plasma conditions; the following references are only representative examples (Lindner, 1930; Yeddanapalli, 1942; Hess, 1986). Many properties of plasma chemistry (Yasuda, 2003) are reminding us of similar observations made in proportional counters, such as the larger polymerization rate of silicone as compared to carbon compounds, and the effects of the electrode material and surface quality. Possible connections between plasma chemistry and gas counters ageing have been made (Va'vra, 2003). The progress in the understanding of the basic underlying chemical and physical processes is often counterbalanced by reports on the dramatic loss of detector systems due to unforeseen

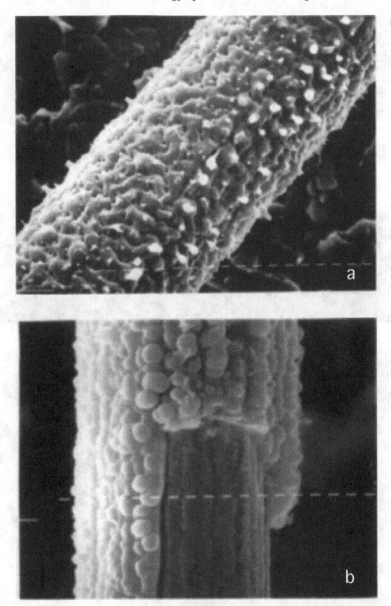

Figure 16.5 Examples of WWPC wires variously coated after irradiation. a: (Ullaland, 1986); b: (Hilke, 1986b); c and d: (Juricic and Kadyk, 1986). By kind permission of LBL.

happenings. Usually obtained in rather specific pressure and charge density conditions, they remain qualitative when extrapolated to the different situations met in gas detectors. Results of systematic investigations on polymer formation in conditions close to those of gas counters have been reported (Kurvinen *et al.*, 2003) and

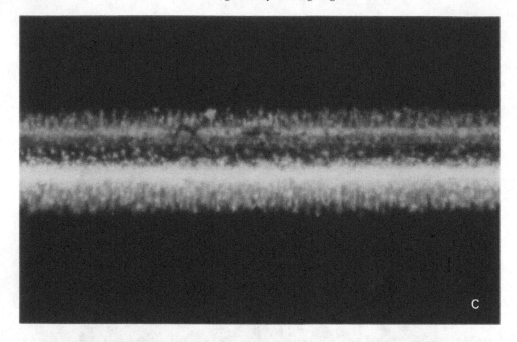

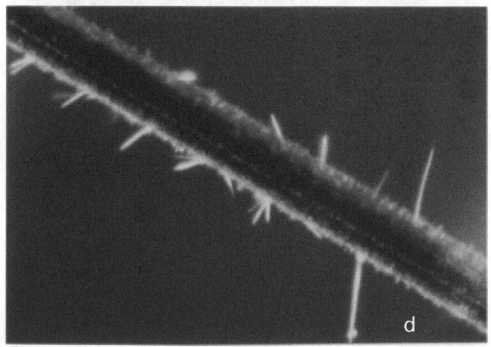

Figure 16.5 *(cont.)*

Figure 16.6   Silicone filament growth in a radiation-damaged chamber (Binkley *et al.*, 2003). By kind permission of Elsevier.

confirm the strong reduction of polymer formation with increasing current density, an effect casting doubt on the validity of results obtained with very high rate exposures.

16.3 Quantitative assessment of the ageing rates

The verification in reasonable time of the survival of a detector for several years of operation needs an accelerated test procedure. Intuitively, if ageing is a consequence of the formation and accumulation of polymers or other chemical processes occurring in the avalanches, the scale invariant should be the total accumulated charge, independently of other factors such as voltage, type and flux of radiation, and gas flow. It has become customary to express the accumulated charge in Coulombs per unit length of wire, a detector being qualified good if it could withstand without gain drops up to several C/cm. A quality factor, the ageing rate R, defined as the percentage gain variation normalized to the total collected charge, should ideally be small and constant for a given detector (Juricic and Kadyk, 1986). However, it appears that the value of R depends, sometimes critically,

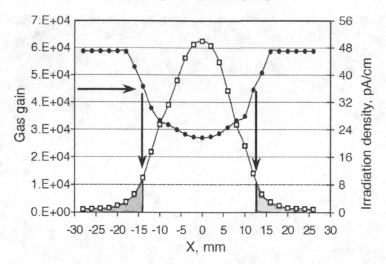

Figure 16.7 Irradiation current density (open squares) and gain reduction (full dots) as a function of position along the wire. Dashed areas correspond to regions of limited streamer formation (Ferguson *et al.*, 2003b). By kind permission of Elsevier.

on test conditions, the main factor being the dose rate, or acceleration factor: data collected at high current densities tend to be optimistic, often by a large factor, compared to those obtained at lower rates (Bouclier *et al.*, 1994b).

Depending on geometry and irradiation conditions, a space charge gain reduction can set in at high current densities, possibly reducing the polymerization efficiency; a reduction of gain by a factor 2 has been measured in the centre of the irradiated area for the measurement shown in Figure 16.7 (Ferguson *et al.*, 2003b). At high operating voltage, the gain on the edges can be large enough to set in a limited streamer regime, thus substantially increasing the collected charge and giving rise to characteristic ring patterns in the damaged irradiated areas.

The appearance of diffused micro-discharges, which produce irreversible local damages or filament growth ('hairs'), can seriously affect performances. Extreme sensitivity to some pollution sources, particularly if containing traces of silicone compounds, has been observed (Ketzer *et al.*, 2001). The interpretation of irradiation results can be seriously biased if these sources, often external to the chamber, are undetected, making comparisons with other observations doubtful.

A correlation between the wire diameter and the rate of ageing has been also found, thinner wires ageing much faster (see for example Va'vra, 1997); this might be simply due to the smaller area available for the deposits, or to an increased polymerization efficiency due to the higher field close to the anode.

Given the dose rate, a larger irradiated area can also increase the ageing rate (Hott, 2003; Hildebrandt, 2003); the gas flow itself affects the ageing rate, larger for lower flows. Moreover, the gain degradation appears to move along the detectors with the gas flow, the damaged area often largely exceeding the irradiated region (Hohlmann *et al.*, 2002; Titov, 2004). A possible explanation of these effects lies in a dependence of the polymerization rate on the time that reactive radicals produced in the avalanches stay in the region of ionization before being carried away by the gas flow.

16.4 Methods of preventing or slowing down the ageing process

As discussed in the introduction, the addition of low ionization potential vapours like methylal, ethyl or iso-propyl alcohols to the main gas mixture considerably slows down the ageing process, aside from improving the performances of the detectors thanks to their added photon-quenching properties (Charpak *et al.*, 1972; Atac, 1987).

The beneficial effect of adding water to detectors has also been known for a long time. Aside from reducing the polymerization rates in plasma discharges, water has also the property of making all surfaces in the detectors slightly more conductive, thus reducing, if not the formation of polymers, at least the accumulation of ions on the thin layers, responsible for the gain degradation and the increase of dark current through the Malter effect.

Results of measurements with water addition to the gas mixture are, however, contradictory, as seen by comparing Figure 16.8 (DeWulf, 1986) and Figure 16.9 (Kotthaus, 1986); while in the first case the addition of a few parts per mille of water substantially reduces the ageing rate, in the second case no effect is observed. This is a representative example of outcomes in a field where the presence of trace impurities may dominate the ageing process. In some cases, the culprit has been identified; Figure 16.10 shows the effect on ageing of the addition in the input gas line of a short section of plastic tubing: the gain loss rate in the aged position is increased by a factor of two.

Extensively used in the early experimental setups, the addition of vapours, that requires the use of temperature-controlled bubblers, is not very practical in large systems, and has encouraged the search for alternative methods of control.

Over the years, experience has shown that even a supposedly robust detector can suffer from severe ageing problems due to the unexpected presence of pollutants. A therapy to cure the disease, even exceptionally, would be very welcome. Procedures to evaporate oily deposits on anodes by heating the wires with an externally supplied current were tested, but can only be used with high resistivity wires (Va'vra, 1997). In most cases, however, detectors are not designed to heat up the wires by DC current, either for lack of connections on both sides of the wires of because of their low resistance. An alternative method consisting of operating the

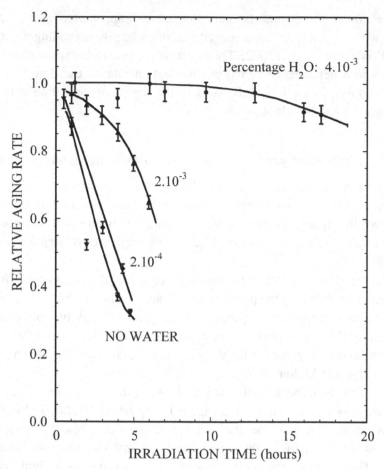

Figure 16.8 An example of the beneficial effect on ageing with the addition of water vapours to the gas mixture (DeWulf, 1986). By kind permission of Elsevier.

detectors in pure argon with a reverse voltage to 'burn out' deposits has been tried with some success (Kollefrath *et al.*, 1998).

A rather extreme method of removing the debris deposited on the anodes has been tested by applying a high voltage pulse to 'zap' the wires with an energy of a few tens of joules (Marshall, 2003): aged chambers cured using the method recovered and operated successfully. This rather extreme treatment can presumably only be used after removal of the electronics, and it is not clear what the fate of the zapped fragments is; moreover, the sublimation yields can affect the efficiency of the detector.

Operation of a damaged detector with the addition of small amounts of oxygen ($\sim$500 ppm), appears to gradually reduce the current, presumably by burning away the organic deposits (Boyarski, 2003). This method, if applicable in other devices, has the advantage of not requiring special connections to the wires.

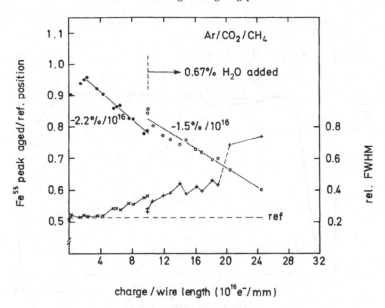

Figure 16.9   In this example, addition of water has no effect on ageing rate (Kotthaus, 1986). By kind permission of Elsevier.

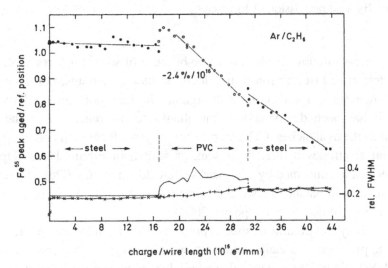

Figure 16.10   Strong ageing under irradiation after introduction in the gas line of a short section of PVC. The degradation continues even after removal of the plastic tube (Kotthaus, 1986). By kind permission of Elsevier.

The presence of silicone deposits on wires, observed when removing all possible sources of organic contamination, has suggested the addition of carbon tetrafluoride to the gas mixture: CF_4 is indeed used in the semiconductor industry to etch silicon in wafer manufacturing. The use of carbon tetrafluoride as a

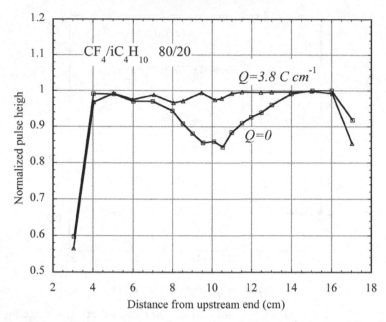

Figure 16.11   Gain scan along a wire of a damaged chamber before ($Q = 0$) and after irradiation with CF_4 in the gas mixture ($Q = 3.8$ C cm^{-1}) (Openshaw *et al.*, 1991). By kind permission of Elsevier.

quencher in gas mixtures is also attractive because of some properties relevant for particle detectors: fast electron drift velocity at moderate fields, non-flammability and low neutron cross section. Non-organic, it does not form polymers, and actually it has been demonstrated that, thanks to the reactivity of the species produced in the avalanches, CF_4 can prevent polymer formation and even remove them from electrodes if already present, as shown in Figure 16.11 (Openshaw *et al.*, 1991), and confirmed by later works (Belostotski *et al.*, 2008). However, this requires a delicate balance between etching and deposition processes, not always easy to achieve in detectors (Capeáns, 2003).

The reactivity and lifetime of species produced by CF_4 in the avalanches, particularly in the presence of organic pollutants, water or oxygen, have been extensively analysed (Schreiner, 2001). It was also found that, when exposing detectors containing CF_4 in the gas mixture, long-lived electro-negative molecules form and propagate with the gas flow (Capeáns *et al.*, 1993); this propagation has been confirmed by other studies (Danilov *et al.*, 2003; Albrecht *et al.*, 2003; Akesson *et al.*, 2002) and can affect the efficiency of large systems having several detectors in series. The design of a re-circulating gas system, required in large experiments, particularly if making use of xenon, is singularly difficult in the presence of species that can react with delicate components such as purifiers and filters.

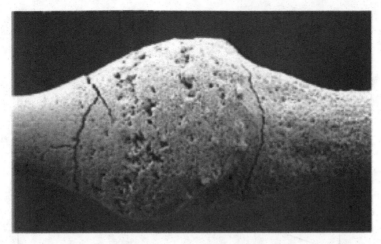

Figure 16.12   Anode wire swelling under the action of fluorinated compounds released by the use of CF_4 in the gas mixture (Akesson *et al.*, 2002). By kind permission of Elsevier.

Damage during operation in CF_4 of the detector's materials, and in particular on the anode wires, has been reported (Akesson *et al.*, 2002; Ferguson *et al.*, 2003a). Under irradiation, presumably because of the penetration of fluorinated compounds into cracks of the wire, the gold-plated tungsten anodes swell and shed metallic flakes; the ill effect of such a happening need not be discussed. The swelling under irradiation has been correlated to the quality of the anode wire surface and to the amount of residual water in the gas, although on the second point the tolerance limits are not clear. The picture in Figure 16.12 gives an extreme example of serious damage to a gold-plated wire in these conditions (Akesson *et al.*, 2002).

The creation in the avalanches of fluorine, that combining with residual water vapours produces hydrofluoric acid (HF), can also have the direst consequences on the constituents of the detector, in particular glass or glass-containing materials, as observed by the ATLAS TRT group: the short glass beads used to join together two segments of long anode wires were seriously degraded under high-flux operation (Akesson *et al.*, 2004b). The use of CF_4 during long-term data taking has been discontinued, in favour of a less aggressive mixture containing oxygen as mild etching agent; short-term runs with carbon tetrafluoride are used to clean the system from silicon containing manufacturing residuals (Akesson *et al.*, 2004b).

16.5 Ageing of resistive plate chambers

The long-term behaviour of resistive plate chambers has been studied extensively in view of their large use in particle physics experiments, in some cases following serious degradation of performance. Problems may arise because of imperfections

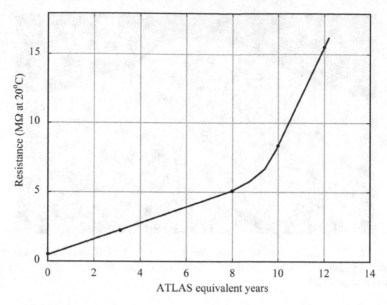

Figure 16.13 Long-term increase of the RPC electrodes' resistivity on exposure to radiation. One Atlas equivalent year corresponds to a particle flux of 100 Hz/cm² (Aielli *et al.*, 2002). By kind permission of Elsevier.

in the detector quality, as observed in the first generation of the BaBar RPCs, where the main source of degradation was identified as the incomplete polymerization of the linseed oil used to improve the surface quality of the Bakelite electrodes (Anulli *et al.*, 2003), or could be due to modifications of the electrodes during operation. A common observation is an increase of resistivity of the electrodes, due either to a depletion of the ionic charge carriers or to 'drying out' of the components during operation (Figure 16.13); this results in a decreased rate capability of the detector, from several kHz/cm² to a few hundred Hz/cm² (Aielli *et al.*, 2002). The intricate chemistry of the linseed oil coating with time and exposure to water has been also studied in detail (Va'vra, 2003). The process can be reversed by adding a small amount of moisture in the gas mixture; a strict control of the ambient temperature and humidity in the experiments is needed to ensure long-term stability of operation (Carboni *et al.*, 2004; Band *et al.*, 2006).

The release of fluorine in the freon-rich gas mixtures used for RPC operation has also been found to be responsible for long-term degradation of glass RPCs exposed to radiation, due to the etching action of hydrofluoric acid on the electrodes; the degradation is particularly fast for detectors operated in the streamer mode.

Observed initially on glass RPCs, this type of chemical damage has been also reported for phenolic laminate-based detectors (Band *et al.*, 2008; Kim *et al.*, 2009).

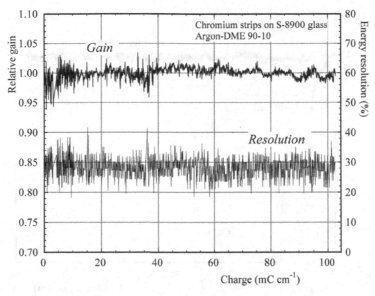

Figure 16.14   Relative gain and energy resolution on 5.9 keV X-rays of a MSGC under long-term irradiation (Bouclier *et al.*, 1995b). By kind permission of Elsevier.

Partial recovery has been achieved by purging the damaged detectors with ammonia-saturated argon (Sakai *et al.*, 2002; Bhide *et al.*, 2006).

16.6 Micro-pattern detectors

Micro-strip gas counters (MSGC) are particularly prone to fast ageing under irradiation, probably because of the small area of the anodes where deposits can easily build up; their radiation tolerance has been extensively studied, in view of the use at high rates and in harsh environments.

While the time evolution of gain is relatively easy to assess with long-term exposures to radiation, the extreme sensitivity of the results to trace pollutants demands a continuous monitoring of the operating conditions and gas purity. In a setup operated at CERN in the framework of an international collaboration to study ageing processes,[2] the gas quality was continuously monitored with a mass chromatograph equipped with a mass spectrometer or an electron capture head to cover a wide range of sensitivity (Bouclier *et al.*, 1996a).

The main results of the measurements, reported in Sauli (1998), indicate that long-term operation of the detector is only possible under strict conditions for the choice of materials and gas purities. Figure 16.14 is an example of long-term gain

---

[2] RD-28 Collaboration: Development of gas microstrip chambers for radiation detection and tracking at high rates, F. Sauli, spokesman.

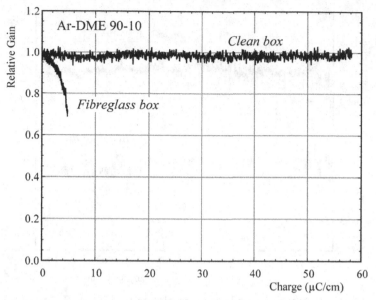

Figure 16.15   Ageing properties of a MSGC plate in a standard and clean assembly (Bouclier *et al.*, 1994a). By kind permission of Elsevier.

and resolution measured with a MSGC operated in argon-DME,[3] under continuous irradiation (Bouclier *et al.*, 1995b); no change is observed up to a total collected charge above 100 mC/cm, corresponding to an integrated minimum ionizing particles flux of about 10^{13} cm^{-2}. It should be noted that, to avoid gain shifts due to insulators charging-up during the irradiation, the measurement can only be performed with detectors manufactured on controlled resistivity substrates, capable of operating at high fluxes (see Section 13.1).

The presence of unwanted pollutants can shorten the detector lifetime by many orders of magnitude. Figure 16.15 is an example, comparing the results obtained with the same type of MSGC plate assembled within a standard fibreglass frame with uncertified epoxies, or mounted in a 'clean box', an all-metal and ceramics envelope chemically cleaned and vacuum-pumped before operation; the difference is striking (note the horizontal scale in µC/cm) (Bouclier *et al.*, 1994a).

Results of the systematic search on the effect on ageing of materials and epoxies used for MSGC detectors construction are collected in Bouclier *et al.* (1996a); extended summary tables suggest the recommended choices.

As discussed in Section 13.1, the use of MSGCs has been discouraged by the discharge problems encountered; however, they remain a very sensitive tool to study ageing properties of gaseous detectors.

---

[3] Dimethyl ether, CH_3OCH_3.

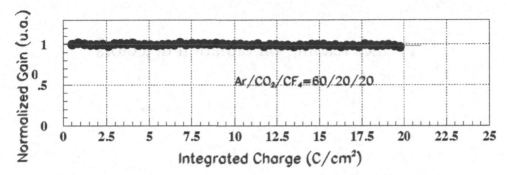

Figure 16.16 Normalized gain as a function of collected charge for a triple-GEM detector (Alfonsi *et al.*, 2004). By kind permission of Elsevier.

The new generation of micro-pattern detectors, Micromegas and GEM in particular, is more tolerant to high radiation fluxes; even though the basic processes of polymer formation in the avalanches are the same as in other counters, their structure does not include thin electrodes, which can be easily coated by deposits with ensuing field modifications. In accelerated ageing tests, realized with continuous exposure to high-rate X-rays, no sign of degradation has been observed with an argon-CO_2 gas filling up to an accumulated charge of 1 C/cm^2 (Guirl *et al.*, 2002; Altunbas *et al.*, 2003).[4]

The addition of carbon tetrafluoride to the mixture, thanks to its etching properties, prevents the most insidious source of ageing due to silicon compound deposits; however, the use of CF_4 in detectors requires special precautions, due to the reactivity of fluorine with water that can generate hydrofluoridric acid, very corrosive for many materials. An example of long-term measurement of normalized gain as a function of accumulated charge for a triple-GEM detector, filled with an Ar-CO_2-CF_4 mixture, is given in Figure 16.16, showing no change up to an accumulated charge of $\sim$20 C cm^{-2}; taking into account the gain of the detector, this corresponds to an integrated flux of about 4×10^{14} minimum ionizing particles per square centimetre (Alfonsi *et al.*, 2004). Similar results have been reported for Micromegas-based detectors (Puill *et al.*, 1999; Kane *et al.*, 2003).

Further reading

Workshop on Radiation Damage to Wire Chambers (1986), J. Kadyk (ed.), LBL-21170.
Ageing Workshop 2001 (2003): International Workshop on Ageing Phenomena in Gaseous Detectors, M. Hohlmann, C. Padilla, N. Tesh and M. Titov (eds.), *Nucl. Instr. and Meth.* **A515**.

---

[4] Unlike wire- and microstrip-based detectors, for which the flux given per unit length of the anodes, for devices having continuous electrodes it is expressed per unit surface.

Further reading on radiation detectors

Curran, S. C. and Craggs, J. D. (1949) *Counting Tubes, Theory and Applications*, London, Butterworth.

Korff, S. A. (1955) *Electron and Nuclear Counters*, New York, Van Nostrand.

Franzen, W. and Cochran, L. W. (1956): Pulse ionization chambers and proportional counters, in *Nuclear Instruments and Their Use*, A. H. Snell (ed.) New York, Wiley.

Price, W (1958) *Nuclear Radiation Detection*, New York, McGraw Hill.

Sharpe, J. (1964) *Nuclear Radiation Detectors*, New York, Methuen.

Shutt, R. P. (ed.) (1967) *Bubble and Spark Chambers*, New York, Academic Press.

Henderson, C. (1970) *Cloud and Bubble Chambers*, London, Methuen.

Ritson, D. M. (1971) *Techniques of High Energy Physics*, New York, Interscience.

Perkins, D. H. (1972) *Introduction to High Energy Physics*, Reading, Addison-Wesley.

Rice-Evans, P. (1974) *Spark, Streamer, Proportional and Drift Chambers*, London, Richelieu.

Fernow, R. C. (1986) *Introduction to Experimental Particle Physics*, Cambridge, Cambridge University Press.

Leo, R. (1987) *Techniques for Nuclear and Particle Physics Experiments*, Berlin, Springer-Verlag.

Knoll, G. F. (1989) *Radiation Detection and Measurements*, New York, Wiley.

Gilmore, R. (1992) *Single Particle Detection and Measurement*, London, Taylor and Francis.

Sauli, F. (ed.) (1992) *Instrumentation in High Energy Physics*, Singapore, Word Scientific.

Grupen, C. (1996) *Particle Detectors* (Cambridge Monographs on Partical Physics) Cambridge, Cambridge University Press.

Kleinknecht, K. (1998) *Detectors for Particle Radiation*, Cambridge, Cambridge University Press.

Sauli, F. (2004) From bubble chambers to electronic systems: 25 years of evolution in particle detectors at CERN (1979–2004). *Phys. Rep.* **404**, 471.

Blum, W. and Rolandi, L. (1993) *Particle Detection with Drift Chambers*, Berlin, Springer-Verlag.

Titov, M. (2012) Gaseous Detectors, in *Handbook of Particle Detection and Imaging*, C. Grupen and I. Buvat (eds.) Berlin, Springer.

Nappi, E. and Peskov, V. (2013) *Imaging Gaseous Detectors and their Applications*, Weinheim, Wiley-VCH Verlag & co.

References

Abbaneo, D., *et al.* (1998) Test of a CMS MSGC tracker prototype in a high-intensity hadron beam. *Nucl. Instr. and Meth.* **A 409**, 37.

Abbiendi, C., *et al.* (2009) The CMS muon barrel drift tubes system commissioning. *Nucl. Instr. and Meth.* **A598**, 192.

Abbon, P., *et al.* (2007) The COMPASS experiment at CERN. *Nucl. Instr. and Meth.* **A 577**, 455.

Abbrescia, M., *et al.* (1997) Effect of the linseed oil treatment on the performance of the Resistive Plate Counters. *Nucl. Instr. and Meth.* **394**, 13.

Abbrescia, M., *et al.* (1999a) The simulation of resistive plate chambers in the avalanche mode: charge spectra and efficiency. *Nucl. Instr. and Meth.* **A 431**, 413.

Abbrescia, M., *et al.* (1999b) Progress in the simulation of resistive plate chambers in avalanche mode. *Nuc. Phys. B (Proc. Suppl.)* **78**, 459.

Abbrescia, M. (2004) The dynamic behaviour of Resistive Plate Chambers. *Nucl. Instr. and Meth.* **A533**, 7.

Abe, K., *et al.* (1998) The performance of the Barrel CRID at the SLD; long-term operational experience. *IEEE Trans. Nucl. Sci.* **45**, 648.

Abgrall, N., *et al.* (2011) Time projection chambers for the T2K near detector. *Nucl. Instr. and Meth.* **A637**, 25.

Ableev, V., *et al.* (2004) TPG construction. *Nucl. Instr. and Meth.* **A535**, 294.

Abrams, G., *et al.* (1989) The Mark II Detector for the SLC. *Nucl. Instr. and Meth.* **A281**, 55.

Ackermann, K. H., *et al.* (2003) The forward time projection chamber in STAR. *Nucl. Instr. and Meth.* **A499**, 713.

Ackerstaff, K., *et al.* (1998) The HERMES Spectrometer. *Nucl. Instr. and Meth.* **A417**, 230.

Acosta, D., *et al.* (2000) Large CMS cathde strip chambers: design and performance. *Nucl. Instr. and Meth.* **A453**, 182.

Acosta, D., *et al.* (2003) Aging tests of full-scale CMS muon cathode strip chambers. *Nucl. Instr. and Meth.* **A515**, 226.

Adam, W., *et al.* (1996) Particle identification algorithms for the DELPHI RICH detector. *Nucl. Instr. and Meth.* **A371**, 240.

Adams, M., *et al.* (1983) $\pi/K/p$ identification with a large-aperture ring-imaging Cherenkov counter. *Nucl. Instr. and Meth.* **217**, 237.

Adeva, B., *et al.* (1990) The construction of the L3 experiment. *Nucl. Instr. and Meth. in Phys. Res.* **A289**, 35.

Adeva, B., *et al.* (1999) The Micro Wire Detector. *Nucl. Instr. and Meth.* **A 435**, 402.

Adeva, B., *et al.* (2001) Performance of the Microwire Detector. *Nucl. Insr. and Meth.* **A461**, 33.

Adorisio, C., *et al.* (2007) A non-invasive technique to replace the anode wires into the drift tube chambers of the muon spectrometer of the ATLAS experiment at the LHC proton–proton collider. *Nucl. Instr. and Meth.* **A575**, 532.

Afanasiev, S., *et al.* (1999) The NA49 large acceptance hadron detector. *Nucl. Instr. and Meth.* **A430**, 210.

Agrawal, P. C. and Ramsey, B. D. (1989) Penning gas mixtures for improving the energy resolution of proportional counters. *IEEE Trans. Nucl. Sci.* **NS-36**, 866.

Agrawal, P. C., *et al.* (1989) Study of argon-based Penning gas mixtures for use in proportional counters. *Nucl. Instr. and Meth.* **277**, 557.

Aguilar-Benitez, M., *et al.* (2002) Construction and test of the final CMS Barel Drift Tube Muon Chamber prototype. *Nucl. Instr. and Meth.* **A480**, 658.

Ahn, S. H., *et al.* (2000) Temperature and humidity dependence of bulk resistivity of bakelite for resistive plate chambers in CMS. *Nucl. Instr. and Meth.* **A451**, 582.

Ahn, S. H., *et al.* (2004) Characteristics of a double gap resistive plate chamber for the endcap region of CMS/LHC: data vs. simulation in avalanche mode. *Nucl. Instr. and Meth.* **A533**, 32.

Aielli, G., *et al.* (2002) RPC aging studies. *Nucl. Instr. and Meth.* **A478**, 271.

Aielli, G., *et al.* (2004) Electrical conduction properties of phenolic-melaminic laminates. *Nucl. Instr. and Meth.* **A533**, 86.

Aielli, G., *et al.* (2006) New results on ATLAS RPCs ageing at CERN's GIF. *IEEE Trans. Nucl. Sci.* **53**, 567.

Aihara, H., *et al.* (1983) Spatial resolution of the PEP-4 Time Projection Chamber. *IEEE Trans. Nucl. Sci.* **NS-30**, 76.

Akbari, H., *et al.* (1992) The L3 vertex detector: design and performance. *Nucl. Instr. and Meth.* **A315**, 161.

Akesson, T., *et al.* (1995) Study of straw proportional tubes for a transition radiation detector/tracker at LHC. *Nucl. Instr. and Meth.* **A361**, 440.

Akesson, T., *et al.* (2002) Aging studies for the ATLAS Transition Radiation Tracker (TRT). *Nucl. Instr. and Meth.* **A 515**, 166.

Akesson, T., *et al.* (2004a) Status of design and construction of the Transition Radiation Tracker (TRT) for the ATLAS experiment at the LHC. *Nucl. Instr. and Meth.* **A522**, 131.

Akesson, T., *et al.* (2004b) Operation of the ATLAS Transition Radiation Tracker under very high irradiation at the CERN LHC. *Nucl. Instr. and Meth.* **A522**, 25.

Akindinov, A., *et al.* (2000) The multigap resistive plate chamber as a time-of-flight detector. *Nucl. Instr. and Meth.* **A456**, 16.

Akindinov, A. N., *et al.* (2004) Latest results on the performance of the multigap resistive plate chamber used for the ALICE TOF. *Nucl. Instr. and Meth.* **A533**, 74.

Akindinov, A., *et al.* (2009) Construction and test of the MRPC detectors for TOF in ALICE. *Nucl. Instr. and Meth.* **A602**, 658.

Akindinov, A., *et al.* (2012) The MRPC-based ALICE Time-Of-Flight detector: Commissioning and first performance. *Nucl. Instr. and Meth.* **A661**, S98.

Albrecht, A., *et al.* (2005) Status and characterization of COMPASS RICH-1. *Nucl. Instr. and Meth.* **A553**, 215.

Albrecht, E., *et al.* (2003) COMPASS RICH-1. *Nucl. Instr. and Meth.* **A502**, 112.

Albrecht, E., *et al.* (1999) Operation, optimisation, and performance of the DELPHI RICH detectors. *Nucl. Instr. and Meth.* **A433**, 47.

Albrecht, H., *et al.* (2003) Aging studies for the large honeycomb drift tube system of the Outer Tracker of HERA-B. *Nucl. Instr. and Meth.* **A515**, 155.

Aleksa, M., *et al.* (2000) Rate effects in high-resolution drift chambers. *Nucl. Instr. and Meth.* **A446**, 435.

Aleksa, M., *et al.* (2002) Limits to drift chamber performance at LHC luminosities. *Nucl. Instr. and Meth.* **A478**, 135.

Alexander, J. P., *et al.* (1989) The Mark-II vertex drift chamber. *Nucl. Instr. and Meth.* **A283**, 519.

Alexeev, A. D., *et al.* (1980) Investigation of self-quencing streamer discharge in a wire chamber. *Nucl. Instr. and Meth.* **177**, 385.

Alexeev, M., *et al.* (2010) TGEM based photon detectors for Cherenkov imaging applications. *Nucl. Instr. and Meth.* **A617**, 396.

Alexeev, M., *et al.* (2011) Progress towards a TGEM-based detector of single photons. *Nucl. Instr. and Meth.* **A 639**, 130.

Alexeev, M., *et al.* (2012) Detection of single photons with ThickGEM-based counters. *JINST* **7**, C02014.

Alexopulos, T., *et al.* (2011) A spark-resistant bulk micromegas chamber for high rate applications. *Nucl. Instr. and Meth.* **A640**, 110.

Alfonsi, M., *et al.* (2004) High-rate particle triggering with triple-GEM detector. *Nucl. Instr. and Meth.* **A518**, 106.

Alfonsi, M., *et al.* (2010) Activity of CERN and LNF groups on large area GEM detectors. *Nucl. Instr. and Meth.* **A 617**, 151.

Alkhazov, G. D. (1969) Mean value and variance of gas amplification in proportional counters. *Nucl. Instr. and Meth.* **75**, 161.

Alkhazov, G. D. (1970) Statistics of electron avalanches and ultimate resolution of proportional counters. *Nucl. Instr. and Meth.* **89**, 155.

Alleyn, H., *et al.* (1968) Wire stretching machine for spark chambers, *CERN NP* 68–34 CERN.

Allison, J., *et al.* (1982) An electrodeless drift chamber. *Nucl. Instr. and Meth.* **201**, 341.

Allison, W. W., *et al.* (1974) The identification of secondary particles by ionization sampling (ISIS). *Nucl. Instr. and Meth.* **119**, 499.

Allison, W. W., *et al.* (1984) Relativistic charged particle identification with ISIS2. *Nucl. Instr. and Meth.* **224**, 396.

Alme, J., *et al.* (2010) The ALICE TPC, a large 3-dimensional tracking device with fast readout for ultra-high multiplicity events. *Nucl. Instr. and Meth.* **A622**, 316.

Alner, G. J., *et al.* (2004) The DRIFT-I dark matter detector at Boulby: design, installation and operation. *Nucl. Instr. and Meth.* **A 535**, 644.

Aloisio, A., *et al.* (2004) The trigger chambers of the ATLAS muon spectrometer: production and tests. *Nucl. Instr. and Meth.* **A 535**, 265.

Altunbas, C., *et al.* (2002) Construction, test and commissioning of the Triple-GEM tracking detector for COMPASS. *Nucl. Instr. and Meth.* **A490**, 177.

Altunbas, C., *et al.* (2003) Aging measurements with the Gas Electron Multiplier (GEM). *Nucl. Instr. and Meth.* **A515**, 249.

Amaldi, U. (1971) *Fisica delle Radiazioni, Boringhieri*, Torino, Boring hieri.

Amaldi, U., *et al.* (2011) Construction, test and operation of a proton range radiography system. *Nucl. Instr. and Meth.* **A629**, 337.

Ambrosio, M., *et al.* (2002) The MACRO detector at Gran Sasso. *Nucl. Instr. and Meth.* **A486**, 663.

Amendolia, S. R., *et al.* (1985a) Ion trapping properties of a Synchronously Gated Time Projection Chamber. *Nucl. Instr. and Meth.* **239**, 192.

Amendolia, S. R., *et al.* (1985b) Influence of the magnetic field on the gating of a Time Projection Chamber. *Nucl. Instr. and Meth.* **A234**, 47.

Amendolia, S. R., *et al.* (1989) The spatial resolution of the ALEPH TPC. *Nucl. Instr. and Meth.* **A283**, 573.

Amerio, S., *et al.* (2004) Design, construction and test of the ICARUS T600 detector. *Nucl. Instr. and Meth.* **A527**, 329.

Ammosov, V., *et al.* (1997) Electric field and currents in resistive plate chambers. *Nucl. Instr. and Meth.* **A401**, 217.

Amram, N., *et al.* (2011) Position resolution and efficiency measurements with large scale Thin Gap Chambers for the super LHC. *Nucl. Instr. and Meth.* **A628**, 177.

An, S., *et al.* (2008) A 20 ps timing device – a multigap Resistive Plate Chamber with 24 gaps. *Nucl. Insr. and Meth.* **A594**, 39.

Anderhub, H., *et al.* (1986) A Time Expansion Chamber as a vertex detector for the experiment mark J at DESY. *Nucl. Instr. and Meth.* **252**, 357.

Anderhub, H., *et al.* (2003) Experience with the L3 vertex drift chamber at LEP. *Nucl. Instr. and Meth. in Phys. Res.* **A515**, 31.

Anderson, D. F. (1980) A xenon gas scintillation proportional counter coupled to a photoionization detector. *Nucl. Instr. and Meth.* **178**, 125.

Anderson, D. F. (1988) Measurement of TMAE and TEA vapor pressure. *Nucl. Instr. and Meth.* **A270**, 416.

Anderson, D. F., *et al.* (1983) Coupling of a BaF2 Scintillator to a TMAE photocathode and a low- pressure wire chamber. *Nucl. Instr. and Meth.* **217**, 217.

Anderson, D. F., *et al.* (1984) Recent developments in BaF2 scintillator coupled to a low - pressure wire chamber. *Nucl. Instr. and Meth.* **225**, 8.

Anderson, D. F., *et al.* (1994) High counting rate resistive plate chamber. *Nucl. Instr. and Meth.* **A348**, 324.

Anderson, M., *et al.* (2003) The STAR time projection chamber: a unique tool for studying high multiplicity events at RHIC. *Nucl. Instr. and Meth.* **A499**, 659.

Anderson, W., *et al.* (2011) Design, construction, operation and performance of a Hadron Blind Detector for the PHENIX experiment. *Nucl. Instr. and Meth.* **A 646**, 35.

Anderson, W. S., *et al.* (1992) Electron attachment, effective ionization coefficient and electron drift velocity for CF_4 gas mixtures. *Nucl. Instr. and Meth.* **A323**, 273.

Andronic, A., *et al.* (2003) Pulse height measurements and electron attachment in drift chambers operated with Xe, CO2 mixtures. *Nucl. Instr. and Meth.* **A498**, 143.

Anelli, M., *et al.* (1991) Glass electrode spark counters. *Nucl. Instr. and Meth.* **300**, 572.

Angelini, F., *et al.* (1993) The Micro-Gap Chamber. *Nucl. Instr. and Meth.* **A335**, 69.

Angelini, F., *et al.* (1994) Behaviour of Microstrip Gas Chamber in strong magnetic field. *Nucl. Instr. and Meth.* **A343**, 441.

Angelini, F., *et al.* (1995) Development of a very large area microstrip gas chamber for the CMS central tracking system. *Nucl. Instr. and Meth.* **A360**, 22.

Angelini, F., *et al.* (1996) Operation of MSGCs with gold strips built on surface-treated thin glass. *Nucl. Instr. and Meth.* **A 382**, 461.

Anghinolfi, F., *et al.* (2004) NINO: an ultra-fast and low-power front-end amplifier/ discriminator ASIC designed for the multigap resistive plate chamber. *Nucl. Instr. and Meth.* **A533**, 183.

Antchev, G., *et al.* (2010) The TOTEM detector at LHC. *Nucl. Instr. and Meth.* **A617**, 62.

Anulli, F., *et al.* (2002) Resistive Plate Chambers performance in the BaBar IFR system. *IEEE Trans. Nucl. Sci.* **49**, 888.

Anulli, F., *et al.* (2003) Mechanisms affecting performance of the BaBar resistive plate chambers and searches for remediation. *Nucl. Instr. and Meth.* **A508**, 128.

Aprile, E., *et al.* (1991) Measurement of the lifetime of conduction electrons in liquid xenon. *Nucl. Instr. and Meth.* **A300**, 343.

Aprile, E., *et al.* (1994) Electron extraction from a CsI photocathode into condensed Xe, Kr and Ar. *Nucl. Instr. and Meth.* **A343**, 129.

Aprile, E., *et al.* (2001) A liquid xenon time projection chamber for gamma ray imaging in astrophysics: present status and future directions. *Nucl. Instr. and Meth.* **A461**, 256.

Aprile, E., *et al.* (2002) Detection of gamma-rays with a 3.5 l xenon ionization chamber triggered by the primary scintillation light. *Nucl. Instr. and Meth.* **A480**, 636.

Aprile, E., *et al.* (2004) Proportional light in a dual-phase xenon chamber. *IEEE Trans. Nucl. Sci.* **NS-51**, 1986.

Aprile, E., *et al.* (2008) Compton imaging of MeV gamma-rays with the Liquid Xenon Gamma-Ray Imaging Telescope (LXeGRIT). *Nucl. Instr. and Meth.* **A593**, 414.

Apsimov, R. J., *et al.* (1986) A Ring Imaging Cherenkov detector for the CERN Omega Spectrometer – the design and recent performances. *Nucl. Instr. and Meth.* **A248**, 76.

Arai, R., *et al.* (1983) Development of a large cylindrical drift chamber for the VENUS detector at TRISAN. *Nucl. Instr. and Meth.* **217**, 181.

Arai, Y., *et al.* (1996) A modular straw drift tube tracking system for the Solenoidal Detector Collaboration experiment. Part I, Design. *Nucl. Instr. and Meth.* **A381**, 355.

Armstrong, T. A., *et al.* (1999) Large straw-tube tracking chamber for AGS experiment E864. *Nucl. Instr. and Meth.* **A425**, 210.

Arnaldi, R., *et al.* (1999) Study of Resistive Plate Chambers for the Alice dimuon spectrometer. *Nucl. Phys. B (Proc. Suppl.)* **78**, 84.

Arnaldi, R., *et al.* (2000) A low-resistivity RPC for the ALICE dimuon arm. *Nucl. Instr. and Meth.* **A451**, 462.

Arnaldi, R., *et al.* (2002) Spatial resolution of RPC in streamer mode. *Nucl. Instr. and Meth.* **A490**, 51.

Arnaldi, R., *et al.* (2009) Final results of the tests on the resistive plate chambers for the ALICE muon arm. *Nucl. Instr. and Meth.* **A602**, 740.

Arneodo, F., *et al.* (2000) First observation of a 140-cm drift ionizing tracks in the ICARUS liquid-argon TPC. *Nucl. Instr. and Meth.* **A449**, 36.

Arnold, R., *et al.* (1986) Photosensitive gas detectors for the ring-imaging Cherenkov (RICH) technique and the DELPHI barrel RICH prototype. *Nucl. Instr. and Meth.* **A252**, 188.

Arnold, R., *et al.* (1988a) A Ring Imaging Cherenkov detector, the DELPHI barrel RICH prototype. Part A: Experimental studies on the detection efficiency and the spatial resolution. *Nucl. Instr. and Meth.* **A270**, 255.

Arnold, R., *et al.* (1988b) A Ring Imaging Cherenkov detector, the DELPHI barrel RICH prototype. Part B: Experimental studies of the detector performance for particle identification. *Nucl. Instr. and Meth.* **A270**, 289.

Arnold, R., *et al.* (1992) A fast cathode-pad photon detector for Cherenkov ring imaging. *Nucl. Instr. and Meth.* **A314**, 465.

Artuso, M., *et al.* (2003) Construction, pattern recognition and performance of the CLEO III Lif-TEA RICH detector. *Nucl. Instr. and Meth.* **A502**, 91.

Ashford, V., *et al.* (1986) Development of the Cherenkov Ring Imaging Detector for the SLD. *IEEE Trans. Nucl. Sci.* **33**, 113.

Atac, M. (1987) Wire chamber ageing and wire material. *IEEE Trans. Nucl. Science* **NS-34**, 475.

Atac, M. and Tollestrup, A. V. (1982) Self-quenching streamers. *IEEE Trans. Nucl. Sci.* **NS-29**, 388.

References

Atencio, L. G., *et al.* (1981) Delay-line readout drift chambers. *Nucl. Instr. and Meth.* **187**, 381.

Atwood, W. B., *et al.* (1991) Performance of the ALEPH Time Projection Chambers. *Nucl. Instr. and Meth.* **A306**, 446.

Au, J. W., *et al.* (1993) The valence shell photoabsorption of the linear alchanes. *Chem. Phys.* **173**, 209.

Azevedo, C. D. R., *et al.* (2010) Towards THGEM UV-photon detectors for RICH: on single-photon detection efficiency in Ne/CH4 and Ne/CF4. *J. Instr.* **5**, P01002.

Bachmann, S., *et al.* (1999) Charge amplification and transfer processes in the Gas Electron Multiplier. *Nucl. Instr. and Meth.* **A438**, 376.

Bachmann, S., *et al.* (2001) Development and applications of the Gas Electron Multiplier. *Nucl. Instr. and Meth.* **A 471**, 115.

Bachmann, S., *et al.* (2002) Discharge studies and prevention in the gas electron multiplier. *Nucl. Instr. and Meth.* **A 479**, 294.

Bachmann, S., *et al.* (2004) Developments for the outer tracking system of the LHCb experiment. *Nucl. Instr. and Meth.* **A518**, 59.

Badertscher, A., *et al.* (2010) Operation of a double-phase pure argon Large Electron Multiplier Time Projection Chamber: comparison of single and double phase operation. *Nucl. Instr. and Meth.* **A 617**, 188.

Badertscher, A., *et al.* (2011) First operation of a double phase LAr Large Electron Multiplier Time Projetion Chamber with a 2D projective readout anode. *Nucl. Instr. and Meth.* **A 641**, 48.

Bagliesi, M. G., *et al.* (2010) The TOTEM T2 telescope based on triple-GEM chambers. *Nucl. Instr. and Meth.* **A 617**, 134.

Baines, J. T. M., *et al.* (1993) The data acquisition system of the OPAL deetctor at LEP. *Nucl. Instr. and Meth.* **A325**, 271.

Baksay, L., *et al.* (1976) Multiwire Proportional Chamber spectrometer for the CERN Intersecting Storage Rings. *Nucl. Instr. and Meth.* **133**, 219.

Balla, A., *et al.* (2011) Status of the cylindrical-GEM project for the KLOE-2 inner tracker. *Nucl. Instr. and Meth.* **A 628**, 194.

Bambynek, W. (1973) On selected problems in the field of proportional counters. *Nucl. Instr. and Meth.* **112**, 103.

Band, H. R., *et al.* (2006) Performance and aging studies of BaBar Resistive Plate Chambers. *Nucl. Phys. (Proc. Suppl.)* **158**, 139.

Band, H. R., *et al.* (2008) Study of HF production in BaBar Resistive Plate Chambers. *Nucl. Instr. and Meth.* **A594**, 33.

Baranco Luque, M., *et al.* (1980) The construction of the central detector for an experiment at the CERN proton–antiproton collider. *Nucl. Instr. and Meth.* **176**, 175.

Barashko, V., *et al.* (2008) Fast algorithm for track segmant and hit reconstruction in the CMS Cathode Strip Chambers. *Nucl. Instr. and Meth.* **A589**, 383.

Baringer, P., *et al.* (1987) A drift chamber constructed of aluminized mylar tubes. *Nucl. Instr. and Meth.* **A254**, 542.

Barr, A., *et al.* (1996) Operation of high rate microstrip gas chambers. *Nucl. Phys. B (Proc. Suppl.)* **61B**, 264.

Barr, A., *et al.* (1998) Construction, test and operation in a high intensity beam of a small system of micro-strip gas chambers. *Nucl. Instr. and Meth.* **A403**, 31.

Barrelet, E., *et al.* (1982) A two-dimensional, single photoelectron drift detector for Cherenkov Ring Imaging. *Nucl. Instr. and Meth.* **200**, 219.

Bartol, F., *et al.* (1996) The C.A.T. pixel proportional gas counter detector. *J. Phys. III France* **6**, 337.

Bateman, J. E. (2003) A general parametric model for the gain of gas avalanche counters with particular attention to non-cylindrical geometries. *Phys. Rep.* **375**, 411.

Bateman, J. E., *et al.* (1976) Spatial resolution in a xenon filled MWPC X-ray imaging detector – a computing physicist approach. *Nucl. Instr. and Meth.* **135**, 235.

Bateman, J. E., *et al.* (1980) The development of the Rutherford Laboratory MWPC positron camera. *Nucl. Instr. and Meth.* **176**, 83.

Bateman, J. E., *et al.* (2010) The FastGas detector. *Nucl. Instr. and Meth.* **A616**, 59.

Bateman, J. E., *et al.* (2012) The OSMOND detector. *Nucl. Instr. and Meth.* **A698**, 168.

Battistoni, G., *et al.* (1978) Detection of induced pulses in proportional wire devices with resistive cathodes. *Nucl. Instr. and Meth.* **152**, 423.

Battistoni, G., *et al.* (1979a) Operation of limited streamer tubes. *Nucl. Instr. and Meth.* **164**, 57.

Battistoni, G., *et al.* (1979b) A cube lattice multiwire detector. *Nucl. Instr. and Meth.* **164**, 453.

Battistoni, G., *et al.* (1982) Resistive cathode transparency. *Nucl. Instr. and Meth.* **202**, 459.

Battistoni, G., *et al.* (1983) Influence of gas mixture and cathode material on limited streamer operation. *Nucl. Instr. and Meth.* **217**, 433.

Battistoni, G., *et al.* (1986) The Nusex Detector. *Nucl. Instr. and Meth.* **A245**, 277.

Bauers, G., *et al.* (1987) Upgraded muon detector system for UA1 based on limited streamer tubes. *Nucl. Instr. and Meth.* **253**, 179.

Baur, R., *et al.* (1994) The CERES RICH detector system. *Nucl. Instr. and Meth.* **A343**, 87.

Bay, A., *et al.* (2002) Study of sparking in micromegas chambers. *Nucl. Instr. and Meth.* **A488**, 162.

Beard, C., *et al.* (1990) Thin, high gain wire chambers for electromagnetic presampling in OPAL. *Nucl. Instr. and Meth.* **A286**, 117.

Becker, U. (1984) Study of resolution for a large array of drift chambers. *Nucl. Instr. and Meth.* **225**, 456.

Becker, U., *et al.* (1975) A comparison of drift chambers. *Nucl. Instr. and Meth.* **128**, 593.

Beckers, T., *et al.* (1994) Optimization of microstrip gas chamber design and operating conditions. *Nucl. Instr. and Meth.* **A346**, 95.

Bedjidian, M., *et al.* (2010) Glass resistive plate chambers for a semi-digital HCAL. *Nucl. Instr. and Meth.* **A623**, 120.

Behnke, T. (2011) Detector concepts at the international linear collider. *Nucl. Instr. and Meth.* **A628**, 19.

Behnke, T., *et al.* (2001) TESLA TDR, DESY-01-011, Humburg, DESY.

Behrends, S. and Melissinos, A. C. (1981) Properties of argon-ethane/methane mixtures for use in proportional counters. *Nucl. Instr. and Meth.* **188**, 521.

Beingessner, S. P., *et al.* (1987) The UA1 Central Detector at present and future luminosity (ACOL). *Nucl. Instr. and Meth.* **A257**, 552.

Bélanger, G., *et al.* (1969) The far-ultraviolet spectra of perfluoro-normal-paraffin. *Chem. Phys. Lett.* **3**, 649.

Bella, G., *et al.* (1986) Development of calorimeters using thin chambers operating in a high gain mode. *Nucl. Instr. and Meth.* **A252**, 503.

Bellazzini, R. and Muleri, F. (2010) X-ray polarimetry: a new window on the high energy sky. *Nucl. Instr. and Meth.* **A 623**, 766.

Bellazzini, R., *et al.* (1998a) The micro-groove detector. *Nucl. Instr. and Meth.* **A 424**, 444.

Bellazzini, R., *et al.* (1998b) Substrate-less, spark-free micro-strip gas counters. *Nucl. Instr. and Meth.* **A409**, 14.

Bellazzini, R., *et al.* (2002) X-ray polarimeter with a micro pattren gas detector with pixel read out. *IEEE Trans. Nucl. Sci.* **NS-49**, 1216.

Bellazzini, R., *et al.* (2004) Reading a GEM with a VLSI pixel ASIC used as direct charge collecting anode. *Nucl. Instr. and Meth.* **A 535**, 477.

Bellazzini, R., *et al.* (2007a) Gas pixel detectors. *Nucl. Instr. and Meth.* **A 572**, 160.

Bellazzini, R., *et al.* (2007b) Imaging with the invisible light. *Nucl. Instr. and Meth.* **581**, 246.

Belostotski, S., *et al.* (2008) Extension of the operational lifetime of the proportional chambers in the HERMES spectrometer. *Nucl. Instr. and Meth.* **A591**, 353.

Bencivenni, G. and Domenici, D. (2007) An ultra-light cylindrical detector as inner tracker at KLOE-2. *Nucl. Instr. and Meth.* **A 581**, 221.

Benlloch, J., *et al.* (1998) Further developments and beam tests of the Gas Electron Multiplier (GEM). *Nucl. Instr. and Meth.* **A419**, 410.

Benot, M., *et al.* (1972) Cherenkov counters for particle identification at high energies. *Nucl. Instr. and Meth.* **105**, 431.

Bensinger, J., *et al.* (2002) Construction of monitored drift tube chambers for ATLAS end-cap muon spectrometer at IHEP (Protvino). *Nucl. Instr. and Meth.* **A494**, 480.

Berger, M. J., *et al.*, (1998), XCOM: photon cross sections database: physics.nist.gov/ PhysRefData/Xcom/Text/XCOM.html.

Berger, M. J., *et al.*, (2011), Stopping-power and range tables for electrons, protons, and helium ions: www.nist.gov/pml/data/star/index.cfm.

Bergnoli, A., *et al.* (2009) Performances of the OPERA RPCs. *Nucl. Instr. and Meth.* **A602**, 635.

Beringer, J. (2012) The review of particle properties. *Phys. Rev.* **D86**, 010001.

Berkowitz, J. (2002) *Atomic and Molecular Photoabsorption*, New York, Academic Press.

Bernard, D. and Delbart, A. (2012) High angular precision gamma-ray astronomy and polarimetry. *Nucl. Instr. and Meth.* **A695**, 71.

Bernet, C., *et al.* (2005) The 40×40 cm^2 gaseous microstrip detector Micromegas for the high-luminosity COMPASS experiment at CERN. *Nucl. Instr. and Meth.* **A536**, 61.

Bernreuther, S., *et al.* (1995) Design and performance of the large HERMES drift chambers. *Nucl. Instr. and Meth.* **367**, 96.

Bertolin, A., *et al.* (2009) The RPC system of the OPERA experiment. *Nucl. Instr. and Meth.* **A602**, 631.

Bertozzi, W., *et al.* (1977) Focal plane instrumentation; a very high resolution MWPC system for inclined tracks. *Nucl. Instr. and Meth.* **141**, 457.

Bettini, A., *et al.* (1991) A study of the factors affecting the electron lifetime in ultra-pure liquid argon. *Nucl. Instr. and Meth.* **A305**, 177.

Bhadra, S., *et al.* (1988) A computer-controlled wire tension measurement system used in the fabrication of the CDF central drift tube array. *Nucl. Instr. and Meth.* **269**, 33.

Bhide, S. S., *et al.* (2006) On aging problem of glass Resistive Plate Chambers. *Nucl. Phys. (Proc. Suppl.)* **158**, 195.

Biagi, S. (1999) Monte Carlo simulation of electron drift and diffusion in counting gases under the influence of electric and magnetic fields. *Nucl. Instr. and Meth.* **A 421**, 234.

Biagi, S. F. and Jones, T. J. (1995) The microdot gas avalanche chamber: an investigation of new geometries. *Nucl. Instr. and Meth.* **A361**, 72.

Biagi, S. and Veenhof, R., (1995a), Electron-molecules cross sections: rjd.web.cern.ch/rjd/ cgi-bin/cross.

Biagi, S. and Veenhof, R., (1995b), MAGBOLTZ: consult.cern.ch/writeup/magboltz/.

Biebel, O., *et al.* (1992) Performance of the OPAL jet chamber. *Nucl. Instr. and Meth.* **323**, 169.

Biino, C., *et al.* (1988) Charge division in a small proportional chamber constucted with aluminized mylar tubes. *Nucl. Instr. and Meth.* **A271**, 417.

Bilevych, Y., *et al.* (2011) Spark protection layers for CMOS pixel anode chips in MPGDs. *Nucl. Instr. and Meth.* **A 629**, 66.

Bindi, M. (2012) Operation and performance tuning of ATLAS RPCs through the detector control system at the startup of 2009 LHC run. *Nucl. Instr. and Meth.* **A661 Suppl. 1**, S10.

Binkley, M., *et al.* (2003) Aging in large CDF tracking chambers. *Nucl. Instr. and Meth.* **A515**, 53.

Birk, M., *et al.* (1976) A simple efficient method of delay-line termination and timing-signals extraction in position-sensitive proportional counters. *Nucl. Instr. and Meth.* **137**, 393.

Biscossa, A., *et al.* (1999) Construction and test of a full-scale prototype of an ATLAS muon spectrometer tracking chamber. *Nucl. Instr. and Meth.* **A525**, 140.

Biswas, S., *et al.* (2009) Performances of linseed oil-free bakelite RPC prototypes with cosmic ray muons. *Nucl. Instr. and Meth.* **A602**, 749.

Bittl, X., *et al.* (1997) Diffusion and drift studies of Ar-DME/CO2/CH4 gas mixtures for a radial TPC in the E/B field. *Nucl. Instr. and Meth.* **A398**, 249.

Bittner, B., *et al.* (2011) Development of fast high-resolution muon drift-tube detectors for high counting rates. *Nucl. Instr. and Meth.* **A628**, 154.

Black, J. K., *et al.* (2003) X-ray polarimeter with an active-matrix pixel proportional counter. *Nucl. Instr. and Meth.* **A 513**, 639.

Blanco, A., *et al.* (2003) Perspectives for positron emission tomography with RPCs. *Nucl. Instr. and Meth.* **A508**, 88.

Blanco, A., *et al.* (2009) Efficiency of RPC detectors for whole-body human TOF-PET. *Nucl. Instr. and Meth.* **A602**, 780.

Bloch, F. and Bradbury, N. E. (1935) On the mechanism of unimolecular electron capture. *Phys. Rev.* **48**, 689.

Blum, W. and Rolandi, G. (1993) *Particle Detection with Drift Chambers*, Berlin, Springer-Verlag.

Blum, W., *et al.* (1986) Measurement of avalanche broadening caused by the wire ExB effect. *Nucl. Instr. and Meth.* **A252**, 407.

Bobkov, S., *et al.* (1984) Drift Precision Imager. *Nucl. Instr. and Meth.* **226**, 376.

Bock, R., *et al.* (1994) A wiring system for mass production of drift and proportional chambers. *Nucl. Instr. and Meth.* **345**, 256.

Bohm, J., *et al.* (1995) High rate operation and lifetime studies with micro-strip gas chambers. *Nucl. Instr. and Meth.* **A360**, 34.

Böhmer, F. V., *et al.* (2013) Simulation of space-charge effects in an ungated GEM-based TPC. *Nucl. Instr. and Meth.* **719**, 101.

Böhmer, V. (1973) New investigations into the detection properties of hybrid chambers. *Nucl. Instr. and Meth.* **107**, 157.

Boie, R. A., *et al.* (1982) High resolution X-ray gas proportional detectors with delay line position sensing for high counting rates. *Nucl. Instr. and Meth.* **201**, 93.

Bologna, G., *et al.* (1979) Electrostatic field in a cylindrical proportional chamber, LNF-79/48, Frascati, INFN.

Bolotnikov, A. and Ramsey, B. (1999) Studies of light and charge produced by alpha-particles in high-pressure xenon. *Nucl. Instr. and Meth.* **A428**, 391.

Bolozdynya, A. I. (1999) Two-phase emission detectors and their applications. *Nucl. Instr. and Meth.* **A422**, 314.

Bondar, A. E., *et al.* (1983) Spatial resolution of induction chambers. *Nucl. Instr. and Meth.* **207**, 379.

Bondar, A. E., *et al.* (2003) Study of ion feedback in multi-GEM structures. *Nucl. Instr. and Meth.* **A496**, 325.

Bondar, A. E., *et al.* (2006) Two-phase argon and xenon avalanche detectors based on Gas Electron Multiplier. *Nucl. Instr. and Meth.* **A 556**, 273.

Borghesi, A. (1978) Tension control device for wires in large Multiwire Proportional Chambers. *Nucl. Instr. and Meth.* **153**, 379.

Borisov, A., *et al.* (2002) ATLAS monitored drift tube assembly and test at IHEP (Protvino). *Nucl. Instr. and Meth.* **A494**, 214.

Botner, O., *et al.* (1990) Production of prompt electrons in the charm Pt region at $s = 630$ GeV. *Phys. Lett. B* **236**, 488.

Bouclier, R., *et al.* (1970) Investigation on some properties of Multiwire Proportional Chambers. *Nucl. Instr. and Meth.* **88**, 149.

Bouclier, R., *et al.* (1974) Proportional Chambers for a 50 000-wire detector. *Nucl. Instr. and Meth.* **115**, 235.

Bouclier, R., *et al.* (1983) Progress in Cherenkov Ring Imaging: Part1. Detection and localization of photons with the multistep proportional chamber. *Nucl. Instr. and Meth.* **205**, 403.

Bouclier, R., *et al.* (1989) Recent developments of the multidrift tube. *Nucl. Instr. and Meth.* **A283**, 509.

Bouclier, R., *et al.* (1992) High flux operation of microstrip gas chambers on glass and plastic supports. *Nucl. Instr. and Meth.* **A323**, 240.

Bouclier, R., *et al.* (1994a) Results of wire chamber ageing tests with CH4 and DME based gas mixtures. *Nucl. Instr. and Meth.* **A346**, 114.

Bouclier, R., *et al.* (1994b) Ageing studies with micro-strip gas chambers. *Nucl. Instr. and Meth.* **A348**, 109.

Bouclier, R., *et al.* (1995a) Optimization of design and beam tests of micro-strip gas chambers. *Nucl. Instr. and Meth.* **A367**, 163.

Bouclier, R., *et al.* (1995b) Development of micro-strip gas chambers for high rate operation. *Nucl. Instr. and Meth.* **367**, 168.

Bouclier, R., *et al.* (1995c) On some factors affecting the discharge conditions in micro-strip gas chambers. *Nucl. Instr. and Meth.* **A365**, 65.

Bouclier, R., *et al.* (1996a) Ageing of Microstrip Gas Chambers: problems and solutions. *Nucl. Instr. and Meth.* **A381**, 289.

Bouclier, R., *et al.* (1996b) High rate operation of micro-strip gas chambers on diamond-coated glass. *Nucl. Instr. and Meth.* **A 369**, 328.

Bouclier, R., *et al.* (1997) The Gas Electron Multiplier (GEM). *IEEE Trans. Nucl. Sci.* **NS-44**, 646.

Boudjemline, K., *et al.* (2007) Spatial resolution of a GEM readout TPC using the charge dispersion signal. *Nucl. Instr. and Meth.* **A 574**, 22.

Boyarski, A. (2003) Additives that prevent or reverse cathode aging in drift chambers with helium-isobutane gas. *Nucl. Instr. and Meth.* **A 515**, 190.

Boyarski, A. M. (2004) Model of high-current breakdown from cathode field emission in aged wire chambers. *Nucl. Instr. and Meth.* **A535**, 632.

Bozzo, M., *et al.* (1980) Development of large planar proportional chambers. *Nucl. Instr. and Meth.* **178**, 77.

Bradamante, F. and Sauli, F. (1967) Magnetostrictive thin-gap wire spark chambers of large dimensions. *Nucl. Instr. and Meth.* **56**, 268.

Bradbury, N. E. and Tatel, H. E. (1934) The formation of negative ions in gases Part II. CO2, N2O, SO2, H2S and H2O. *J. Chem. Phys.* **2**, 835.

Braem, A., *et al.* (2003) Technology of photocathode production. *Nucl. Instr. and Meth.* **A502**, 205.

Brand, C., *et al.* (1976) Development of a 150 m^2 proportional chamber system with a 1 million bit buffer: the EMI for BEBC. *Nucl. Instr. and Meth.* **136**, 485.

Brand, C., *et al.* (1986) Results on space measurement accuracy from tests of a half-scale DELPHI TPC Prototype. *Nucl. Instr. and Meth.* **A252**, 413.

Brand, C., *et al.* (1989) The DELPHI Time Projection Chamber. *Nucl. Instr. and Meth.* **A283**, 567.

Brehin, S., *et al.* (1975) Some observations concerning the construction of proportional chambers with thick sense wires. *Nucl. Instr. and Meth.* **123**, 225.

Breidenbach, M., *et al.* (1973) Time properties of MWPC using electronegative gases, and their efficiency of detection for normal and abnormal particles (quarks). *Nucl. Instr. and Meth.* **108**, 23.

Breskin, A. (2000) Advances in gas avalanche radiation detectors for biomedical applications. *Nucl. Instr. and Meth.* **A454**, 26.

Breskin, A. and Zwang, N. (1977) Timing properties of Parallel Plate Avalanche Counters with light particles. *Nucl. Instr. and Meth.* **144**, 609.

Breskin, A., *et al.* (1974a) Further results on the operation of high-accuracy drift chambers. *Nucl. Instr. and Meth.* **119**, 9.

Breskin, A., *et al.* (1974b) Recent observations and measurements with High-Accuracy Drift Chambers. *Nucl. Instr. and Meth.* **124**, 189.

Breskin, A., *et al.* (1974c) Two-dimensional drift chambers. *Nucl. Instr. and Meth.* **119**, 1.

Breskin, A., *et al.* (1977) High-accuracy, bidimensional read-out of proportional chambers with short resolution times. *Nucl. Instr. and Meth.* **143**, 29.

Breskin, A., *et al.* (1978) A fast, bidimensional, position-sensitive detection system for heavy ions. *Nucl. Instr. and Meth.* **148**, 275.

Breskin, A., *et al.* (1988) A high efficiency low-pressure UV-RICH detector with optical avalanche recording. *Nucl. Instr. and Meth.* **A273**, 798.

Breskin, A., *et al.* (1995) New ideas in CsI-based photon detectors: wire photomultipliers and protection of the photocathodes. *IEEE Trans. Nucl. Sci.* **NS-42**, 298.

Breskin, A., *et al.* (2001) Sealed gas UV-photon detector with a multi-GEM multiplier. *IEEE Trans. Nucl. Sci.* **NS-48**, 417.

Breskin, A., *et al.* (2002) GEM photomultiplier operation in CF4. *Nucl. Instr. and Meth.* **A 483**, 670.

Breskin, A., *et al.* (2009) A concise review of THGEM detectors. *Nucl. Instr. and Meth.* **A 598**, 107.

Breskin, A., *et al.* (2010) Progress in gaseous photomultipliers for the visible spectral range. *Nucl. Instr. and Meth.* **A 623**, 318.

Bressan, A., *et al.* (1999a) High rate behavior and discharge limits in micro-pattern detectors. *Nucl. Instr. and Meth.* **A424**, 321.

Bressan, A., *et al.* (1999b) Beam tests of the gas electron multiplier. *Nucl. Instr. and Meth.* **A425**, 262.

Bressan, A., *et al.* (1999c) Two-dimensional readout in GEM detectors. *Nucl. Instr. and Meth.* **A425**, 254.

Breyer, B. (1973) Pulse height distribution in low energy proportional counter measurements. *Nucl. Instr. and Meth.* **112**, 91.

Brinkmann, D., *et al.* (1995) Image data analysis for the NA35 steamer chamber. *Nucl. Instr. and Meth.* **A354**, 419.

Brown, S. C. (1959) *Basic Data on Plasma Physics*, New York, Wiley.

Broyles, C. D., *et al.* (1953) The measurement and interpretation of the K auger intensities of Sn113, Cs137, and Au198. *Phys. Rev.* **89**, 715.

Bucciantonio, M., *et al.* (2013) Development of a fast proton range radiography system for quality assurance in hadrontherapy. *Nucl. Instr. and Meth.* **A718**, 160.

Buckley, E., *et al.* (1989) A study of ionization electrons drifting over large distances in liquid argon. *Nucl. Instr. and Meth.* **A275**, 364.

Budagov, Y. A., *et al.* (1987) How to use electrodeless drift chambers in experiments at accelerators. *Nucl. Instr. and Meth.* **A255**, 493.

Buffet, J. C., *et al.* (2005) Advances in detectors for single crystal neutron diffraction. *Nucl. Instr. and Meth.* **A554**, 392.

Bunemann, O., *et al.* (1949) Design of grid ionization chambers. *Canadian J. of Res.* **27A**, 191.

Burckhart, H. J., *et al.* (1986) Investigation of very long jet chambers. *Nucl. Instr. and Meth.* **A244**, 416.

Buskens, J., *et al.* (1983) Small high-precision wire chambers for the measurement of proton-antiproton elastic scattering at the CERN collider. *Nucl. Instr. and Meth.* **207**, 365.

Büttner, C., *et al.* (1998) Progress with the Gas Electron Multiplier. *Nucl. Instr. and Meth.* **A 409**, 79.

Buys, A. (1996) RICH in operating experiments. *Nucl. Instr. and Meth.* **A371**, 1.

Buzulutskov, A., *et al.* (2000) The GEM photomultiplier operated with noble gas mixtures. *Nucl. Instr. and Meth.* **A 443**, 164.

Bychkov, V. N., *et al.* (2006) The large size straw drift chambers of the COMPASS experiment. *Nucl. Instr. and Meth.* **A556**, 66.

Byrne, J. (1969) Single electron detection in proportional gas counters. *Nucl. Instr. and Meth.* **74**, 291.

Byrne, J. (2002) Electron avalanches in inhomogeneous media. *Nucl. Instr. and Meth.* **A491**, 122.

Calcaterra, A., *et al.* (2004) Test of large area glass RPCs at the DAΦNE Test Beam Facility (BTF). *Nucl. Instr. and Meth.* **A533**, 154.

Calvert, J. C. and Pitts, J. N. (1966) *Photochemistry*, New York, Wiley.

Camarri, P. (2009) Operation and performance of RPCs in the ARGO-YBJ experiment. *Nucl. Instr. and Meth.* **A602**, 668.

Camarri, P., *et al.* (1998) Streamer suppression with SF6 in RPCs operated in avalanche mode. *Nucl. Instr. and Meth.* **A 414**, 317.

Candela, A., *et al.* (2004) Ageing and recovering of glass RPC. *Nucl. Instr. and Meth.* **A533**, 116.

Capeáns, M. (2003) Aging and materials: lessons for detectors and gas systems. *Nucl. Instr. and Meth.* **A515**, 73.

Capeáns, M., *et al.* (1993) Aging properties of straw proportional tubes with a $Xe-Co_2-CF_4$ gas mixture. *Nucl. Instr. and Meth.* **A337**, 122.

Carboni, G., *et al.* (2004) Final results from an extensive ageing test of bakelite Resistive Plate Chambers. *Nucl. Instr. and Meth.* **A533**, 107.

Cardarelli, R., *et al.* (1988) Progress in resistive plate counters. *Nucl. Instr. and Meth.* **A263**, 20.

Cardarelli, R., *et al.* (1993) Performance of a resistive plate chamber operating with pure CF3Br. *Nuc. Phys. B (Proc. Suppl.)* **A333**, 399.

Cardarelli, R., *et al.* (1996) Avalanche and streamer mode operation of resistive plate chambers *Nucl. Instr. and Meth.* **A382**, 470.

Carr, J. and Kagan, H. (1986) Wire stability studies for an SSC central drift tracker. *Proc. 1986 Summer Study on the Physics of the Superconducting Supercollider.* Stanford. p. 396.

Carrillo, C. (2012) The CMS project, results from 2009 cosmic-ray data. *Nucl. Instr. and Meth.* **A661**, S19.

Cavalli-Sforza, M., *et al.* (1975) A system of multiwire proportional chambers for a large aperture spectrometer. *Nucl. Instr. and Meth.* **124**, 73.

Cennini, P., *et al.* (1999) Detection of scintillation light in coincidence with ionizing tracks in a liquid argon time projection chamber. *Nucl. Instr. and Meth.* **A432**, 240.

Cerminara, G. (2010) Commissioning, operation and performance of the CMS drift tube chambers. *Nucl. Instr. and Meth.* **A617**, 144.

Cerrito, L., *et al.* (1999) Particle identification using cluster counting in a large drift chamber at normal conditions. *Nucl. Instr. and Meth.* **A 434**, 261.

Cerron-Zeballos, E., *et al.* (1996) A new type of resistive plate chamber: the multigap RPC. *Nucl. Instr. and Meth.* **A 374**, 132.

Cerutti, F. (2004) Performance studies of the monitored drift-tube chambers of the ATLAS muon spectrometer. *Nucl. Instr. and Meth.* **A535**, 175.

Chan, W. F., *et al.* (1993a) The electronic spectrum of carbon dioxide. Discrete and continuum photoabsorption oscillator strengths (6–203 eV) *Chem. Phys.* **178**, 401.

Chan, W. F., *et al.* (1993b) Absolute optical oscillator strengths for discrete and continuum photoabsorption of molecular nitrogen (11–200 eV). *Chem. Phys.* **170**, 81.

Chaplier, G., *et al.* (1999) Preliminary results of the experimental and simulated intrinsic properties of the Compteur A Trou (CAT) detector: behavior with synchrotron radiation. *Nucl. Instr. and Meth.* **A426**, 339.

Charles, G., *et al.* (2011) Discharge studies in Micromegas detectors in low energy hadron beams. *Nucl. Instr. and Meth.* **A468**, 174.

Charpak, G. (1970) Evolution of the automatic spark chambers. *Ann. Rev. Nucl. Sci.* **20**, 195.

Charpak, G. and Sauli, F. (1971) Multiwire chambers operating in the Geiger–Müller mode; new simple method of particle localization. *Nucl. Instr. and Meth.* **96**, 363.

Charpak, G. and Sauli, F. (1974) High-accuracy, two-dimensional readout in Multiwire Proportional Chambers. *Nucl. Instr. and Meth.* **113**, 235.

Charpak, G. and Sauli, F. (1978) The multistep avalanche chamber: a new high rate, high accuracy gaseous detector. *Phys. Letters* **78 B**, 523–528.

Charpak, G., *et al.* (1963) A new method for determining the position of a spark in a spark chamber by measurement of currents. *Nucl. Instr. and Meth.* **24**, 501.

Charpak, G., *et al.* (1968) The use of Multiwire Proportional Counters to select and localize charged particles. *Nucl. Instr. and Meth.* **62**, 262.

Charpak, G., *et al.* (1970) Some developments in the operation of Multiwire Proportional Chambers. *Nucl. Instr. and Meth.* **80**, 13.

Charpak, G., *et al.* (1971) Some features of large Multiwire Proportional Chambers. *Nucl. Instr. and Meth.* **97**, 377.

Charpak, G., *et al.* (1972) Time degeneracy of Multiwire Proportional Chambers. *Nucl. Instr. and Meth.* **99**, 279.

Charpak, G., *et al.* (1973) High-Accuracy Drift Chambers and their use in strong magnetic fields. *Nucl. Instr. and Meth.* **108**, 413.

Charpak, G., *et al.* (1978) Progress in high-accuracy proportional chambers. *Nucl. Instr. and Meth.* **148**, 471.

Charpak, G., *et al.* (1979a) Detection of far ultraviolet photons with the Multistep Avalanche Chamber. Applications to Cherenkov light imaging and to some problems in high-energy physics. *Nucl. Instr. and Meth.* **164**, 419.

Charpak, G., *et al.* (1979b) High-accuracy localization of minimum ionizing particles using the cathode-induced charge centre-of-gravity read-out. *Nucl. Instr. and Meth.* **167**, 455.

Charpak, G., *et al.* (1987) An optical, proportional, continuously operating avalanche chamber. *Nucl. Instr. and Meth.* **A258**, 177.

Charpak, G., *et al.* (1988) Studies of light emission by continuously sensitive avalanche chambers. *Nucl. Instr. and Meth.* **A269**, 142.

Chechik, R., *et al.* (1996) Real-time Secondary Electron Emission Detector for high-rate X-ray crystallography. *IEEE Trans. Nucl. Sci.* **NS-43**, 1248.

Chechik, R., *et al.* (2004) Thick GEM-like hole multipliers: properties and possible applications. *Nucl. Instr. and Meth.* **A535**, 303.

Chefdeville, M., *et al.* (2006) An electron-multiplying "Micromegas" grid made in silicon wafer post-processing technology. *Nucl. Instr. and Meth.* **A556**, 490.

Cheng, D. C., *et al.* (1974) Very large proportional drift chambers with high spatial and time resolution. *Nucl. Instr. and Meth.* **117**, 157.

Chepel, V. and Araujo, H. (2013) Liquid noble gas detectors for low energy particle physics. *JINST* **8**, R04001.

Chesnokov, Y. A., *et al.* (2013) Bent crystal channeling applications for beam splitting, extraction and collimation in the U-70 accelerator of IHEP. *Nucl. Instr. and Meth.* **B 309**, 105.

Chiavassa, E., *et al.* (1978) Multiwire and drift chambers for the OMICRON spectrometer. *Nucl. Instr. and Meth.* **156**, 187.

Chiba, J., *et al.* (1983) Study of position resolution for cathode readout MPWC with measurement of induced charge distribution. *Nucl. Instr. and Meth.* **206**, 451.

Chiba, Y., *et al.* (1988) Electron attachment of oxygen in a drift chamber filled with xenon+10% methane. *Nucl. Instr. and Meth.* **A269**, 171.

Chirihov-Zorin, I. E. and Pukhov, O. E. (1996) On sensitivity of gas-discharge detectors to light. *Nucl. Instr. and Meth.* **A371**, 375.

Christodoulides, A. A. and Christophorou, L. G. (1971) Electron attachment to brominated aliphatic hydrocarbons of the form $nCNH2N+1Br$. *J. Chem. Phys.* **54**, 4691.

Christophorou, L. G. (1971) *Atomic and Molecular Radiation Physics*, London, Wiley.

Christophorou, L. G., *et al.* (1966) Interaction of thermal electrons with polarizable and polar molecules. *J. Chem. Phys.* **44**, 3506.

Christophorou, L. G., *et al.* (1979) Fast gas mixtures for gas-filled particle detectors. *Nucl. Instr. and Meth.* **163**, 141.

Chtchetkovski, A. I., *et al.* (2000) A planar avalanche counter with a thin resistive cathode for light ions. *Nucl. Instr. and Meth.* **A451**, 449.

Cisbani, F. C., *et al.* (2003) Quantum efficiency measurement system for large area CsI photodetectors *Nucl. Instr. and Meth.* **A 502**, 251.

Clark, A. R., *et al.* (1976) Proposal for a PEP facility based on the Time Projection Chamber, PUB-5012, Berkeley.

Clergeau, J. F., *et al.* (2001) Operation of sealed microstrip gas chambers at the ILL. *Nucl. Instr. and Meth.* **A471**, 60.

Cockroft, A. L. and Curran, S. C. (1951) The elimination of the end effects in counters. *Rev. Sci. Instrum.* **22**, 37.

Coelho, L. C. C., *et al.* (2007a) Xenon GPSC high-pressure operation with large-area avalanche photodiode readout. *Nucl. Instr. and Meth.* **A575**, 444.

Coelho, L. C. C., *et al.* (2007b) Measurement of the photoelectron-collection efficiency in noble gases and methane. *Nucl. Instr. and Meth.* **A 581**, 190.

Colaleo, A., *et al.* (2009) The compact muon solenoid RPC barrel detector. *Nucl. Instr. and Meth.* **A602**, 674.

Colas, P. (2004) Ion backflow in the Micromegas TPC for the future linear collider. *Nucl. Instr. and Meth.* **A535**, 226.

Colas, P., *et al.* (2001) Electron drift velocity measured at high electric fields. *Nucl. Instr. and Meth.* **A478**, 215.

Colli, L. and Facchini, U. (1952) Drift velocity of electrons in argon. *Rev. Sci. Instrum.* **23**, 39.

Commichau, V., *et al.* (1985) Test of a high resolution drift chamber prototype. *Nucl. Instr. and Meth.* **A235**, 267.

Cozza, D., *et al.* (2003) The CsI-based RICH detector array for the identification of high momentum particles in ALICE. *Nucl. Instr. and Meth.* **A502**, 101.

Crittenden, R. R., *et al.* (1981) A design for one mm pitch multiwire proportional chamber operating at high rates. *Nucl. Instr. and Meth.* **185**, 75.

Curran, S. C. and Craggs, J. D. (1949) *Counting Tubes Theory and Applications*, London, Butterworth.

Cwiok, M., *et al.* (2005) Optical Time Projection Chamber for imaging of two-proton decay of 45Fe nucleus. *IEEE Trans. Nucl. Sci.* **NS-52**, 2895.

Dagendorf, V., *et al.* (1994) Thermal neutron imaging detectors combining novel composite foil convertors and gaseous electron multipliers. *Nucl. Instr. and Meth.* **A350**, 503.

Dalla Torre, S. (2011) Status and perspectives of gaseous photon detectors. *Nucl. Instr. and Meth.* **A639**, 111.

Danilov, M., *et al.* (2003) Aging studies for the muon detector of HERA-B. *Nucl. Instr. and Meth.* **A515**, 202.

Davies-White, W., *et al.* (1972) A large cylindrical drift chamber for the Mark II detector at SPEAR. *Nucl. Instr. and Meth.* **160**, 227.

Davisson, C. M. and Evans, R. D. (1952) Gamma-ray absorption coefficients. *Rev. Modern Phys.* **24**, 79.

De Boer, W., *et al.* (1978) Performance of drift chambers without field shaping in high magnetic fields. *Nucl. Instr. and Meth.* **156**, 249.

De Cataldo, G., *et al.* (2011) The ALICE-HMPID detector control system: its evolution towards an expert and adaptive system. *Nucl. Instr. and Meth.* **A639**, 211.

De Graaf, E. J., *et al.* (1979) Construction and application of a delay line for position read-out of wire chambers. *Nucl. Instr. and Meth.* **166**, 139.

De Lima, E. P., *et al.* (1982) Fano factors of rare gases and their mixtures. *Nucl. Instr. and Meth.* **192**, 575.

De Lima, E. P., *et al.* (1985) The gain divergence in the transition to the Self-Quenching Streamers. *IEEE Trans. Nucl. Sci.* **NS-32**, 510.

De Palma, M., *et al.* (1983) A system of large multiwire proportional chambers for a high intensity experiment. *Nucl. Instr. and Meth.* **217**, 135.

Decamp, D., *et al.* (1990) ALEPH: a detector for electron–positron annihilations at LEP. *Nucl. Instr. and Meth.* **A294**, 121.

Delbart, A. (2010) Production and calibration of 9 m^2 of bulk micromegas detectors for the readout of the ND280/TPCs of the T2K experiment. *Nucl. Instr. and Meth.* **A623**, 105.

Delbart, A., *et al.* (2001) New developments of micromegas detector. *Nucl. Instr. and Meth.* **A461**, 84.

Delbart, A., *et al.* (2002) Performance of micromegas with preamplification at high intensity hadron beams. *Nucl. Instr. and Meth.* **A478**, 205.

den Boggende, A. J. F., *et al.* (1969) Comments on the ageing effect of gas-filled proportional counters. *J. Phys. E: Sci. Instrum.* Ser.2 Vol. **2**, 701.

Derenzo, S. E., *et al.* (1974) Electron avalanche in liquid xenon. *Phys. Rev. A* **9**, 2582.

Derré, J., *et al.* (2000) Spatial resolution in Micromegas detectors. *Nucl. Instr. and Meth.* **A459**, 523.

DeWulf, J. P. (1986) The results of the streamer-tube system of the CHARM II neutrino detector. *Nucl. Instr. and Meth.* **A252**, 443.

Di Mauro, A. (2004) Recent CsI-RICH developments. *Nucl. Instr. and Meth.* **A525**, 173.

Di Mauro, A., *et al.* (2011) The VHMPID RICH upgrade project for ALICE at LHC. *Nucl. Instr. and Meth.* **A639**, 274.

Dias, T. H. V. T., *et al.* (1991) The Fano factor in gaseous xenon: a Monte Carlo calculation for X-rays in the 0.1 to 25 keV energy range. *Nucl. Instr. and Meth.* **A307**, 341.

Dick, L., *et al.* (2004) FGLD: a novel and compact micro-pattern detector. *Nucl. Instr. and Meth.* **A535**, 347.

Dighe, P. M., *et al.* (2003) Boron-lined proportional counters with improved neutron sensitivity. *Nucl. Instr. and Meth.* **A496**, 154.

Dixit, M. S., *et al.* (2004) Position sensing from charge dispersion in micro-pattern gas detectors with a resistive anode. *Nucl. Instr. and Meth.* **A 518**, 721.

do Carmo, S. J. C., *et al.* (2008) Experimental study of the w-values and Fano factors of gaseous xenon and Ar-Xe mixtures for X-rays. *IEEE Trans. Nucl. Sci.* **NS55**, 2637.

Doke, T. (1982) Recent developments of liquid xenon detectos. *Nucl. Instr. and Meth.* **196**, 87.

Doke, T. and Masuda, K. (1999) Present status of liquid rare gas scintillation detectors and their new application to gamma-ray calorimeters. *Nucl. Instr. and Meth.* **A420**, 62.

Doke, T., *et al.* (1992) Fano factors in rare gases and their mixtures. *Nucl. Instr. and Meth.* **B 63**, 373.

Doll, P., *et al.* (1988) Large area Multiwire Chamber DE-E telescope for (n,Z) studies in a continuous energy neutron beam. *Nucl. Instr. and Meth.* **A270**, 437.

Dominik, W., *et al.* (1989) A gaseous detector for high-accuracy autoradiography of radioactive compounds with optical readout of avalanche positions. *Nucl. Instr. and Meth.* **A278**, 779.

Doroud, K., *et al.* (2009a) Recombination: an important effect in multigap resistive plate chambers. *Nucl. Instr. and Meth.* **A610**, 649.

Doroud, K., *et al.* (2009b) Simulation of temperature dependence of RPC operation. *Nucl. Instr. and Meth.* **A602**, 723.

Dörr, R., *et al.* (1985) Characteristics of a multiwire circular electrodeless drift chamber. *Nucl. Instr. and Meth.* **A238**, 238.

Drumm, H., *et al.* (1980) Experience with the Jet-Chamber of the JADE detector at PETRA. *Nucl. Instr. and Meth.* **176**, 333.

Druyvesteyn, M. J. and Penning, F. M. (1940) The mechanism of electrical discharges in gases of low pressure. *Rev. Mod. Phys.* **12**, 87.

Dubbert, J., *et al.* (2007a) Integration, commissioning and installation of monitored drift tube chambers for the ATLAS barrel muon spectrometer. *Nucl. Instr. and Meth.* **A572**, 53.

Dubbert, J., *et al.* (2007b) Modelling of the space-to-drift-time relationship of the ATLAS monitored drift-tube chambers in the presence of magnetic fields. *Nucl. Instr. and Meth.* **A572**, 50.

Duchazeaubeneix, J. C., *et al.* (1980) Nuclear scattering radiography. *J. Comp. Ass. Tomography*, **4**, 803.

Duerdoth, P., *et al.* (1975) Measurements of the time resolution and rate capability of Multiwire Proportional Chambers. *Nucl. Instr. and Meth.* **129**, 461.

Durand, E. (1966) *Electrostatique*, Paris, Masson et Cie.

Ekelöf, T. (1984) The experimental method of ring-imaging Cherenkov (RICH) counters, CERN-EP/84–168, CERN.

Ekelöf, T., *et al.* (1981) The Cherenkov Ring Imaging detector: recent progress and new developments. *Phys. Scripta* **23**, 718.

Engelfried, J. (2011) Cherenkov light imaging – fundamentals and recent developments. *Nucl. Instr. and Meth.* **A639**, 1.

Erskine, G. A. (1972) Electrostatic problems in Multiwire Proportional Chambers. *Nucl. Instr. and Meth.* **105**, 565.

Erskine, G. A. (1982) Charges and current induction induced by moving ions in multiwire chambers. *Nucl. Instr. and Meth.* **198**, 325.

Erzen, B., *et al.* (1999) Analysis of the DELPHI RICH data at LEP II. *Nucl. Instr. and Meth.* **A433**, 247.

Esbensen, H., *et al.* (1977) Channeling of protons, pions and deuterons in the GeV region. *Nucl. Phys. B* **127**, 281.

Evans, C. J. (1969) The development of incined sparks in a track-following spark chamber. *Nucl. Instr. and Meth.* **69**, 61.

Evans, R. D. (1958) Compton effect, *in* Handbook der Physik, Flügge, J. (ed.) p. 218, Berlin, Springer Verlag.

Fabbietti, L., *et al.* (2003) Photon detection efficiency in the CsI based HADES RICH. *Nucl. Instr. and Meth.* **A502**, 256.

Fano, U. (1963) Penetration of protons, alpha particles and mesons. *Ann. Rev. Nuclear Science*, **13**, 1.

Farmer, E. C., Brown, S. C. (1948) A study of the deterioration of methane-filled Geiger–Müller counters. *Phys. Rev.* **74**, 902.

Farr, F., *et al.* (1978) Space resolution of drift chambers operated at high gas pressure. *Nucl. Instr. and Meth.* **154**, 175.

Ferguson, T., *et al.* (2003a) Swelling phenomena in anode wires aging under a high accumulated dose. *Nucl. Instr. and Meth.* **A515**, 266.

Ferguson, T., *et al.* (2003b) Gas gain and space charge effects in aging tests of gaseous detectors. *Nucl. Instr. and Meth.* **A515**, 283.

Ferguson, T., *et al.* (2005) Anode front-end electronics for the cathode strip chambers of the CMS Endcap Muon detector. *Nucl. Instr. and Meth.* **A539**, 386.

Ferroni, F. (2009) The second generation BaBar RPCs: final evaluation of performances. *Nucl. Instr. and Meth.* **A602**, 649.

Filatova, N. A., *et al.* (1977) Study of drift chamber system for a K-e scattering experiment at the Fermi National Accelerator. *Nucl. Instr. and Meth.* **143**, 17.

Fischer, G. and Plch, J. (1972) The high voltage read-out for Multiwire Proportional Chambers. *Nucl. Instr. and Meth.* **100**, 515.

Fischer, H. M., *et al.* (1989) The OPAL Jet Chamber. *Nucl. Instr. and Meth.* **A283**, 492.

Fischer, J. and Shibata, S. (1972) The hybrid chamber: a proportional chamber with gated spark readout. *Nucl. Instr. and Meth.* **101**, 401.

Fischer, J., *et al.* (1976) Large proportional multiwire chambers for transition radiation detection with unambiguous position readout. *Nucl. Instr. and Meth.* **136**, 19.

Fischer, J., *et al.* (1978) Spatial distribution of the avalanches in proportional counters. *Nucl. Instr. and Meth.* **151**, 451.

Fischle, H., *et al.* (1991) Experimental determination of ionization cluster size distributions in counting gases. *Nucl. Instr. and Meth.* **A301**, 202.

Florent, J. J., *et al.* (1993) The electrostatic field in microstrip chambers and its influence on detector performances. *Nucl. Instr. and Meth.* **A329**, 125.

Foeth, H., *et al.* (1973) On the localization of the position of the particle along the wire of a multiwire proportional chamber. *Nucl. Instr. and Meth.* **109**, 521.

Fonte, P. (1996) A model of breakdown in parallel-plate detectors. *IEEE Trans. Nucl. Sci.* **NS-43**, 2135.

Fonte, P., et al. (1991) Feedback and breakdown in parallel-plate chambers. *Nucl. Instr. and Meth.* **A305**, 91.

Fonte, P., et al. (1999) A spark-protected high-rate detector. *Nucl. Instr. and Meth.* **A431**, 154.

Fonte, P., et al. (2000) A new high-resolution TOF technology. *Nucl. Instr. and Meth.* **A 443**, 201.

Ford, J. L. C. (1979) Position sensitive proportional counters as focal plane detectors. *Nucl. Instr. and Meth.* **162**, 277.

Ford, W. T., et al. (1987) Trigger drift chambers for the upgraded MARK II detector at PEP. *Nucl. Instr. and Meth.* **A255**, 486.

Fourletov, S. (2004) Straw tube tracking detector (STT) for ZEUS. *Nucl. Instr. and Meth.* **A535**, 191.

Fowler, I. L. (1973) Very large boron trifluoride proportional counters. *Rev. Sci. Instrum.* **34**, 731.

Fowler, R. H. (1931) The analysis of photoelectric sensitivity curves for clean metals at various temperatures. *Phys. Rev.* **38**, 45.

Fraenkel, Z., et al. (2005) A hadron blind detector for the PHENIX experiment at RHIC. *Nucl. Instr. and Meth.* **A 546**, 466.

Fraga, F. A. F., et al. (2002) CCD readout of GEM-based neutron detectors. *Nucl. Instr. and Meth.* **A478**, 357.

Fraga, M. M., et al. (1992) Fragments and radicals in gaseous detectors. *Nucl. Instr. and Meth.* **A323**, 284.

Fraga, M. M. R., et al. (2001) Pressure dependence of secondary NIR scintillation in Ar and Ar/CF4. *IEEE Trans. Nucl. Sci.* **48**, 330.

Franzen, W. and Cochran, L. W. (1956) Pulse Ionization Chambers and Proportional Counters, in *Nuclear Instruments and their Use*, Snell, A. H. (ed.) p. 3–81. New York, Wiley.

Friedman, H. (1960) The Sun's ionizing radiation, in *Physics of the Upper Atmosphere*, Ratcliffe, J. (ed.), New York, Academic Press.

Friedrich, D., et al. (1979) Positive ion effects in large-volume drift chambers. *Nucl. Instr. and Meth.* **158**, 81.

Friese, J., et al. (1999) Enhanced quantum efficiency for CsI grown on a graphite-based substrate coating. *Nucl. Instr. and Meth.* **A438**, 86.

Frieze, W., et al. (1976) A high resolution Multiwire Proportional Chamber System. *Nucl. Instr. and Meth.* **136**, 93.

Frolov, A. R., et al. (1991) Position resolution of the spark counter with a localized discharge. *Nucl. Instr. and Meth.* **A307**, 497.

Fuchs, M. (1995) Very high multiplicity tracking in heavy ion collisions with the NA49 TPC detector. *Nucl. Instr. and Meth.* **A367**, 349.

Fujii, K., et al. (1984) Study of limited streamer drift tube performance. *Nucl. Instr. and Meth.* **A225**, 23.

Fujimoto, J., et al. (1986) Cathode readout of limited streamer tubes with conductive plastic walls. *Nucl. Instr. and Meth.* **A252**, 53.

Fujita, K., et al. (2007) A high-resolution two-dimensional 3He neutron MSGC with pads for neutron scattering experiments. *Nucl. Instr. and Meth.* **A580**, 1027.

Fukui, S. and Miyamoto, A. (1959) A new type of particle detector: the discharge chamber. *Nuovo Cimento* **11**, 113.

Fulbright, H. W. (1958) Ionization chambers in nuclear physics, in *Encyclopedia of Physics*, Flügge, S. (ed.), Berlin, Springer Verlag. p. 1–51.

Garber, D. I. and Kinsey, R. R. (1976) Neutron cross sections, BNL Report No. 325 Vol II, BNL.

Gatti, E., *et al.* (1979) Optimum geometry for strip cathodes or grids in MWPC for avalanche localization along the anode wires. *Nucl. Instr. and Meth.* **163**, 83.

Geiger, H. and Müller, W. (1928) Das Elektronenzählrohr. *Phys. Zeits.* **29**, 839.

Gilmore, R. S. (1980) Particle identification by Cherenkov and transition radiation, SLAC-PUB-2606, SLAC.

Gilmore, R. S., *et al.* (1983) An optical readout for accurate positioning of UV photons in an Avalanche Chamber. *Nucl. Instr. and Meth.* **206**, 189.

Giomataris, I., *et al.* (1996) Micromegas: a high-granularity, position-sensitive gaseous detector for high particle flux environments. *Nucl. Instr. and Meth.* **A376**, 29.

Giomataris, I., *et al.* (2006) Micromegas in a bulk. *Nucl. Instr. and Meth.* **A560**, 405.

Giubellino, P., *et al.* (1986) Investigation of breakdown conditions in drift chambers. *Nucl. Instr. and Meth.* **A245**, 155.

Golovatyuk, V. M., *et al.* (1985) Some characteristics of plastic streamer tubes. *Nucl. Instr. and Meth.* **A236**, 300.

Gordon, J. S. and Mathieson, E. (1984) Cathode charge distributions in multiwire chambers: I. Measurement and theory. *Nucl. Instr. and Meth.* **227**, 267.

Gray, D. E. (ed.) (1963) *The American Institute of Physics Handbook*, New York, McGraw Hill.

Grishin, V. M., *et al.* (1991) Ionization energy loss in very thin absorbers. *Nucl. Instr. and Meth.* **A309**, 476.

Grunberg, C., *et al.* (1970) Multiwire Proportional and Semiproportional Counter with a variable sensitive volume. *Nucl. Instr. and Meth.* **78**, 102.

Guirl, L., *et al.* (2002) An aging study of triple GEMs in Ar-CO2. *Nucl. Instr. and Meth.* **A 478**, 263.

Gustavino, C., *et al.* (2001) A glass resistive plate chamber for large experiments. *Nucl. Instr. and Meth.* **A457**, 558.

Haddad, Y., *et al.* (2013) High rate resistive plate chamber for the LHC detector upgrades. *Nucl. Instr. and Meth.* **A718**, 424.

Hallewell, G. D. (1994) Long-term, efficient RICH detector operation with TMAE. *Nucl. Instr. and Meth.* **A343**, 250.

Hallewell, G. (2011) Aspects of the use of saturated fluorocarbon fluids in high energy physics. *Nucl. Instr. and Meth.* **A639**, 207.

Hammarström, R., *et al.* (1980a) Large multiwire proportional chambers for experiment NA3 at the CERN PS. *Nucl. Instr. and Meth.* **174**, 45.

Hammarström, R., *et al.* (1980b) Multitube proportional chambers for high counting rates. *Nucl. Instr. and Meth.* **176**, 181.

Hansen, T. C., *et al.* (2008) The D20 instrument at the ILL: a versatile high-intensity two-axis neutron diffractometer. *Meas. Sci. Technol.* **19**, 034001.

Hanson, G. (1986) The new drift chamber for the Mark II detector at the SLAC linear collider. *Nucl. Instr. and Meth.* **A252**, 343.

Hargrove, C. K., *et al.* (1984) The spatial resolution of the time projection chamber at Triumf. *Nucl. Instr. and Meth.* **219**, 461.

Hasted, J. B. (1964) *Physics of Atomic Collisions*, London, Butterworths.

Heintze, J. (1978) Drift chambers and recent developments. *Nucl. Instr. and Meth.* **156**, 227.

Heintze, J. (1982) The jet-chamber of the JADE experiment. *Nucl. Instr. and Meth.* **196**, 333.

Hempel, G., *et al.* (1975) Development of Parallel-Plate Avalanche Counters for the detection of fission fragments. *Nucl. Instr. and Meth.* **131**, 445.

Hendricks, R. W. (1969) Space charge effects in proportional counters. *Rev. Sci. Instrum.* **40**, 1216.

Henke, B. L., *et al.*, (1993), X-Ray interactions: photoabsorption, scattering, transmission and reflection E = 50–300000 eV, Z = 1–92: www.cxro.lbl.gov/optical_constants/.

Hess, D. W. (1986) Plasma chemistry in wire coatings. *Proc. Workshop on Radiation Damage to Wire Chambers* **LBL-21170**, 15.

Hildebrandt, M. (2003) Aging tests with GEM-MSGCs. *Nucl. Instr. and Meth.* **A515**, 255.

Hilke, H. J. (1986a) Detector calibration with lasers: a review. *Nucl. Instr. and Meth.* **A252**, 169.

Hilke, H. J. (1986b) Summary of aging studies in wire chambers by AFS, DELPHI and EMC groups. *Proc. Workshop on Radiation Damage to Wire Chambers* **LBL21170**, 153.

Hoedlmoser, H., *et al.* (2006) Production technique and quality evaluation of CsI photocathodes for the ALICE/HMPD detector. *Nucl. Instr. and Meth.* **566**, 338.

Hoedlmoser, H., *et al.* (2007) Long term performance and ageing of CsI photocayhodes for the ALICE/HMPID detector. *Nucl. Instr. and Meth.* **A574**, 28.

Hohlmann, M., *et al.* (2002) Aging phenomena in gaseous detectors–perspectives from the 2001 workshop. *Nucl. Instr. and Meth.* **A 494**, 179.

Holder, M., *et al.* (1978) A detector for high-energy neutrino interactions. *Nucl. Instr. and Meth.* **148**, 235.

Holroyd, R., *et al.* (1987) Measurement of the quantum efficiency of TMAE and TEA from threshold to 120 nm. *Nucl. Instr. and Meth.* **A261**, 440.

Hott, T. (2003) Aging problems of the inner tracker at HERA-B. *Nucl. Instr. and Meth.* **515**, 242.

Hu, J., *et al.* (2006) Time expansion chamber system for characterization of TWIST low-energy muon beams. *Nucl. Instr. and Meth.* **A566**, 563.

Huffman, R. E., *et al.* (1955) Near ultraviolet absorption cross sections of argon, krypton and xenon. *Bull. Am. Phys. Soc.* **7**, 457.

Hurst, G. S. and Klots, C. E. (1976) Elementary processes in irradiated noble gases. *Adv. Rad. Chem.* **5**, 1.

Huxley, L. G. and Crompton, R. W. (1974) *The Diffusion and Drift of Electrons in Gases*, New York, Wiley.

Iacobaeus, C., *et al.* (2000) A novel portal imaging device for advanced radiation therapy. *IEEE Trans. Nucl. Sci.* **NS-48**, 1496.

Iacobaeus, C., *et al.* (2001) Sporadic electron jets from cathodes – the main breakdown-triggering mechanism in gaseous detectors. *IEEE Trans. Nucl. Sci.* **49**, 1622.

Iarocci, E. (1983) Plastic streamer tubes and their applications in high energy physics. *Nucl. Instr. and Meth.* **217**, 30.

Igo, G. J., *et al.* (1952) Statistical fluctuations in ionization by 3.5 MeV protons. *Phys. Rev.* **89**, 879.

Isobe, T., *et al.* (2006) Development of a Time Projection Chamber using CF4 gas for relativistic heavy ion experments. *Nucl. Instr. and Meth.* **A 564**, 190.

Ivanouchenkov, Y., *et al.* (1999) Breakdown limit studies in high rate gaseous detectors. *Nucl. Instr. and Meth.* **A422**, 300.

Jamil, M., *et al.* (2012) GEANT4 Monte Carlo simulation response of parallel plate avalanche counter for fast neutron detection. *Rad. Meas.* **47**, 277.

Janssen, M. E., *et al.* (2006) R&D studies ongoing at DESY on a Time Projection Chamber for a detector at the International Linear Collider. *Nucl. Instr. and Meth.* **A 566**, 75.

Järvinen, M. L. and Sipilä, H. (1982) Effects of pressure and admixture of neon Penning mixtures on proportional counters resolution. *Nucl. Instr. and Meth.* **193**, 53.

Jean-Marie, B., *et al.* (1979) Systematic measurements of electron drift velocity and study of some properties of four gas mixtures: A-CH4, A-C2H4, A-C2H6, A-C3H8. *Nucl. Instr. and Meth.* **159**, 213.

Jeavons, A. P. (1978) The high-density proportional chamber and its applications. *Nucl. Instr. and Meth.* **156**, 41.

Jeavons, A. P. and Cate, C. (1976) The proportional Chamber Gamma Camera. *IEEE Trans. Nucl. Sci.* **NS-23**, 640.

Jeavons, A. P., *et al.* (1975) The high-density multiwire drift chamber. *Nucl. Instr. and Meth.* **124**, 491.

Jelenak, Z. M., *et al.* (1993) Electronic excitation of the 750- and 811-nm lines in argon. *Phys. Rev. E* **47**, 3566.

Jones, A. R. and Holford, R. M. (1981) Application of Geiger-Müller counters over a wide range of counting rates. *Nucl. Instr. and Meth.* **189**, 503.

Juricic, I. and Kadyk, J. A. (1986) Results from some anode wire aging tests, *Proc. Workshop on Radiation Damage to Wire Chambers*. Berkeley. LBL-21170 141.

Kadyk, J. (1986) *Workshop on Radiation Damage to Wire Chambers*. Berkeley. LBL-21170.

Kageyama, K., *et al.* (1996) Space charge effects in a thin rectangular proportional counter. *Nucl. Instr. and Meth.* **A369**, 151.

Kaminski, J., *et al.* (2006) *GEM TPC Anode Pad Shape Studies TPC Application Workshop*. Berkeley, CA (USA).

Kamyshov, Y., *et al.* (1987) The self-quenching streamer discharge in Ar-CO2 mixtures. *Nucl. Instr. and Meth.* **A257**, 125.

Kane, S., *et al.* (2003) An aging study of a Micromegas with GEM preaplification. *Nucl. Instr. and Meth.* **A515**, 261.

Kanter, K. (1961) Electron scattering by thin foils for energies below 10 kev. *Phys. Rev.* **121**, 461.

Kappler, S., *et al.* (2004) Design and construction of a GEM-TPC prototype for R&D purposes. *IEEE Trans. Nucl. Sci.* **NS-51**, 1524.

Keirim-Markus, I. B. (1972) The relative luminescence yield of gas proportional counters. *Instruments and Experimental Techniques* **15**, 1337.

Ketzer, B., *et al.* (2001) GEM detectors for COMPASS. *IEEE Trans. Nucl. Sci.* **NS-48**, 1065.

Ketzer, B., *et al.* (2004) Performance of triple GEM tracking detectors in the COMPASS experiment. *Nucl. Instr. and Meth.* **A535**, 314.

Khorashad, L. K., *et al.* (2011) Simulation of resistive plate chamber in streamer mode operation. *Nucl. Instr. and Meth.* **A628**, 470.

Killenberg, M., *et al.* (2004) Charge transfer and charge broadening of GEM structures in high magnetic fields. *Nucl. Instr. and Meth.* **A530**, 251.

Kim, H. C., *et al.* (2009) Quantitative aging study with intense irradiation tests for the CMS. *Nucl. Instr. and Meth.* **A602**, 771.

Kiselev, O., *et al.* (1995) Measurement of electron drift velocities and Lorentz angles in fast gas mixtures. *Nucl. Instr. and Meth.* **A367**, 306.

Klein, M. and Schmidt, C. J. (2011) CASCADE neutron detectors for highest count rates in combination with ASIC/FPGA based readout electronics. *Nucl. Instr. and Meth.* **A 628**, 9.

Klyancho, A. V., *et al.* (2012) A GEM-based dose imaging detector with optical readout for proton radiotherapy. *Nucl. Instr. and Meth.* **A694**, 271.

Knoll, G. (1989) *Radiation Detection and Measurements*, New York, Wiley & Sons.

Kobayashi, M., *et al.* (2011) Cosmic ray test of a GEM-based TPC prototype operated in Ar-CF4-isobutane gas mixtures. *Nucl. Instr. and Meth.* **A 641**, 37.

Kobetich, E. J. and Katz, R. (1968) Energy deposition by electron beams and δ ray *Phys. Rev.* **170**, 391.

Koike, T., *et al.* (2011) A new gamma-ray detector with gold-plated gas electron multiplier. *Nucl. Instr. and Meth.* **A 648**, 180.

Kollefrath, M., *et al.* (1998) Ageing studies for the ATLAS-monitored drift tubes. *Nucl. Instr. and Meth.* **A419**, 351.

Koori, N., *et al.* (1984) Other magic gas mixtures for multiwire proportional chambers. *Nucl. Instr. and Meth.* **220**, 453.

Koori, N., *et al.* (1986a) Self-quenching streamers in magic gas mixtures. *Nucl. Instr. and Meth.* **A243**, 486.

Koori, N., *et al.* (1986b) Self-quenching streamer transitions induced by α- and β-rays in a gas counter. *Jap. J. Appl. Phys.* **25**, 986.

Koori, N., *et al.* (1989) Comparison of self-quenching streamer transitions in Xe-, Kr-, and Ar-mixtures *IEEE Trans. Nucl. Sci.* **NS-36**, 223.

Koori, N., *et al.* (1990) A self-quenched streamer tube operated with Ne-and He-mixtures. *Nucl. Instr. and Meth.* **A299**, 80.

Koori, N., *et al.* (1991) On the mechanism of self-quenching streamer formation. *Nucl. Instr. and Meth.* **A307**, 581.

Korff, S. A. (1955) *Electron and Nuclear Counters*, New York, Van Nostrand.

Kotthaus, R. (1986) A laboratory study of radiation damage to drift chambers. *Nucl. Instr. and Meth.* **A252**, 531.

Krämer, M., *et al.* (2008) First results of the PixelGEM central tracking system for COMPASS. *IEEE Nucl. Sci. Symp. Conf. Rec.* p. 2920.

Krienen, F. (1962) Digitized spark chambers. *Nucl. Instr. and Meth.* **16**, 262.

Kruithof, A. A. and Penning, F. M. (1937) Determination of the Towns end ionization coefficient α for mixtures of neon and argon. *Physica* **4**, 430.

Krusche, A., *et al.* (1965) Nanosecond lifetime measurements with a fast gaseous counter. *Nucl. Instr. and Meth.* **33**, 177.

Kubo, H., *et al.* (2003) Development of a time projection chamber with micro-pixel electrodes. *Nucl. Instr. and Meth.* **A513**, 94.

Kubo, T., *et al.* (2003) Study of the effect of water vapor on a glass RPC with and without freon. *Nucl. Instr. and Meth.* **A508**, 50.

Kuhlmann, W. R., *et al.* (1966) Ortsempfindliche Zählrohre. *Nucl. Instr. and Meth.* **40**, 118.

Kurvinen, K., *et al.* (2003) Analysis of organic compounds formed in electron avalanches in a proportional counter filled with Ar/C2H4 gas mixture. *Nucl. Instr. and Meth.* **A515**, 118.

Lami, S., *et al.* (2006) A triple-GEM telescope for the TOTEM experiment. arXiv:physics/0611178v1.

Landi, G. (2003) Properties of the center of gravity as an algorithm for position measurements: two-dimensional geometry. *Nucl. Instr. and Meth.* **A497**, 511.

Langhoff, P. W. and Karplus, M. (1969) Padé summations of the Cauchy dispersion equation. *J. Opt. Soc. Am.* **59**, 863.

Lapique, F. and Piuz, F. (1980) Simulation of the measurement by primary cluster counting of the energy lost by a relativistic ionizing particle in argon. *Nucl. Instr. and Meth.* **175**, 297.

Ledermann, B., *et al.* (2007) Development studies for the ILC: measurements and simulations for a Time Projection Chamber with GEM technology. *Nucl. Instr. and Meth.* **A 581**, 232.

Lehraus, I., *et al.* (1981) Resolution limits of ionization sampling in high pressure drift detectors. *Phys. Scripta* **23**, 727.

Lehraus, I., *et al.* (1982) dE/dx measurements in Ne, Ar, Xe and pure hydrocarbons. *Nucl. Instr. and Meth.* **200**, 199.

Lehraus, I., *et al.* (1984) *Dependence of Drift Velocity and Signal Amplitude on Gas Mixture Pressure and Purity*, CERN-EF Division N-TPC Note 84–6, CERN, Geneva.

Lightfoot, P. K., *et al.* (2005) Developmet of a double-phase Xenon cell using micromegas charge readout for applications in dark matter physics. *Nucl. Instr. and Meth.* **A 554**, 266.

Lindner, E. G. (1930) Organic reactions in gaseous electrical discharge I. Normal paraffin hydrocarbons. *Phys. Rev.* **36**, 1375.

Lintereur, A., *et al.* (2011) 3He and BF3 neutron detectors pressure effect and model comparison. *Nucl. Instr. and Meth.* **A652**, 347.

Lippmann, C. and Riegler, W. (2004) Space charge effects in Resistive Plate Chambers. *Nucl. Instr. and Meth.* **A517**, 54.

Litt, J. and Meunier, R. (1973) Cherenkov counter techniques in high-energy physics. *Ann. Rev. Nucl. Sci.* **23**, 1.

Llopart, X., *et al.* (2002) Medipix2: A 64-k pixel readout chip with 55-μm square elements working in single photon counting mode. *IEEE Trans. Nucl. Sci.* **NS-49**, 2279.

Llopart, X., *et al.* (2007) Timepix, a 65k programmable pixel readout chip for arrival time, energy and/or photon counting measurements. *Nucl. Instr. and Meth.* **A581**, 485.

Loeb, L. B. (1961) *Basic Processes of Gaseous Electronics*, Berkeley, University of California Press.

Lorents, D. C. (1976) The physics of electron beam excited rare gases at high densities. *Physica* **82C**, 19.

Lowke, J. J. and Parker, J. J. (1969) Theory of electron diffusion parallel to electric fields. II: Application to real gases. *Phys. Rev.* **181**, 302.

Lu, C. and McDonald, K. T. (1994) Properties of reflective and semitransparent CsI photocathodes. *Nucl. Instr. and Meth.* **A343**, 135.

LUX luxdarkmatter.org/

LXCAT Electron Scattering Database, www.lxcat.laplace.univ-tlse.fr/

Mahesh, L. (1976) On the counting rate-dependent amplitude shifts in proportional counters. *Nucl. Instr. and Meth.* **133**, 57.

Majewski, S. (1986) Results on ageing and stability with pure DME and isobutane-methylal mixture in thin high-rate multiwire chambers. *Proc. Workshop on Radiation Damage to Wire Chambers* **LBL-21170**, 239.

Majewski, S. and Sauli, F. (1975) *Support lines and beam killers for large size multiwire proportional chambers*, CERN NP Int. Rep. 75–14, CERN.

Majewski, S., *et al.* (1983) A thin multiwire chamber operating in the high multiplication mode. *Nucl. Instr. and Meth.* **217**, 265.

Malai, J., *et al.* (2011) An integrated Micromegas UV-photon detector. *Nucl. Instr. and Meth.* **A633**, 5194.

Malter, L. (1936) Thin film emission. *Phys. Rev.* **50**, 48.

Mangiarotti, A., *et al.* (2004) Exactly solvable model for the time response function of RPC. *Nucl. Instr. and Meth.* **A533**, 16.

Marel, G., *et al.* (1977) Large Planar Drift Chambers. *Nucl. Instr. and Meth.* **141**, 43.

Margato, L. M. S., *et al.* (2004) Performance of an optical readout GEM-based TPC. *Nucl. Instr. and Meth.* **A535**, 231.

Marr, G. V. (1967) *Photoionization Processes in Gases*, New York, Academic Press.

Marsh, J. B., *et al.* (1979) Glow discharge, its sensitivity to infra-red radiation, RL-79-038, Didcot, Rutherford Laboratory

Marshall, T. (2003) Restoring contaminated wires, removing gas contaminants, and aging studies of drift tube chambers. *Nucl. Instr. and Meth.* **A515**, 50.

Martin, F. (2007) The ATLAS transition radiation tracker (TRT) from construction to installation. *Nucl. Instr. and Meth.* **A581**, 535.

Martinez, J. C., *et al.* (2007) Automatic method to manufacture 2D Multiwire Proportional Counter frames. *Nucl. Instr. and Meth.* **A573**, 41.

Martoff, C. J., *et al.* (2000) Suppressing drift chamber diffusion without magnetic field. *Nucl. Instr. and Meth.* **A440**, 355.

Martoff, C. J., *et al.* (2005) Negative ion drift and diffusion in a TPC near 1 bar. *Nucl. Instr. and Meth.* **A555**, 55.

Masaoka, S., *et al.* (2003) Optimization of a micro-strip gas chamber as a two-dimensional neutron detector using gadolinium converter. *Nucl. Instr. and Meth.* **A513**, 538.

Massey, H. S. W., *et al.* (1969) *Electronic and Ionic Impact Phenomena*, Vols. I to IV, Oxford, Oxford University Press.

Mathieson, E. (1986) Dependence of gain on count rate due to space charge in coaxial and multiwire proportional counters. *Nucl. Instr. and Meth.* **A249**, 413.

Mathieson, E. and Harris, T. J. (1978) Induced charges in a multi-wire proportional chamber. *Nucl. Instr. and Meth.* **154**, 189.

Mathieson, E. and Harris, T. J. (1979) Evaluation of the initial angular width of the avalanche in a proportional chamber. *Nucl. Instr. and Meth.* **159**, 483.

Mathieson, E. and Smith, G. (1988) Charge distribution in Parallel Plate Avalanche Chambers. *Nucl. Instr. and Meth.* **A273**, 518.

McCarty, R., *et al.* (1986) Identification of large transverse momentum hadrons using a Ring-Imaging Cherenkov Counter. *Nucl. Instr. and Meth.* **A248**, 69.

McDaniel, E. W. and Mason, E. A. (1973) *The Mobility and Diffusion of Ions in Gases*, New York, Wiley & Sons.

McGregor, D. S., *et al.* (2003) Design considerations for thin-film coated semiconductor thermal neutron detectors – I: basics regarding alpha particle emitting neutron films. *Nucl. Instr. and Meth.* **A500**, 272.

McMaster, W. H. (1969) Compilation of X-ray cross sections, UCRL-50174, Liver more.

Meek, J. and Cragg, J. D. (1953) *Electrical Breakdown of Gases*, Oxford, Clarendon Press.

Meinschad, T., *et al.* (2004) GEM-based photon detector for RICH applications. *Nucl. Instr. and Meth.* **A535**, 324.

Melchart, G., *et al.* (1981) The multistep avalanche chamber as detector for thermal neutrons. *Nucl. Instr. and Meth.* **186**, 613.

Miernik, K., *et al.* (2007) Optical Time Projection Chamber for imaging nuclear decays. *Nucl. Instr. and Meth.* **A 581**, 194.

Miller, G. L., *et al.* (1971) A position sensitive detector for a magnetic spectrograph. *Nucl. Instr. and Meth.* **91**, 389.

Miller, L. S., *et al.* (1968) Charge transport in solid and liquid Ar, Kr and Xe. *Phys. Rev.* **166**, 871.

Mills, W. R., *et al.* (1962) Low voltage He3-filled proportional counters for efficient detection of thermal and epithermal neutrons. *Rev. Sci. Instrum.* **33**, 866.

Miné, P., *et al.* (1988) A BaF2-TMAE detector for positron emission tomography. *Nucl. Instr. and Meth.* **A273**, 881.

Miśkowiec, D. and Braun-Munzinger, P. (2008) Laser calibration systems for the CERES Time Projection Chamber. *Nucl. Instr. and Meth.* **A593**, 188.

Mitchell, J. H. and Ridler, K. E. W. (1934) The speed of positive ions in nitrogen. *Proc. Royal Soc. London* **A146**, 911.

Montgomery, G. C. and Montgomery, D. D. (1941) Geiger–Mueller Counters. *J. Franklin Inst.* **231**, 447.

Morii, H., *et al.* (2004) Quenching effects in nitrogen gas scintillation. *Nucl. Instr. and Meth.* **A526**, 399.

Mörmann, D., *et al.* (2003) GEM-based gaseous photomultipliers for UV and visible photon imaging. *Nucl. Instr. and Meth.* **A504**, 93.

Morozov, A., *et al.* (2011) Effect of electric field on the primary scintillation from CF4. *Nucl. Instr. and Meth.* **A628**, 360.

Morse, P. M. and Feshbach, H. (1953) *Methods of Theoretical Physics*, New York, McGraw-Hill.

Morse, P. M., *et al.* (1935) Velocity distributions for elastically colliding electrons. *Phys. Rev.* **48**, 412.

Moshaii, A. and Doroud, K. (2009) Study on the effect of humidity on the RPC performance. *Nucl. Instr. and Meth.* **A602**, 727.

Mota, B., *et al.* (2004) Performance of the ALTRO chip on data acquired on an ALICE TPC next term prototype. *Nucl. Instr. and Meth.* **A535**, 500.

Mountain, R. J., *et al.* (1999) The CLEO III ring imaging Cherenkov detector. *Nucl. Instr. and Meth.* **A433**, 77.

Mueller, E. R. (1989) A simple method to measure wire tension in multiwire chambers. *Nucl. Instr. and Meth.* **281**, 652.

Müller, T. (1986) A microvertex detector for experiment UA1 at the CERN proton–antiproton collider. *Nucl. Instr. and Meth.* **A252**, 387.

Nappi, E. and Seguinot, J. (2005) Ring Imaging Cherenkov Detectors: the state of the art and perspectives. *Riv. Nuovo Cimento* **28**, 1.

Nishio, K., *et al.* (1997) A system for correlation measurement of fission fragments and prompt neutrons for thermal neutron induced fission. *Nucl. Instr. and Meth.* **A385**, 171.

Nohtomi, A., *et al.* (1994) Count-loss mechanism of self-quenching streamer (SQS) tubes. *Nucl. Instr. and Meth.* **A342**, 538.

Northcliffe, L. C. (1963) Passage of heavy ions through matter. *Ann. Rev. Nuclear Science* **13**, 67.

Nygren, D. and Marx, J. (1978) The Time Projection Chamber. *Physics Today* **31**, 46.

Ochi, A., *et al.* (2002) Development of micro pixel chamber. *Nucl. Instr. and Meth.* **A478**, 196.

Oda, S. X., *et al.* (2006) Development of a time projection chamber using gas electron multipliers (GEM–TPC). *Nucl. Instr. and Meth.* **A 566**, 312.

Oed, A. (1988) Position-sensitive detector with micro-strip anode for electron multiplication with gases. *Nucl. Instr. and Meth.* **A263**, 351.

Oed, A. (2004) Detectors for thermal neutrons. *Nucl. Instr. and Meth.* **A525**, 62.

Ogren, H. (1995) The straw tracker for the SDC detector. *Nucl. Instr. and Meth.* **A367**, 133.

Oh, S. H., *et al.* (1991) Construction and performance of a 2.7 m long straw drift tube prototype chamber for the SSC. *Nucl. Instr. and Meth.* **A309**, 368.

Ohgaki, H., *et al.* (1990) Space charge effect in SQS transitions in a gas counter. *Nucl. Instr. and Meth.* **A295**, 411.

Ohnuki, T., *et al.* (2001) Measurement of carbon disulfide anion diffusion in a TPC. *Nucl. Instr. and Meth.* **A 463**, 142.

Okuno, H., *et al.* (1979) Azimuthal spread of the avalanche in proportional chambers. *IEEE Trans. Nucl. Sci.* **NS-26**, 160.

Openshaw, R., *et al.* (1991) Etching of anode wire deposits with CF4/isobutane (80:20) avalanches. *Nucl. Instr. and Meth.* **A307**, 298.

Östling, J., *et al.* (2000) Novel detector for portal imaging in radiation therapy. *Progr. Biomedical Optics and Imaging* **P7 3977**, 84.

Östling, J., *et al.* (2003) Study of hole-type gas multiplication structures for portal imaging and other high counting rate applications. *IEEE Trans. Nucl. Sci.* **NS-50**, 809.

Pacella, D., *et al.* (2001) Ultrafast soft X ray 2D plasma imaging system based on Gas Electron Multiplier detector with pixel read-out. *Rev. Scient. Instrum.* **72**, 1372.

Pacella, D., *et al.* (2003) X-VUV spectroscopic imaging with a micropattern gas detector. *Nucl. Instr. and Meth.* **A 508**, 414.

Palladino, V. and Sadoulet, B. (1974) Applications of the classical theory of electrons in gases to multiwire proportional and drift chambers, LBL-3013, Berkeley.

Palladino, V. and Sadoulet, B. (1975) Applications of classical theory of electrons in gases to drift proportional chambers. *Nucl. Instr. and Meth.* **128**, 323.

Pansky, A., *et al.* (1993) Detection of X-ray fluorescence of light elements by electron counting in a low-pressure gaseous electron multiplier. *Nucl. Instr. and Meth.* **A330**, 150.

Pansky, A., *et al.* (1995) The scintillation yield of CF_4 and its relevance to detection science. *Nucl. Instr. and Meth.* **A354**, 262.

Parker, J. D., *et al.* (2013) Neutron imaging detector based on the µPIC micro-pixel chamber. *Nucl. Instr. and Meth.* **A697**, 23.

Parker, J. H. and Lowke, J. J. (1969) Theory of electron diffusion parallel to electric fields. I: Theory. *Phys. Rev.* **181**, 290.

Peisert, A. (1983) The Parallel Plate Avalanche Chamber as end-cap detector for Time Projection Chambers. *Nucl. Instr. and Meth.* **217**, 229.

Peisert, A. and Sauli, F. (1984) Drift and diffusion of electrons in gases: a compilation, CERN 84–08, Geneva.

Perez-Mendez, V. and Pfab, J. M. (1965) Magnetostrictive readout for "wire spark chambers". *Nucl. Instr. and Meth.* **33**, 141.

Periale, L., *et al.* (2002) Detection of the primary scintillation light from dense Ar, Kr and Xe with novel photosensitive gaseous detectors. *Nucl. Instr. and Meth.* **A478**, 377.

Periale, L., *et al.* (2004) The developemnt of gaseous detectors with solid photocathodes for low-temperature applications. *Nucl. Instr. and Meth.* **A535**, 517.

Peskov, V. and Fonte, P. (2009) Research on discharges in micropattern and small gap gaseous detectors, arXiv:0911.0463.

Pestov, Y. N. (1982) Status and future developments of spark counters with a localized discharge. *Nucl. Instr. and Meth.* **196**, 45.

Pestov, Y. N. (2002) Review on counters with localized discharge. *Nuc. Phys. B (Proc. Suppl.)* **A494**, 447.

Pestov, Y. N., *et al.* (2000) Timing performance of spark counters and photon feedback. *Nucl. Instr. and Meth.* **A456**, 11.

Petersen, G., *et al.* (1980) The multistep avalanche chamber as detector in radiochromatography imaging. *Nucl. Instr. and Meth.* **176**, 239.

Piuz, F. (2003) Ring Imaging Cherenkov systems based on gaseous photo-detectors: trends and limits around particle physics accelerators. *Nucl. Instr. and Meth.* **A502**, 76.

Piuz, F., *et al.* (1982) Evaluation of systematic errors in the avalanche localization along the wire with cathode strips readout MWPC. *Nucl. Instr. and Meth.* **196**, 451.

Platzman, R. L. (1967) Energy spectrum of primary activations in the action of ionizing radiation, in *Radiation Research*, G. Silini (ed.), Amsterdam, North Holland. p. 20.

Policarpo, A. J. L. P. (1977) The Gas Scintillation Proportional Counter. *Space Sci. Instr.* **3**, 77.

Policarpo, A. J. P. L., *et al.* (1970) The gas proportional scintillation counter under X-ray bombardment: resolution and pulse correlations. *Nucl. Instr. and Meth.* **77**, 309.

Policarpo, A. J. P. L., *et al.* (1972) Improved resolution for low energies with gas proportional scintillation counters. *Nucl. Instr. and Meth.* **102**, 337.

Policarpo, A. J. P. L., *et al.* (1974) Detection of soft X-rays with a xenon Proportional Scintilation Counter. *Nucl. Instr. and Meth.* **118**, 221.

Policarpo, A. P. J. (1981) Light production and gaseous detectors. *Phys. Scripta* **23**, 539.

Prete, G. and Viesti, G. (1985) Discrimination capability of avalanche counters detecting different ionizing particles. *Nucl. Instr. and Meth.* **A234**, 276.

Price, L. E., *et al.* (1982) Investigation of Long Drift Chambers for a Nucleon Decay Detector. *IEEE Trans. Nucl. Sci.* **29**, 383.

Price, W. J. (1958) *Nuclear Radiation Detectors*, New York, McGraw Hill.

Procureur, S., *et al.* (2011) Discharge studies in micromegas detectors in a 150 GeV/c pion beam. *Nucl. Instr. and Meth.* **A 659**, 91.

Puill, G., *et al.* (1999) MICROMEGAS: high rate and radiation hardness results. *IEEE Trans. Nucl. Sci.* **NS-46**, 1894.

Rabus, H., *et al.* (1999) Quantum efficiency of caesium iodide photocathodes in the 120–220 nm spectral range traceable to a primary detector standard. *Nucl. Instr. and Meth.* **A438**, 94.

Radicioni, E. (2007) Design, construction and performance of a large GEM-TPC prototype. *Nucl. Instr. and Meth.* **A572**, 195.

Raether, H. (1964) *Electron Avalanches and Breakdown in Gases*, London, Butterworth.

Rahman, M., *et al.* (1981) A multitrack drift chamber with 60 cm drift space. *Nucl. Instr. and Meth.* **188**, 159.

Ramo, S. (1939) Currents induced by electron motion. *Proc. of IRE* **27**, 584.

Ramsey, B. D. and Agrawal, P. C. (1989) Xenon-based Penning mixtures for proportional counters. *Nucl. Instr. and Meth.* **278**, 576.

Rapp, D. and Englander-Golden, P. (1965) Total cross sections for ionization and attachment in gases by electron inpact. I. Positive ionization. *J. Chem. Phys.* **43**, 1464.

Ravazzani, A., *et al.* (2006) Characterization of 3He proportional counters. *Rad. Meas.* **41**, 582.

Regan, T. (1984) A sensitive instrument for measuring wire tension in multiwire proportional and drift chambers. *Nucl. Instr. and Meth.* **219**, 100.

Rehak, P., *et al.* (2000) First results of the Micro Pin Array detector (MIPA). *IEEE Trans. Nuc. Sci.* **NS-47**, 1426.

Rhodes, C. K. (1979) *Excimer Lasers*, Berlin, Springer-Verlag.

Riegler, W. (2002a) High accuracy wire chambers. *Nucl. Instr. and Meth.* **A494**, 173.

Riegler, W. (2002b) Induced signals in resistive plate chambers. *Nucl. Instr. and Meth.* **A491**, 258.

Riegler, W. (2004) Extended theorems for signal induction in particle detectors VCI 2004. *Nucl. Instr. and Meth.* **A535**, 287.

Riegler, W. and Lippmann, C. (2004) The physics of Resistive Plate Chambers. *Nucl. Instr. and Meth.* **A518**, 86.

Riegler, W., *et al.* (2000) Resolution limits of drift tubes. *Nucl. Instr. and Meth.* **A443**, 156.

Riegler, W., *et al.* (2007) Gain reduction due to space charge in wire chambers. *Nucl. Instr. and Meth.* **A582**, 469.

Robin, M. B. (1974) *Higher Excited States of Polyatomic Molecules*, New York, Academic Press.

Roessler, D. M. and Walker, C. (1967) Optical constants of magnesium oxide and lithium fluoride in the far ultraviolet. *J. Opt. Soc. Am.* **57**, 835.

Rohrbach, F. (1988) Streamer chambers at CERN during the past decade and visual techniques of the future, CERN / EF 88–17, CERN.

Rossegger, S. and Riegler, W. (2010) Signal shape in a TPC wire chamber. *Nucl. Instr. and Meth.* **A623**, 927.

Rossegger, S., *et al.* (2010) An analytical approach to space charge distortions for time projection chambers. *Nucl. Instr. and Meth.* **A617**, 193.

Rossegger, S., *et al.* (2011) Analytical solutions for space charge fields in TPC drift volumes. *Nucl. Instr. and Meth.* **A632**, 52.

Rossi, B. and Staub, H. (1949) *Ionization Chambers and Counters*, New York, McGraw-Hill.

Rotherburg, E. and Walsh, S. (1993) Mechanism of wire breaking due to sparks in proportional or drift chambers. *Nucl. Instr. and Meth.* **A333**, 316.

RPC2005 (2006) 8th International Workshop on Resistive Plate Chambers. *Nucl. Phys. B (Proc. Suppl.)* **158**.

RPC2007 (2009) IX International Workshop on Resistive Plate Chambers. *Nucl. Instr. and Meth.* **A602**.

Rutherford, E. and Geiger, H. (1908) An electrical method of counting the number of α-particles from radio-active substances. *Proc. Royal Soc. A* **81**, 141.

Saito, M., *et al.* (2008) Fluctuation of ionization, scintillation and proportional scintillation yields due to apha-particles in gaseous xenon under normal pressures. *Nucl. Instr. and Meth.* **A593**, 407.

Sakai, H., *et al.* (2002) Study of the effect of water vapor on a resistive plate chamber with glass electrodes. *Nucl. Instr. and Meth.* **A484**, 153.

Salete Leite, M. S. C. P. (1980) Radioluminescence of rare gases. *Portugal Phys.* **11**, 53.

Santonico, R. (2009) An overview of RPC at the LHC startup. *Nucl. Instr. and Meth.* **A602**, 627.

Santonico, R. (2012) RPC impact in today's physics and perspectives for a new R&D phase. *Nucl. Instr. and Meth.* **A661**, S2.

Santonico, R. and Cardarelli, R. (1981) Development of resistive plate counters. *Nucl. Instr. and Meth.* **187**, 377.

Saquin, Y. (1992) The DELPHI time projection chamber. *Nucl. Instr. and Meth.* **A323**, 209.

Saudinos, J., *et al.* (1973) Localisation de particules par compteur à migration. *Nucl. Instr. and Meth.* **111**, 77.

Saudinos, J., *et al.* (1975) Nuclear scattering applied to radiography. *Phys. Med. Biol.* **20**, 890.

Sauli, F. (1977) Principles of operation of multiwire proportional and drift chambers, CERN 77–09, Geneva.

Sauli, F. (1978) Limiting accuracies in multiwire proportional and drift chambers. *Nucl. Instr. and Meth.* **156**, 147.

Sauli, F. (1992) Applications of gaseous detectors in astrophysics, medicine and biology. *Nucl. Instr. and Meth.* **A323**, 1.

Sauli, F. (1997) GEM: a new concept for electron amplification in gas detectors. *Nucl. Instr. and Meth.* **A386**, 531.

Sauli, F. (1998) Development of high rate MSGCs: overview of results from RD-28. *Nucl. Phys. B (Proc. Suppl.)* **61B**, 236.

Sauli, F. (2003) Fundamental understanding of aging processes: review of the workshop results. *Nucl. Instr. and Meth.* **A515**, 358.

Sauli, F. (2005) Novel Cherenkov photon detectors. *Nucl. Instr. and Meth.* **A 533**, 18.

Sauli, F. (2014) Gas Electron Multiplier (GEM) detectors: principles of operation and applications, in *Comprehensive Biomedical Physics*, Brahme, A. (ed.), Elsevier, in press.

Sauli, F. and Sharma, A. (1999) Micro pattern gaseous detectors. *Ann. Rev. Nucl. Part. Sci.* **49**, 341.

Saxon, D. H. (1988) Multicell Drift Chambers. *Nucl. Instr. and Meth.* **A265**, 20.

Schilly, P., *et al.* (1970) Construction and performance of large multiwire proportional chambers. *Nucl. Instr. and Meth.* **91**, 221.

Schindler, H., *et al.* (2010) Calculation of gas gain fluctutions in uniform fields. *Nucl. Instr. and Meth.* **A624**, 78.

Schins, E. (1996) *Thesis at Wuppertal University*, WUB-DIS 96–22.

Schlumbohm, H. (1958) Zur Statistik der Elektronenlawinen im ebenen Field. *Zeit. Phys.* **151**, 563.

Schmidt, B. (1998) Microstrip Gas Chambers: recent developments, radiation damage and long term behaviour. *Nucl. Instr. and Meth.* **A 419**, 230.

Schmidt, H. R. (1999) Pestov spark counters: work principle and applications. *Nuc. Phys. B (Proc. Suppl.)* **78**, 372.

Schreiner, A. (2001) Humidity dependence of anode corrosion in HERA-B Outer Tracker Chambers operated with Ar/CF4/CO2. *Nucl. Instr. and Meth.* **A515**, 146.

Schuh, S., *et al.* (2004) A high precision X-ray tomograph for quality control of the ATLAS Muon Monitored Drift Tube Chambers. *Nucl. Instr. and Meth.* **A518**, 73.

Schultz, G. (1976) Etude d'un Détecteur de Particules a très haute précision spatiale. *Thesis at Université Louis Pasteur*, Strasbourg,

Schultz, G. and Gresser, J. (1978) A study of transport coefficients of electrons in some gases used in proportional and drift chambers. *Nucl. Instr. and Meth.* **151**, 413.

Schultz, G., *et al.* (1977) Mobilities of positive ions in some gas mixtures used in proportional and drift chambers. *Rev. Physique Appliquée* **12**, 67.

Schyns, E. (2002) Status of large area CsI photocathode developments. *Nucl. Instr. and Meth.* **A494**, 441.

Seguinot, J. and Ypsilantis, T. (1977) Photo-ionization and Cherenkov ring imaging. *Nucl. Instr. and Meth.* **142**, 377.

Seguinot, J. and Ypsilantis, T. (1994) A historical survey of ring imaging Cherenkov counters. *Nucl. Instr. and Meth.* **A343**, 1.

Séguinot, J., *et al.* (1980) Imaging Cherenkov detectors: photo-ionization of tri-ethyl-amine. *Nucl. Instr. and Meth.* **173**, 283.

Séguinot, J., *et al.* (1990) Reflective UV photocathodes with gas-phase electron extraction: solid, liquid, and adsorbed thin films. *Nucl. Instr. and Meth.* **A297**, 133.

Seravalli, E., *et al.* (2009) 2D dosimetry in a proton beam with a scintillating GEM detector. *Phys. Med. Biol.* **54**, 3755.

Sernicki, J. (1983) A versatile large area Parallel Plate Avalanche Counter (PPAC) for broad-range magnetic spectrographs. *Nucl. Instr. and Meth.* **212**, 195.

Sharma, A. (1996) Study and optimization of the tracking detector for the FINUDA experiment. *Thesis at Université de Genève*,

Sharma, A. (2009) Summary of RPC2007 the IX International Workshop. *Nucl. Instr. and Meth.* **A602**, 854.

Sharma, A. (2012) Muon tracking and triggering with gaseous detectors and some applications. *Nucl. Instr. and Meth.* **A666**, 98.

Sharma, A. and Sauli, F. (1992) A measurement of the first Townsend coefficient in argon based mixtures at high fields. *Nucl. Instr. and Meth.* **A323**, 280.

Shockley, W. (1938) Currents to conductors induced by a moving point charge. *J. Appl. Phys.* **9**, 635.

Shotanus, P., *et al.* (1986) Development study of a new gamma camera for positron emission tomography. *Nucl. Instr. and Meth.* **A252**, 255.

Siegmund, O. H. W., *et al.* (1983) Improved energy resolution capability of an imaging proportional counter using electron counting techniques. *IEEE Trans. Nucl. Sci.* **NS-30**, 350.

Sipilå, H. (1976) Energy resolution of the proportional counter. *Nucl. Instr. and Meth.* **133**, 351.

Sipilä, H. and Kiuru, E. (1978) On energy dispersive properties of the proportional counter channel, in *Advances in X-ray analysis*, Newkirk, J. B. and Ruud, C. O. (eds.), New York, Plenum Publishing.

Sipilä, H., *et al.* (1980) Mathematical treatment of space charge effects in proportional counters. *Nucl. Instr. and Meth.* **176**, 381.

Smirnov, I. B. (2005) Modeling of ionization produced by fast charged particles in gases. *Nucl. Instr. and Meth.* **A554**, 474.

Smirnov, I. B., (2012), HEED: consult.cern.ch/writeup/heed/.

Smith, G. C., *et al.* (1990) A low pressure, parallel plate avalanche chamber for detection of soft X-ray fluorescence. *Nucl. Instr. and Meth.* **A291**, 135.

Smith, G. C., *et al.* (1992) High rate, high resolution, two-dimensional gas proportional detectors for X-ray synchrotron radiation experiments. *Nucl. Instr. and Meth.* **A323**, 78.

Sorensen, P., *et al.* (2009) The scintillation and ionization yield of liquid xenon for nuclear recoils. *Nucl. Instr. and Meth.* **A601**, 339.

Spielberg, N. and Tsarnas, D. I. (1975) Counting rate dependent gain shifts in flow proportional counters. *Rev. Sci. Instrum.* **46**, 1085.

STAR drupal.star.bnl.gov/STAR/public/img/.

Staric, M. (1989) Analytical approach to the development of charge signals in multiwire proportional chambers. *Nucl. Instr. and Meth.* **A283**, 744.

Staric, M., *et al.* (1983) Some simple approaches used in the construction of multiwire proportional chambers. *Nucl. Instr. and Meth.* **216**, 67.

Staub, H. (1953) Detection methods, in *Experimental Nuclear Physics*, Segré, E. (ed.), New York, Wiley. p. 1.

Steinhaeuser, P., *et al.* (1997) Simulation of the electromagnetic properties of a Pestov spark counter. *Nucl. Instr. and Meth.* **A390**, 86.

Stelzer, H. (1976) A large area parallel plate avalanche counter. *Nucl. Instr. and Meth.* **133**, 409.

Suzuki, M. (1983) Recombination luminescence from ionization tracks produced by alpha particles in high-pressure argon, krypton and xenon gases. *Nucl. Instr. and Meth.* **215**, 345.

Suzuki, M. and Kubota, S. (1979) Mechnism of proportional scintillation in argon, krypton and xenon. *Nucl. Instrum. and Meth.* **164**, 197.

Suzuki, M., *et al.* (1987) The emission spectra of Ar, Kr and Xe+TEA. *Nucl. Instr. and Meth.* **A254**, 556.

Suzuki, M., *et al.* (1988) On the optical readout of Gas Avalanche Chambers and its applications. *Nucl. Instr. and Meth.* **A263**, 237.

Szarka, J. and Povinec, P. (1979) Electrostatic problems in multi-element proportional counters. *Nucl. Instr. and Meth.* **164**, 463.

Takahashi, H., *et al.* (2004) Development of a two-dimensional multi-grid-type microstrip gas chamber for spallation neutron source. *Nucl. Instr. and Meth.* **A529**, 348.

Takahashi, T., *et al.* (1983) Emission spectra from Ar-Xe, Ar-Kr, Ar-N_2, Ar-CH_4, Ar-CO_2 and Xe-N_2 gas scintillation proportional counters. *Nucl. Instr. and Meth.* **205**, 591.

Tanaka, Y., *et al.* (1953) Absorption coefficients of gases in the vacuum ultraviolet. Part IV. Ozone. *J. Chem. Phys.* **21**, 1651.

Teyssier, J. L., *et al.* (1963) Les scintillateurs gazeux: structure, mécanisme et applications. *J. Phys. Rad.* **24**, 55.

Thiess, P. E. and Miley, G. H. (1975) New near-infrared and ultraviolet gas-proportional scintillation counters. *IEEE Trans. Nucl. Sci.* **NS-21**, 125.

Thompson, A. C., (2004), X-Ray data booklet: xdb.lbl.gov.

Thompson, J. R., *et al.* (1985) Cathode charge distribution in multiwire chambers 3: distribution in anode wire direction. *Nucl. Instr. and Meth.* **A234**, 505.

Thomson, G. M., *et al.* (1973) Mobility, diffusion and clustering of K^+ ions in gases. *J. Chem. Phys.* **58**, 2402.

Titov, M. (2004) Radiation damage and long-term aging in gas detectors. arXiv:physics/0403055.

Titov, M. (2007) New developments and future perspectives of gaseous detectors. *Nucl. Instr. and Meth.* **A581**, 25.

Titov, M. (2012) Gaseous detectors, in *Handbook of Particle Detection and Imaging*, Grupen, C. and Buvat, I. (eds.), Berlin, Springer Verlag.

Titt, U., *et al.* (1998) A Time Projection Chamber with optical readout for charged particle track structure imaging. *Nucl. Instr. and Meth.* **A416**, 85.

Tomitani, T. (1972) Analysis of potential distribution in a gaseous counter of rectangular cross-section. *Nucl. Instr. and Meth.* **100**, 179.

Townsend, J. (1947) *Electrons in Gases*, London, Hutchinson.

TPC (1983) *The Time Projection Chamber.* AIP Conf. Proc. 108, Vancouver American Institute of Physics.

Trinchero, G. C., *et al.* (2003) A study of new techniques for large-scale glass RPC production. *Nuc. Phys. B (Proc. Suppl.)* **A508**, 102.

Trippe, T. (1969) *Minimum Tension Requirements for Charpak Chambers' Wires*, CERN NP Int. Rep. 69–18, CERN.

Trower, W. P. (1966) Range–energy end dE/dx plots of charged particles in matter, UCRL–2426, Livermore.

Tsyganov, E. N. (1976) Estimates of cooling and bending processes for charged particle penetration through a monocrystal, Report TM-684, FERMILAB.

Ullaland, O. (1986) The OMEGA and SFMD experience in intense beams. *Proc. Workshop on Radiation Damage to Wire Chambers.* Berkeley. LBL-21170 107.

Ullaland, O. (2005) Fluid systems for RICH detectors. *Nucl. Instr. and Meth.* **A553**, 107.

Urban, M., *et al.* (1981) A position sensitive parallel plate avalanche counter to detect minimum ionizing particles. *Nucl. Instr. and Meth.* **188**, 47.

Va'vra, J. (1986a) High Resolution Drift Chambers. *Nucl. Instr. and Meth.* **A244**, 391.

Va'vra, J. (1986b) *Review of Wire Chamber Aging Workshop on Radiation Damage to Wire Chambers Berkeley.* LBL-21170263.

Va'vra, J. (1992) Wire chamber gases. *Nucl. Instr. and Meth.* **A323**, 34.

Va'vra, J. (1997) Wine ageing with the TEA photocathode. *Nucl. Instr. and Meth.* **A 387**, 183.

Va'vra, J. (2003) Physics and chemistry of aging-early developments. *Nucl. Instr. and Meth.* **A515**, 1.

van der Graaf, H. (2007) GridPix: an integrated readout system for gaseous detectors with a pixel chip as anode. *Nucl. Instr. and Meth.* **A580**, 1023.

van der Graaf, H. (2011) Gaseous detectors. *Nucl. Instr. and Meth.* **A628**, 27.

Veenhof, R. (1998) GARFIELD, recent developments. *Nucl. Instr. and Meth.* **A419**, 726.

Veenhof, R. (2002) GARFIELD, rjd.home.cern.ch/rjd/garfield/.

Vellettaz, N., *et al.* (2004) Two-dimensional gaseous microstrip detector for thermal neutrons. *Nucl. Instr. and Meth.* **392**, 73.

Veress, I., Montvai, A. (1978) Survey on Multiwire Proportional Chambers. *Nucl. Instr. and Meth.* **156**, 73.

Viertel, G. M. (1995) The L3 forward muon detector. *Nucl. Instr. and Meth.* **367**, 115.

Villa, F. (1983) Dimethylether: a low velocity, low diffusion drift chamber gas. *Nucl. Instr. and Meth.* **217**, 273.

Villa, M., *et al.* (2011) Progress on large area GEMs. *Nucl. Instr. and Meth.* **A 628**, 182.

Volpe, G., *et al.* (2008) Gas Cherenkov detectors for high momentum charged particle identification in the ALICE experiment at LHC. *Nucl. Instr. and Meth.* **A595**, 40.

Wagner, E. B., *et al.* (1967) Time-of-Flight Investigations of Electron Transport in Some Atomic and Molecular Gases. *J. Chem. Phys.* **47**, 3138.

Walenta, A. H. (1973) A system of large Multiwire Drift Chambers. *Nucl. Instr. and Meth.* **111**, 467.

Walenta, A. H. (1978) Left-right assignment in drift chambers and MWPCs using induced signals. *Nucl. Instr. and Meth.* **151**, 461.

Walenta, A. H. (1979) The Time Expansion Chamber and single ionization cluster measurement. *IEEE Trans. Nucl. Sci.* **NS-26**, 73.

Walenta, A. H. (1981) Performance and development of dE/dx counters. *Phys. Scripta* **23**, 354.

Walenta, A. H. (1983) State-of-the art and applications of Wire Chambers. *Nucl. Instr. and Meth.* **217**, 65.

Walenta, A. H., *et al.* (1971) The Multiwire Drift Chamber: a new type of proportional wire chamber. *Nucl. Instr. and Meth.* **92**, 373.

Walkowiak, W. (2000) Drift velocity of free electrons in liquid argon. *Nucl. Instr. and Meth.* **A449**, 288.

Walz, C. (2010) Optimization and characterization of Gas Electron Multipliers. *Thesis at Hochschule Offenburg,*

Wang, J. G. (2003) RPC performance at KLM/BELLE. *Nucl. Instr. and Meth.* **A508**, 133.

Warren, R.W. and Parker, J.H. (1962) Ratio of the Diffusion Coefficient to the Mobility Coefficient for Electrons in He, Ar, N2, H2, D2, CO and CO2. *Phys. Rev.* **128**, 2661.

Watanabe, K. and Zelikoff, M. (1953) Absorption coefficients of water vapor in the vacuum ultraviolet. *J. Opt. Soc. Am.* **43**, 753.

Watanabe, T. (1999) A computational analysis of intrinsic detection efficiency of Geiger–Mueller tubes for photons. *Nucl. Instr. and Meth.* **A438**, 439.

White-Grodstein, G. (1957) X-ray attenuation coefficients from 10 keV to 100 MeV, *National Bureau of Standards Circular* **583**, NBS.

Wiedner, D., *et al.* (2008) The use of n-perfluorocarbons as RICH radiators. *Nucl. Instr. and Meth.* **A595**, 216.

Wilkinson, D. H. (1950) *Ionization Chambers and Counters*, Cambridge, Cambridge University Press.

Wilkinson, D. H. (1992) The Geiger discharge revisited Part I. The charge generated. *Nucl. Instr. and Meth.* **A321**, 195.

Wilkinson, D. H. (1996a) The Geiger discharge revisited Part II. Propagation. *Nucl. Instr. and Meth.* **A383**, 516.

Wilkinson, D. H. (1996b) The Geiger discharge revisited Part III. Convergence. *Nucl. Instr. and Meth.* **A383**, 523.

Wilkinson, D. H. (1999) The Geiger discharge revisited Part IV. The fast component. *Nucl. Instr. and Meth.* **A435**, 446.

Williamson, C. and Boujot, J. P. (1962) Tables of range and rates of energy loss of charged particles of energy 0.5 to 150 MeV, Saclay, CEN.

XENON: xenon.astro.columbia.edu.

Yamashita, T., *et al.* (1992) Measurements of electron drift velocities and positive ion mobilities for gases containing CF4 II. *Nucl. Instr. and Meth.* **A317**, 213.

Yasuda, H. (2003) New insights into aging phenomena from plasma chemistry. *Nucl. Instr. and Meth.* **A515**, 15.

Ye, J., *et al.* (2008) Study of position resolution of resistive plate chambers. *Nucl. Instr. and Meth.* **A591**, 411.

Yeddanapalli, L. M. (1942) The decomposition of methane in glow discharge at liquid-air temperature. *J. Chem. Phys.* **10**, 249.

You, W. (2012) Particle identification of the ALICE TPC via dE.dx. *Nucl. Instr. and Meth.* **A706**, 55.

Ypsilantis, T. (1981) Cherenkov Ring Imaging. *Phys. Scripta* **23**, 371.

Ypsilantis, T. and Seguinot, J. (1994) Theory of ring imaging Cherenkov counters. *Nucl. Instr. and Meth.* **A343**, 30.

Zeballos, E. C., *et al.* (1996) A comparison of the wide gap and narrow gap resistive plate chamber. *Nucl. Instr. and Meth.* **A373**, 35.

Zech, G. (1983) Electrodeless drift chambers. *Nucl. Instr. and Meth.* **217**, 209.

Zeuner, T. (2000) The MSGC-GEM inner tracker for HERA-B. *Nucl. Instr. and Meth.* **A 446**, 324.

Zhang, J., *et al.* (2010) The BESIII muon identification system. *Nucl. Instr. and Meth.* **A614**, 196.

Zhang, W., *et al.* (1989) Excitation and ionization of freon molecules 1. *Chem. Phys.* **137**, 391.

Ziegler, J. F. (1977) *The Stopping Power of Ions in Matter*, New York, Pergamon.

Ziegler, M., *et al.* (2001) A triple GEM detector with two dimensional readout. *Nucl. Instr. and Meth.* **A 471**, 260.

Index

Printed in the United States
by Baker & Taylor Publisher Services